A study guide containing numerous worked examples and practice problems has been prepared to assist the student in mastering the various statistical concepts that are presented in this text. This *Student Study Guide—Introduction to Statistics, 3rd edition*, by William D. Ergle and Ronald E. Walpole may be purchased from your local bookstore.

Frequently Used Formulas

Sample mean
$$\bar{x} = \frac{\Sigma x}{n}$$

Sample variance
$$s^2 = \frac{n\Sigma x^2 - (\Sigma x)^2}{n(n-1)}$$

Additive rule of probability
$$P(A \cup B) = P(A) + P(B) - P(A \cap B)$$

Multiplicative rule of probability
$$P(A \cap B) = P(A)P(B|A)$$

Conditional probability
$$P(B|A) = \frac{P(A \cap B)}{P(A)}$$

Mean of a discrete random variable
$$\mu = \Sigma x f(x)$$

Variance of a discrete random variable
$$\sigma^2 = \Sigma(x - \mu)^2 f(x)$$

Binomial distribution
$$b(x; n, p) = \binom{n}{x}p^x q^{n-x}$$

Binomial mean	$\mu = np$
Binomial variance	$\sigma^2 = npq$
Hypergeometric distribution	$h(x; N, p, k) = \dfrac{\binom{k}{x}\binom{N-k}{n-x}}{\binom{N}{n}}$
Poisson distribution	$p(x; \mu) = \dfrac{e^{-\mu}\mu^x}{x!}$
Confidence interval for μ (σ known)	$\bar{x} \pm z_{\alpha/2}\dfrac{\sigma}{\sqrt{n}}$
Confidence interval for μ (σ unknown)	$\bar{x} \pm t_{\alpha/2}\dfrac{s}{\sqrt{n}}$
Sample size for estimating μ	$n = \left(\dfrac{z_{\alpha/2}\sigma}{e}\right)^2$
Confidence interval for $\mu_1 - \mu_2$ (σ_1 and σ_2 known)	$(\bar{x}_1 - \bar{x}_2) \pm z_{\alpha/2}\sqrt{\dfrac{\sigma_1^2}{n_1} + \dfrac{\sigma_2^2}{n_2}}$
Pooled estimate of σ^2	$s_p^2 = \dfrac{(n_1 - 1)s_1^2 + (n_2 - 1)s_2^2}{n_1 + n_2 - 2}$
Confidence interval for $\mu_1 - \mu_2$ ($\sigma_1 = \sigma_2$ but unknown)	$(\bar{x}_1 - \bar{x}_2) \pm t_{\alpha/2}s_p\sqrt{\dfrac{1}{n_1} + \dfrac{1}{n_2}}$
Confidence interval for μ_D (paired observations)	$\bar{d} \pm t_{\alpha/2}\dfrac{s_d}{\sqrt{n}}$

Continued on rear endpapers

Introduction to Statistics

3 rd edition Introduction to Statistics

Ronald E. Walpole

Professor of Mathematics and Statistics, Roanoke College

Macmillan Publishing Co., Inc.
NEW YORK

Collier Macmillan Publishers
LONDON

Macmillan Publishing Co., Inc.
866 Third Avenue, New York, New York 10022

Collier Macmillan Canada, Inc.

Library of Congress Cataloging in Publication Data

Walpole, Ronald E.
  Introduction to statistics.

  Bibliography: p.
  Includes index.
  1. Statistics. I. Title.
QA276.12.W36 1982   519.5   81-8189
ISBN 0-02-424150-4 (Hardbound)   AACR2
ISBN 0-02-977650-3 (International Edition)

Printing: 1 2 3 4 5 6 7 8   Year: 2 3 4 5 6 7 8

Dedicated to the memory of my parents

Preface

Like the previous editions, this third edition of *Introduction to Statistics* has been written to serve as an introductory statistics text for students majoring in any of the academic disciplines, whether it be sociology, psychology, economics, business administration, or one of the sciences. A knowledge of high school algebra is sufficient to comprehend the basic concepts of descriptive and inferential statistics that are presented, although experience seems to indicate that a student benefits more from such a course when it is preceded by either a semester course in elementary functions or finite mathematics.

In preparing this third edition, all examples and exercises dealing with numerical measurements have been converted to metric units. Unlike the first two editions in which the exercises were placed only at the end of each chapter, now the numerous illustrative exercises, many of which are new, have been placed immediately following appropriate sections within each chapter. Optional statistical concepts and procedures defined throughout many of the exercise sets may be included at the discretion of the professor. A *Student Study Guide,* which provides additional insight and practice in solving the various types of problems presented throughout the text, accompanies this edition.

The concepts of descriptive statistics, which previously were scattered throughout the text, have now been placed in the first three chapters. To complement this area, the treatment of percentiles, quartiles, and deciles has been expanded to cover both grouped and ungrouped data, the Pearsonian coefficient of skewness is defined, and new material has been added to illustrate the use of a random number table in selecting a random sample. The field of modern statistics, with an increased emphasis on statistical inference, is based primarily on the theory of probability. An introduction to the basic concepts of probability theory using set notation is, therefore, presented in Chapter 4. The material in Chapter 5, introducing random variables and their mathematical expectations, and discrete and continuous probability distributions and the various properties describing these distributions, has been retained from the second edition, although the discussion on mathematical expectations has been completely revised.

An introduction to several discrete probability distributions is presented in Chapter 6, followed by a discussion of the normal distribution in Chapter 7. This then naturally leads to the treatment of sampling theory, estimation theory, and hypothesis testing in Chapters 8, 9, and 10, respectively. Chapter 8 has been revised somewhat by including a discussion of simulated experiments as well as a brief treatment of systematic, stratified, and cluster sampling procedures, while delaying our study of the chi-square and F distributions until Chapter 9. A new section on Bayesian methods of estimating the binomial parameter and the normal mean has also been included in Chapter 9. Chapter 11 presents the general techniques of curve fitting followed by a discussion of regression theory and an expanded treatment of correlation analysis that includes tests of significance and a new section on partial and multiple correlation. An introduction to analysis of variance with an expanded treatment of experimental designs covering randomized block and Latin square designs is presented in Chapter 12. Although some authors prefer to use a regression approach to analysis of variance, no attempt has been made in this text to relate the two chapters. Either chapter may be considered without a knowledge of the other. For greater emphasis, the material on nonparametric tests has been greatly expanded and now constitutes the final chapter of the text.

The text contains sufficient material to allow for flexibility in the length of the course and in the selection of topics. A semester course, meeting three hours a week, should include most of the material in Chapters 1 through 5; Sections 6.1, 6.2, and 6.3 of Chapter 6; Chapters 7 through 10, perhaps excluding Sections 9.9 and 9.10 of Chapter 9 and Sections 10.8, 10.9, and 10.10 of Chapter 10; and Sections 11.1, 11.2, 11.3, 11.4, and 11.8 of Chapter 11.

I wish to acknowledge my appreciation to all those who assisted in the preparation of this textbook. I am particularly grateful to Rhonda Haga and Kim Marzke for typing and proofreading this revised third edition and the accompanying *Solutions Manual;* to Dr. William D. Ergle for writing the major part of the *Student Study Guide;* to the Macmillan Publishing Co. Inc., for their editorial assistance; and to the many teachers and reviewers of the first two editions for their helpful suggestions in preparing this third edition.

I am indebted to the Literary Executor of the late Sir Ronald A. Fisher, F.R.S., Cambridge, and to Oliver & Boyd Ltd., Edinburgh, for their permission to reprint Table A.5 from their book *Statistical Methods for Research Workers;* to Professor E. S. Pearson and the Biometrika Trustees for permission to reprint in abridged form, as Tables A.6 and A.7, Tables 8 and 18 from *Biometrika Tables for Statisticians,* Vol. I; to D. Van Nostrand Company, Inc., for permission to reproduce in Table A.3 material from E. C. Molina's *Poisson's Exponential Binomial Limit.* I wish also to express my appreciation for permission to reproduce Table A.8 from a publication by the American Cyanamid Company, Table A.9 from the *Bulletin of the Educational Research at Indiana University,* Tables A.10 and A.14 from the *Annals of Mathematical Statistics,* Table A.11 from *Biometrics,* and Table A.13 from the *Journal of the American Statistical Association.*

R.E.W.

Contents

Introduction to Statistics

1 Introduction

The processing of statistical information has a history that extends back to the beginning of mankind. In early biblical times nations compiled statistical data to provide descriptive information relative to all sorts of things, such as taxes, wars, agricultural crops, and even athletic events. Today, with the development of probability theory, we are able to use statistical methods that not only describe important features of the data but methods that allow us to proceed beyond the collected data into the area of decision making through generalizations and predictions.

Descriptive and Inferential Statistics

In the study of statistics we are basically concerned with the presentation and interpretation of **chance outcomes** that occur in a planned or scientific investigation. For example, we may record the number of accidents that occur monthly at the intersection of Driftwood Lane and Royal Oak Drive, hoping to justify the installation of a traffic light, we might classify responses in an opinion poll as "yes" or "no," or we may be interested in the amount of residue deposited in a chemical reaction when the concentration of the catalyst is varied. Hence the statistician is usually dealing with either **numerical data** representing **counts** or **measurements,** or perhaps with **categorical data** that can be classified according to some criterion.

We shall refer to any recording of information, whether it be numerical or categorical, as an **observation.** Thus the numbers 3, 1, 0, and 2, representing the number of accidents that occurred for each month from January through April during the past year at the intersection of Driftwood Lane and Royal Oak Drive, constitute a set of observations. Similarly, the measurements 2.5, 3.1, and 1.8 grams of residue deposited in the chemistry experiment are recorded as observations.

Example 1. The students at Hidden Valley High are given blood tests to determine their type of blood. A person's blood can be classified in 8 ways. It must be AB, A, B, or O, with a plus or minus sign, depending on the presence or absence of the Rh antigen. Of course, this results in a descriptive or categorical representation of the data rather than a numerical measurement or count.

Statistical methods are those procedures used in the *collection, presentation, analysis, and interpretation of data.* We shall categorize these methods as belonging to one of two major areas called **descriptive statistics** and **statistical inference.** First, let us consider the field of descriptive statistics.

DEFINITION

Descriptive Statistics. Descriptive statistics comprises those methods concerned with collecting and describing a set of data so as to yield meaningful information.

Let it be clearly understood that descriptive statistics provides information only about the collected data and in no way draws inferences or conclusions concerning a larger set of data. The construction of tables, charts, graphs, and other relevant computations in various newspapers and magazines usually fall in the area categorized as descriptive statistics.

Example 2. The data in Table 1.1 appeared in a Sunday edition of the *Roanoke Times and World News.*

TABLE 1.1
National League Batting

Team	AB	R	H	HR	RBI	Pct
St. Louis	2720	362	775	54	343	.285
Chicago	2542	337	693	74	313	.273
Pittsburgh	2646	346	715	73	313	.270
Philadelphia	2761	356	742	62	335	.269
Cincinnati	2841	361	756	67	339	.266
Montreal	2518	325	660	62	298	.262
Los Angeles	2821	364	725	96	356	.257
Atlanta	2758	364	707	72	334	.256
San Francisco	2711	389	680	72	357	.251
Houston	2773	311	691	32	294	.249
New York	2640	294	652	41	281	.247
San Diego	2825	315	679	53	293	.240

Such a table allows one to summarize a large amount of data collected from the beginning of the baseball season so as to provide immediate meaningful information concerning the batting accomplishments of each team. By scanning the table, it is seen that Cincinnati has the most times "at bat" but St. Louis has collected the most "hits." However, San Francisco leads the league in "runs" scored and also in "runs batted in." The "home-run" lead at this time was held by Los Angeles.

Example 3. The **bar chart** of Figure 1.1 shows, for women relative to men, the 1978 starting salaries offered to new bachelor's graduates in a variety of scientific professions. Clearly, women scientists start their careers at a lower salary than do men in almost every field, with the exception of new graduates in engineering, a field in which women constitute less than 2% of all engineers in the United States.

Example 4. The **pie chart** shown in Figure 1.2 represents an investor's monthly distribution of funds paid into the stock market. To obtain the size of each wedge in the pie chart, one needs to compute the fraction of the monthly payment invested in each area and multiply that fraction by 360°.

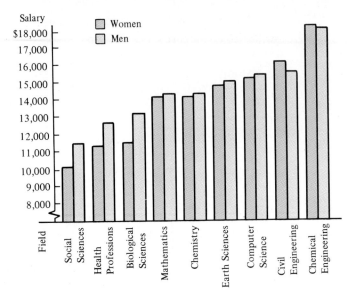

FIGURE 1.1

Beginning annual salary offers to bachelor's degree candidates, 1977–1978. (From The College Placement Council, *A Study of 1977–78 Beginning Offers by Business and Industry,* Bethlehem, Pa.)

> Hence, for a monthly payment of \$100, \$25 or $\frac{1}{4}$ is invested in oils. Therefore, a wedge that has a central angle of $(\frac{1}{4})(360°) = 90°$ represents pictorially the proportion or percentage of the investor's funds used to purchase oil stocks.

Although the presentation of statistical data in tables and charts as illustrated in Examples 2 through 4 has many useful purposes, our ultimate objective in most statistical studies is to make decisions and draw conclusions about a large set of data about which we have only partial knowledge based

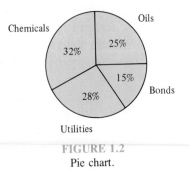

FIGURE 1.2
Pie chart.

on a smaller subset of the data. This takes us into the field of statistical inference.

DEFINITION

_____

Statistical Inference. Statistical inference comprises those methods concerned with the analysis of a subset of data leading to predictions or inferences about the entire set of data.

_____

The distinction between descriptive statistics and statistical inference becomes quite clear if we consider the following examples.

Example 5. Suppose that a set of measurements representing the total precipitation in the resort area of Lake Placid during the month of July has been recorded for the past 30 years. Any value describing the data, such as the average precipitation for July during the past 30 years or the driest July in the past 30 years, is a value in the field of descriptive statistics. We are not attempting to say anything about the precipitation of any other years except the 30 years from which the information was obtained. However, if the average precipitation for July in this area was 3.3 centimeters during the past 30 years and we make the statement that next July we can expect between 3.2 and 3.4 centimeters of rain, we are generalizing and thereby placing ourselves in the field of statistical inference.

Example 6. Academic records of the graduating classes during the past 5 years at a nearby state university show that 72% of the entering freshmen eventually graduated. The numerical value, 72%, is a descriptive statistic. If you are a member of the present freshman class and conclude from this study that your chances of graduating are better than 70%, you have made a statistical inference that is subject to uncertainty.

The generalizations associated with statistical inferences are always subject to uncertainties, since we are dealing only with partial information obtained from a subset of the data of interest. To cope with uncertainties, an understanding of probability theory is essential. In this text we shall keep this theory to a minimum. By omitting certain proofs that require a knowledge of differential and integral calculus, one is able to learn a great deal about statistical procedures using only elementary concepts of high school algebra.

Every effort is made in this text to present material in such a manner that all areas of learning may benefit, whether it be the field of science, psychology, business, agriculture, or medicine. The basic techniques for collecting and analyzing data are the same no matter what the field of application may be.

For example, the chemist runs an experiment using 3 variables and measures the amount of desired product. The results are then analyzed by statistical procedures. These same procedures are used to analyze the results obtained by measuring the yield of grain when 3 fertilizers are tested or to analyze the data representing the number of defectives produced by 3 similar machines. Many statistical methods that were derived primarily for agricultural applications have proved to be equally valuable for applications in other areas.

Statisticians are employed today by every progressive industry to direct their quality-control process and to assist in the establishment of good advertising and sales programs for their products. In business the statistician is responsible for decision making, for the analysis of time series, and for the formation of index numbers. Indeed, statistics is a very powerful tool if properly used. The abuse of statistical procedures will frequently lead to erroneous results. One should be careful to apply the correct and most efficient procedure for the given conditions to obtain maximum information from the data available.

The procedures used to analyze a set of data depend to a large degree on the method used to collect the information. For this reason it is desirable in any investigation to consult with the statistician from the time the project is planned until the final results are analyzed and interpreted.

1.2
Populations and Samples

The statistician is primarily concerned with numerical data. For the classification of blood types in Example 1 on page 2, it may be convenient to use numbers from 1 to 8 to represent each blood type and then record the appropriate number for each student. In any particular study, the number of possible observations may be small, large but finite, or infinite. In the classification of blood types we can only have as many observations as there are students at Hidden Valley High. The project therefore results in a finite number of observations. On the other hand, if we could toss a pair of dice indefinitely and record the totals that occur, we would obtain an infinite set of values, each value representing the result of a single toss of a pair of dice.

The totality of observations with which we are concerned, whether their number be finite or infinite, constitutes what we call a **population.** In past years the word "population" referred to observations obtained from statistical studies involving people. Today, the statistician uses the term to refer to observations relevant to anything of interest, whether it be groups of people, animals, or objects.

DEFINITION

Population. A *population* consists of the totality of the observations with which we are concerned.

The number of observations in the population is defined to be the **size** of the population. If there are 600 students in the school that we classified according to blood type, we say that we have a population of size 600. The numbers on the cards in a deck, the heights of residents in a certain city, and the lengths of fish in a particular lake are examples of populations with finite size. In each case the total number of observations is a finite number. The die-tossing experiment generates a population whose size is infinite. Similarly, the observations obtained by measuring the atmospheric pressure every day from the past on into the future, or all measurements on the depth of a lake from any conceivable position, are examples of populations whose sizes are infinite. Some finite populations are so large that in theory we assume them to be infinite. This is true if you consider the population of all possible lifetimes of a certain type of storage battery being manufactured for mass distribution throughout the country.

In the field of statistical inference the statistician is interested in arriving at conclusions concerning a population when it is impossible or impractical to observe the entire set of observations that make up the population. For example, in attempting to determine the average length of life of a certain brand of light bulb, it would be impossible to test all such bulbs if we are to have any left to sell. Exorbitant costs can also be a prohibitive factor in studying the entire population. Therefore, we must depend on a subset of observations from the population to help us make inferences concerning that same population. This takes us into the theory of sampling.

DEFINITION

Sample. A *sample* is a subset of a population.

If our inferences from the sample to the population are to be valid, we must obtain samples that are representative of the population. All too often we are tempted to choose a sample by selecting the most convenient members of the population. Such a procedure may lead to erroneous inferences concerning the population. Any sampling procedure that produces inferences that consistently overestimate or consistently underestimate some characteristic of the population is said to be **biased.** To eliminate any possibility of a bias in the sampling procedure, it is desirable to choose a **simple random sample,** or stated more briefly, a **random sample.**

Simple Random Sample. A *simple random sample* of *n* observations is a sample that is chosen in such a way that every subset of *n* observations of the population has the same probability of being selected.

The process for selecting a simple random sample is relatively easy for small finite populations but considerably more complicated for large and infinite populations. For a small finite population we might write the observations constituting the population on cards, place them in a large container, and then, for a simple random sample of size 5, select 5 cards in succession from the container. If the tags were thoroughly mixed, every subset of 5 tags in the population has the same chance of being selected.

The method described here for choosing a simple random sample can become very involved and time-consuming for large finite populations. There may also be some question as to whether the tags have been thoroughly mixed before selecting the sample. A better and perhaps more efficient method for selecting a simple random sample is achieved by using tables of **random numbers** (or **random digits**) such as Table A.12 of the Appendix. Such a table is constructed by generating in some random order a lengthy string of the digits 0, 1, 2, ..., 9. This could be done by repeatedly spinning a wheel, such as a roulette wheel, on which we have divided the face into 10 equal sectors labeled 0, 1, 2, ..., 9. Today, we would be more inclined to use a computer program to generate such a table.

The randomly produced digits can be grouped in any convenient way. In Table A.12 we have grouped them into sets of five digits and numbered the rows for ease of use. However, even in this form, a digit in any position in the table has the same chance of being any one of the digits 0, 1, 2, ..., 9, and digits in different positions are independent of each other. In using Table A.12, we choose the starting point at random and read off the digits in groups containing either one, two, three, or more of the digits in any predetermined direction. To illustrate the use of a table of random numbers in selecting a simple random sample, consider the following example.

Example 7. Use Table A.12 to choose a simple random sample of size 7 from a group of 80 mice to be used in studying the growth rate of tumors in a cancer research experiment.

Solution. First we label the mice 01, 02, 03, ..., 80 in any order. Next, we arbitrarily enter Table A.12 on line 28 and read out the pairs of digits in the 16th and 17th columns, going down the page. If we ignore numbers as they recur and all numbers greater than 80, our simple random sample of size 7 consists of the growth rates for tumors in the mice assigned labels

$$19 \quad 48 \quad 73 \quad 79 \quad 26 \quad 60 \quad 40$$

Suppose now that our population is large, as is the case if we wish to arrive at a conclusion concerning the proportion of coffee-drinking people in the United States who prefer Blue Mountain coffee imported from Jamaica over any other brand. It would be practically impossible to assign a number or label to every coffee-drinking American in order to select a random sample as outlined in Example 7. Instead, one could proceed in a haphazard fashion and select a large group of Americans from many different sections of the country, being careful not to favor any easily accessible region or to ignore areas that may be some distance away. The method suggested here for choosing a large sample to estimate the proportion of Americans favoring Blue Mountain coffee does not strictly satisfy our definition of a simple random sample, and consequently our estimate may well be biased. A correct procedure for estimating the population proportion when the population has been stratified into various regions will be demonstrated in Section 8.6, where some alternative sampling procedures are discussed. However, for most of the statistical procedures discussed throughout this text, the samples, whether obtained by random number tables or in some haphazard manner, will be treated as simple random samples.

1.3

Problems for the Statistician

In this age of advanced technology and productive research, one has only to read a newspaper or listen to a commercial on television to realize that there are a broad class of problems that can be solved by means of statistical procedures. Some of these problems we shall be able to solve using the methods of this text, whereas others must be left to more advanced techniques. In many of these problems we are attempting to estimate an unknown property of the population. Such is the case when one estimates the number of people who watched a rerun on television of _The Wizard of Oz,_ or when a biologist attempts to estimate the number of fish in Lake Louise. Other problems are solved by using a variety of testing procedures. Statistical tests have been developed to prove or disprove claims that one product is superior to another, or to test the effectiveness of a new vaccine or drug in arresting or eradicating a disease.

In the text *Statistics: A Guide to the Unknown* by Judith M. Tanur et al. (see the References), statistical techniques have been applied to a variety of fascinating problems that occur in the biological, political, social, and physical world. Among these one finds a discussion of the large-scale testing of the Salk polio vaccine in 1954, the development of a mathematical model to deal with epidemics, controlled experiments to test the effect of speed limits on traffic-accident rates, rapid procedures for projecting winners on election night, the use of statistical analysis to determine the authenticity of authorship of historical documents, cloud seeding and rainmaking, and numerous other applications.

We conclude this section with a more detailed description of a variety of problems to be solved later in the text.

Problem 1. In a certain federal prison it is known that $\frac{2}{3}$ of the inmates are under 25 years of age. It is also known that $\frac{3}{5}$ of the inmates are male and that $\frac{5}{8}$ of the inmates are female or 25 years of age or older. We use probability theory to determine the likelihood of selecting a prisoner at random from this prison who is female and at least 25 years old.

Problem 2. A soft-drink machine is regulated so that the amount of drink dispensed averages 200 milliliters per cup. Periodically, the machine is checked by taking a sample of 9 drinks and computing the average content. What action should one take if the sample of 9 drinks has an average content of 219 milliliters? Statistical procedures are used to determine if one is likely to obtain an average of 219 milliliters for a random sample of 9 drinks when the true average for the machine is indeed 200 milliliters per cup. If so, the machine is thought to be operating satisfactorily; otherwise, adjustments are made.

Problem 3. A study is made to estimate the percentage of citizens in a town who favor the construction of a new civic center. By using statistical methods, we can determine how large a sample is needed if one wishes to be highly confident that our estimate be within, say, 1% of the true percentage.

Problem 4. It is claimed that an automobile is driven on the average less than 20,000 kilometers per year. To test this claim, a random sample of 100 automobile owners kept a record of the kilometers they traveled and reported an average of 23,500 kilometers. Using statistical procedures, we can decide whether the higher observed average distance driven by the 100 drivers is significantly larger than the claimed value, resulting in a rejection of the claim, or whether one could expect such an experimental result even though the claim is in fact valid.

Problem 5. An argument has developed between two companions concerning the relationship between the size of a family and the educational level attained by the father. Archie claims that the more a man is educated, the less likely he is to have a large family. Harry says that this is nonsense. To settle the argument, a random sample of 200 married men, all retired, were classified according to education and number of children. Following statistical analysis of the data by appropriate techniques, it was shown that Harry was correct. In other words, based on this particular study, the size of a family is independent of the level of education attained by the father.

Problem 6. The following data represent the memberships at a local country club during the past 7 years:

x (years)	1	2	3	4	5	6	7
y (enrollment)	304	341	393	457	548	670	882

Using statistical procedures, we can derive the "best-fitting" growth curve relating the values of the two variables x and y and then, assuming that this growth pattern will continue, predict the size of the membership at some future date.

Problem 7. Two judges at a college homecoming parade ranked 8 floats in the following order:

Judge	Float							
	1	2	3	4	5	6	7	8
A	5	8	4	3	6	2	7	1
B	7	5	4	2	8	1	6	3

Can we conclude from these data that the two judges are in general agreement, or are they substantially apart? Perhaps one of the judges is incompetent. A question such as this is answered by calculating a measure of correlation between the rankings of the two judges.

1.4

**Summation
Notation**

In statistics it is frequently necessary to work with sums of numerical values. For example, we may wish to compute the average cost of a certain brand of toothpaste sold at 10 different stores. Perhaps we would like to know the total number of heads that occur when 3 coins are tossed several times.

Consider a controlled experiment in which the decreases in weight over a 6-month period were 15, 10, 18, and 6 kilograms, respectively. If we designate the first recorded value x_1, the second x_2, and so on, then we can write $x_1 = 15$, $x_2 = 10$, $x_3 = 18$, and $x_4 = 6$.

Using the Greek letter Σ (capital sigma) to indicate "summation of," we can write the sum of the 4 weights as

$$\sum_{i=1}^{4} x_i,$$

where we read "summation of x_i, i going from 1 to 4." The numbers 1 and 4 are called the *lower* and *upper limits* of summation. Hence

$$\sum_{i=1}^{4} x_i = x_1 + x_2 + x_3 + x_4$$

$$= 15 + 10 + 18 + 6 = 49.$$

Also,

$$\sum_{i=2}^{3} x_i = x_2 + x_3 = 10 + 18 = 28.$$

In general, the symbol $\sum_{i=1}^{n}$ means that we replace i wherever it appears after the summation symbol by 1, then by 2, and so on up to n, and then add up the terms. Therefore, we can write

$$\sum_{i=1}^{3} x_i^2 = x_1^2 + x_2^2 + x_3^2,$$

$$\sum_{j=2}^{5} x_j y_j = x_2 y_2 + x_3 y_3 + x_4 y_4 + x_5 y_5.$$

The subscript may be any letter, although i, j, and k seem to be preferred by statisticians. Obviously,

$$\sum_{i=1}^{n} x_i = \sum_{j=1}^{n} x_j.$$

The lower limit of summation is not necessarily a subscript. For instance, the sum of the natural numbers from 1 to 9 may be written

$$\sum_{x=1}^{9} x = 1 + 2 + \cdots + 9 = 45.$$

When we are summing over all the values of x_i that are available, the limits of summation are often omitted and we simply write $\Sigma\, x_i$. If in the diet experiment only 4 people were involved, then $\Sigma\, x_i = x_1 + x_2 + x_3 + x_4$. In fact, some authors even drop the subscript and let $\Sigma\, x$ represent the sum of all available data.

Example 8. If $x_1 = 3$, $x_2 = 5$, and $x_3 = 7$, find

(a) $\displaystyle\sum x_i$;   **(b)** $\displaystyle\sum_{i=1}^{3} 2x_i^2$;   **(c)** $\displaystyle\sum_{i=2}^{3} (x_i - i)$.

Solution

(a) $\displaystyle\sum x_i = x_1 + x_2 + x_3 = 3 + 5 + 7 = 15.$

(b) $\displaystyle\sum_{i=1}^{3} 2x_i^2 = 2x_1^2 + 2x_2^2 + 2x_3^2 = 18 + 50 + 98 = 166.$

(c) $\displaystyle\sum_{i=2}^{3} (x_i - i) = (x_2 - 2) + (x_3 - 3) = 3 + 4 = 7.$

Example 9. Given $x_1 = 2$, $x_2 = -3$, $x_3 = 1$, $y_1 = 4$, $y_2 = 2$, and $y_3 = 5$, evaluate

(a) $\displaystyle\sum_{i=1}^{3} x_i y_i.$ **(b)** $\displaystyle\left(\sum_{i=2}^{3} x_i\right)\left(\sum_{j=1}^{2} y_j^2\right).$

Solution

(a) $\sum_{i=1}^{3} x_i y_i = x_1 y_1 + x_2 y_2 + x_3 y_3$

$\qquad = (2)(4) + (-3)(2) + (1)(5) = 7.$

(b) $\left(\sum_{i=2}^{3} x_i \right) \left(\sum_{j=1}^{2} y_j^2 \right) = (x_2 + x_3)(y_1^2 + y_2^2)$

$\qquad\qquad = (-2)(20) = -40.$

Three theorems that provide basic rules in dealing with summation notation are given below.

THEOREM 1.1

The summation of the sum of two or more variables is the sum of their summations. Thus

$$\sum_{i=1}^{n} (x_i + y_i + z_i) = \sum_{i=1}^{n} x_i + \sum_{i=1}^{n} y_i + \sum_{i=1}^{n} z_i.$$

Proof. Expanding the left side and regrouping, we have

$$\sum_{i=1}^{n} (x_i + y_i + z_i)$$

$$= (x_1 + y_1 + z_1) + (x_2 + y_2 + z_2)$$
$$+ \cdots + (x_n + y_n + z_n)$$

$$= (x_1 + x_2 + \cdots + x_n) + (y_1 + y_2 + \cdots + y_n)$$
$$+ (z_1 + z_2 + \cdots + z_n)$$

$$= \sum_{i=1}^{n} x_i + \sum_{i=1}^{n} y_i + \sum_{i=1}^{n} z_i.$$

THEOREM 1.2

If c is a constant, then

$$\sum_{i=1}^{n} c x_i = c \sum_{i=1}^{n} x_i.$$

Proof. Expanding the left side and factoring, we get

$$\sum_{i=1}^{n} cx_i = cx_1 + cx_2 + \cdots + cx_n$$

$$= c(x_1 + x_2 + \cdots + x_n)$$

$$= c\sum_{i=1}^{n} x_i.$$

THEOREM 1.3

_____

If c is a constant, then

$$\sum_{i=1}^{n} c = nc.$$

_____

Proof. If in Theorem 1.2 all the x_i are equal to 1, then

$$\sum_{i=1}^{n} c = \underbrace{c + c + \cdots + c}_{n \text{ terms}} = nc.$$

The use of Theorems 1.1 through 1.3 in simplifying summation problems is illustrated in the following examples.

Example 10. If $x_1 = 2$, $x_2 = 4$, $y_1 = 3$, $y_2 = -1$, find the value of

$$\sum_{i=1}^{2} (3x_i - y_i + 4).$$

Solution

$$\sum_{i=1}^{2} (3x_i - y_i + 4) = \sum_{i=1}^{2} 3x_i - \sum_{i=1}^{2} y_i + \sum_{i=1}^{2} 4$$

$$= 3\sum_{i=1}^{2} x_i - \sum_{i=1}^{2} y_i + (2)(4)$$

$$= (3)(2 + 4) - (3 - 1) + 8$$

$$= 24.$$

Example 11. Simplify

$$\sum_{i=1}^{3} (x - i)^2.$$

Solution

$$\sum_{i=1}^{3} (x - i)^2 = \sum_{i=1}^{3} (x^2 - 2xi + i^2)$$

$$= \sum_{i=1}^{3} x^2 - \sum_{i=1}^{3} 2xi + \sum_{i=1}^{3} i^2$$

$$= 3x^2 - 2x \sum_{i=1}^{3} i + \sum_{i=1}^{3} i^2$$

$$= 3x^2 - 2x(1 + 2 + 3) + (1 + 4 + 9)$$

$$= 3x^2 - 12x + 14.$$

Often our data may be classified according to two criteria. For example, x_{ij} might represent the amount of gas released when a chemical experiment is run at the ith temperature level and the jth pressure level. To sum such observations, it is convenient to adopt a double-summation notation. The symbol $\sum_{i=1}^{m} \sum_{j=1}^{n}$ means that we first sum over the subscript j, using the theory for single summation, and then perform a second summation by allowing i to assume values from l to m. Hence, for the data in the following table,

		Pressure			
		1	*2*	*3*	*4*
Temperature	1	x_{11}	x_{12}	x_{13}	x_{14}
	2	x_{21}	x_{22}	x_{23}	x_{24}

we have

$$\sum_{i=1}^{2} \sum_{j=2}^{4} x_{ij} = \sum_{i=1}^{2} (x_{i2} + x_{i3} + x_{i4})$$

$$= (x_{12} + x_{13} + x_{14}) + (x_{22} + x_{23} + x_{24}).$$

Similarly, if $f(x_i, y_j)$ represents the textbook sales for publisher x_i at university y_j, then

$$\sum_{i=1}^{3} \sum_{j=1}^{2} f(x_i, y_j) = \sum_{i=1}^{3} [f(x_i, y_1) + f(x_i, y_2)]$$

$$= [f(x_1, y_1) + f(x_1, y_2)] + [f(x_2, y_1) + f(x_2, y_2)]$$

$$+ [f(x_3, y_1) + f(x_3, y_2)].$$

gives the total sales of a certain three publishers at two specific universities.

Some useful theorems that simplify the use of double-summation notation are found in Exercises 12 through 15.

EXERCISES

1. Classify the following statements as belonging to the area of descriptive statistics or statistical inference:
   (a) As a result of recent cutbacks by the oil-producing nations, we can expect the price of gasoline to double in the next year.
   (b) At least 5% of all fires reported last year in a certain city were deliberately set by arsonists.
   (c) Of all patients who have received this particular type of drug at a local clinic, 60% later developed significant side effects.
   (d) Assuming that less than 20% of the Columbian coffee beans were destroyed by frost this past winter, we should expect an increase of no more than 30 cents for a kilogram of coffee by the end of the year.
   (e) As a result of a recent poll, most Americans are in favor of building additional nuclear power plants.

2. In a new subdivision it was noted that 12 of the homes were colonials, 4 were Tudors, 5 were French provincials, and 9 were of a contemporary design. Classify the following conclusions that are drawn from these numbers as being in the field of descriptive statistics or statistical inference:
   (a) More colonial homes are constructed in new subdivisions than most other designs.
   (b) Next to colonial homes, more residents in this subdivision preferred a contemporary design.
   (c) The colonial designs outnumbered the Tudor designs by 3 to 1.
   (d) At least 30% of all new homes being built today are of a contemporary design.
   (e) If the present trend continues, builders will construct more contemporary homes than colonials in the next 5 years.

3. Define suitable populations from which the following samples are selected:
   (a) Persons in 200 homes are called by telephone in the city of Richmond and asked to name the candidate that they favor for election to the school board.

(b) A coin is tossed 100 times and 34 tails are recorded.

(c) Two hundred pairs of a new type of tennis shoe are tested on the professional tour and, on the average, lasted 4 months.

(d) On 5 different occasions it took a lawyer 21, 26, 24, 22, and 21 minutes to drive from her suburban home to her midtown office.

4. The number of tickets issued for traffic violations by 8 state troopers during the Memorial holiday weekend are 5, 4, 7, 7, 6, 3, 8, and 6.

(a) If these values represent the number of tickets issued by a random sample of 8 state troopers from Montgomery County in Virginia, define a suitable population.

(b) If the values represent the number of tickets issued by a random sample of 8 state troopers from South Carolina, define a suitable population.

5. By entering Table A.12 on line 2 and reading down the page using the 39th and 40th columns, make a random selection of 4 physicians from the following 30 physicians:

Atkinson	Davis	Hagan	Moorman	Snead
Baetz	Donnelly	Henretta	Nolan	Taylor
Bivens	Erwin	Hurt	Pierce	Vance
Bockner	Farley	Jennings	Renich	Walton
Clark	Grayson	Little	Roth	Wright
Crum	Greer	Meyer	Shumate	Yates

Assign the numbers from 1 through 30 to the physicians alphabetically.

6. The American Automobile Association wants to select a simple random sample of 15 of the 730 service stations along the nation's interstate highways to solicit information concerning their operating hours during the tourist season. Assigning numbers from 1 through 730 to the various stations, which ones will be selected if the sample is chosen by entering Table A.12 at the 13th, 14th, and 15th columns, beginning with line 16 and reading down successive three-digit columns?

7. A random selection of 5 scouts is to be chosen from 300 scouts at a jamboree to serve on an activities committee.

(a) By labeling the scouts from 1 to 300, which ones will be selected for the committee if the sample is chosen by entering Table A.12 at the 3rd, 4th, and 5th columns, beginning with line 42 and reading down successive three-digit columns?

(b) A more efficient method for making our random selection is achieved by labeling the scouts from 1 to 300 and then assigning the numbers 001, 301, and 601 to the first scout, the numbers 002, 302, and 602 to the second scout, ..., and the numbers 300, 600, and 900 to the last scout. Repeat part (a) using this numbering scheme.

8. Write each of the following expressions in full:

(a) $\sum\limits_{i=6}^{10} w_i^2$;   (b) $\sum\limits_{h=2}^{4} (x_h + h)$;   (c) $\sum\limits_{j=1}^{5} 3(v_j - 2)$.

Simplify, leaving your answer as a polynomial:

(a) $\sum\limits_{i=2}^{4} (2x + i)^2$;   (b) $\sum\limits_{y=0}^{3} (x - y + 3)^3$.

10. If $x_1 = 4$, $x_2 = -3$, $x_3 = 6$, and $x_4 = -1$, evaluate the following:

(a) $\sum\limits_{i=1}^{4} x_i^2(x_i - 3)$;   (b) $\sum\limits_{i=2}^{4} (x_i + 1)^2$;   (c) $\sum\limits_{i=2}^{3} (x_i + 2)/x_i$.

11. Given $x_1 = -2$, $x_2 = 3$, $x_3 = 1$, $y_1 = 4$, $y_2 = 0$, and $y_3 = -5$, find the value of the following:

(a) $\sum x_i y_i^2$;   (b) $\sum\limits_{i=2}^{3} (2x_i + y_i - 3)$;   (c) $\left(\sum x^2 \right)\left(\sum y \right)$.

12. Show that

$$\sum_{i=1}^{m} \sum_{j=1}^{n} (x_{ij} + y_{ij} + z_{ij}) = \sum_{i=1}^{m} \sum_{j=1}^{n} x_{ij} + \sum_{i=1}^{m} \sum_{j=1}^{n} y_{ij} + \sum_{i=1}^{m} \sum_{j=1}^{n} z_{ij}.$$

13. Show that

$$\sum_{i=1}^{m} \sum_{j=1}^{n} cx_{ij} = c \sum_{i=1}^{m} \sum_{j=1}^{n} x_{ij}.$$

14. Show that

$$\sum_{i=1}^{m} \sum_{j=1}^{n} c = mnc.$$

15. Verify that

$$\sum_{i=1}^{m} \sum_{j=1}^{n} x_i y_j = \left(\sum_{i=1}^{m} x_i \right)\left(\sum_{j=1}^{n} y_j \right).$$

2 Statistical Measures of Data

A variety of statistical measures are employed to summarize and describe sets of data. Some of these statistical measures define, in some sense, the center of a set of data and consequently are called measures of central location or measures of central tendency. Others provide a measure of variability among the observations and are classified as measures of variation. Together these two kinds of statistical measures are very useful in describing the actual distribution of the observations that constitute our set of data.

2.1
Parameters and Statistics

The terminology and notation adopted by statisticians in their treatment of statistical data depend entirely on whether the data set constitutes a population or a sample selected from a population. Consider, for example, the following set of data representing the number of typing errors made by a secretary on 10 different pages of a document: 1, 2, 1, 2, 3, 1, 1, 4, 0, and 2. First, let us assume that the document contains exactly 10 pages so that the data constitute a small finite population. A quick study of this population could lead to a number of conclusions. For instance, we could make the statement that the largest number of typing errors on any single page was 4, or we might state that the arithmetic mean (average) of the 10 numbers is 1.5. The numbers 4 and 1.5 are descriptive properties of our population. We refer to such values as **parameters** of the population.

DEFINITION

Parameter. Any numerical value describing a characteristic of a population is called a *parameter*.

It is customary to represent parameters by Greek letters. By tradition the arithmetic mean of a population is denoted by μ. Hence, for our population of typing errors, $\mu = 1.5$. Note that a parameter is a *constant* value describing the population.

Now let us suppose that the data representing the number of typing errors constitute a sample obtained by counting the number of errors on 10 pages randomly selected from a large manuscript. Clearly, the population is now a much larger set of data about which we only have partial information provided by the sample. The numbers 4 and 1.5 are now descriptive measures of the sample and are not to be considered parameters of the population. A value computed from a sample is called a **statistic.**

DEFINITION

Statistic. Any numerical value describing a characteristic of a sample is called a *statistic*.

A statistic is usually represented by ordinary letters of the English alphabet. If the statistic happens to be the sample mean, we shall denote it by $\bar{x}$. For our random sample of typing errors we have $\bar{x} = 1.5$. Since many random samples are possible from the same population, we would expect the statistic to vary from sample to sample. In other words, if a second random sample of 10 pages is selected from the manuscript and the number of typing errors per page tabulated, the largest value might turn out to be 5 rather than 4, and the arithmetic mean would probably be close to 1.5, but almost certainly different.

In our study of statistical inference we shall use the value of a statistic as an estimate of the corresponding population parameter. The size of the populations will be, for the most part, large or infinite. To know how accurate the statistic estimates the parameter, we must first investigate the distribution of the values of the statistic obtained in repeated sampling, a subject to be considered later.

2.2
Measures of Central Location

To investigate a set of quantitative data, it is useful to define numerical measures that describe important features of the data. One of the important ways of describing a group of measurements, whether it be a sample or a population, is by the use of an *average*.

An average is a measure of the center of a set of data when the data are arranged in an increasing or decreasing order of magnitude. For example, if an automobile averages 14.5 kilometers to 1 liter of gasoline, this can be considered a value indicating the center of several more values. In the country 1 liter of gasoline may give considerably more kilometers per liter than in the congested traffic of a large city. The number 4.5 in some sense defines a center value.

Any measure indicating the center of a set of data, arranged in an increasing or decreasing order of magnitude, is called a **measure of central location** or a **measure of central tendency.** The most commonly used measures of central location are the **mean, median,** and **mode.** The most important of these and the one we shall consider first is the mean.

DEFINITION

Population Mean. If the set of data $x_1, x_2, \ldots, x_N$, not necessarily all distinct, represents a finite population of size N, then the *population mean* is

$$\mu = \frac{\sum_{i=1}^{N} x_i}{N}.$$

Example 1. The number of employees at 5 different drugstores are 3, 5, 6, 4, and 6. Treating the data as a population, find the mean number of employees for the 5 stores.

Solution. Since the data are considered to be a finite population,

$$\mu = \frac{3 + 5 + 6 + 4 + 6}{5} = 4.8.$$

DEFINITION

Sample Mean. If the set of data x_1, x_2, ..., x_n, not necessarily all distinct, represents a finite sample of size n, then the _sample mean_ is

$$\bar{x} = \frac{\sum\limits_{i=1}^{n} x_i}{n}.$$

Example 2. A food inspector examined a random sample of 7 cans of a certain brand of tuna to determine the percent of foreign impurities. The following data were recorded: 1.8, 2.1, 1.7, 1.6, 0.9, 2.7, and 1.8. Compute the sample mean.

Solution. This being a sample, we have

$$\bar{x} = \frac{1.8 + 2.1 + 1.7 + 1.6 + 0.9 + 2.7 + 1.8}{7} = 1.8\%.$$

Often, it is possible to simplify the work in computing a mean by using **coding** techniques. For example, it is sometimes convenient to add (or subtract) a constant to all our observations and then compute the mean. How is this new mean related to the mean of the original set of observations? If we let $y_i = x_i + a$, then

$$\bar{y} = \frac{\sum\limits_{i=1}^{n} y_i}{n} = \frac{\sum\limits_{i=1}^{n} (x_i + a)}{n} = \bar{x} + a.$$

Therefore, the addition (or subtraction) of a constant to all observations changes the mean by the same amount. To find the mean of the numbers -5, -3, 1, 4, and 6, we might add 5 first to give the set of all positive values 0, 2, 6, 9, and 11 that have a mean of 5.6. Therefore, the original numbers have a mean of $5.6 - 5 = 0.6$.

Now suppose that we let $y_i = ax_i$. It follows that

$$\bar{y} = \frac{\sum\limits_{i=1}^{n} y_i}{n} = \frac{\sum\limits_{i=1}^{n} ax_i}{n} = a\bar{x}.$$

Therefore, if all observations are multiplied or divided by a constant, the new observations will have a mean that is the same constant multiple of the original mean. The mean of the numbers 4, 6, 14 is equal to 8, and therefore, after dividing by 2, the mean of the set 2, 3, and 7 must be $\frac{8}{2} = 4$.

The second most useful measure of central location is the median. For a population we designate the median by $\tilde{\mu}$ and for a sample we write $\tilde{x}$.

DEFINITION

Median. The *median* of a set of observations arranged in an increasing or decreasing order of magnitude is the middle value when the number of observations is odd or the arithmetic mean of the two middle values when the number of observations is even.

Example 3. On 5 term tests in sociology a student has made grades of 82, 93, 86, 92, and 79. Find the median for this population of grades.

Solution. Arranging the grades in an increasing order of magnitude, we get

$$79 \quad 82 \quad 86 \quad 92 \quad 93$$

and hence $\tilde{\mu} = 86$.

Example 4. The nicotine contents for a random sample of 6 cigarettes of a certain brand are found to be 2.3, 2.7, 2.5, 2.9, 3.1, and 1.9 milligrams. Find the median.

Solution. If we arrange these nicotine contents in an increasing order of magnitude, we get

$$1.9 \quad 2.3 \quad 2.5 \quad 2.7 \quad 2.9 \quad 3.1$$

and the median is then the mean of 2.5 and 2.7. Therefore,

$$\tilde{x} = \frac{2.5 + 2.7}{2} = 2.6 \text{ milligrams.}$$

The third and final measure of central location that we shall discuss is the mode.

DEFINITION

Mode. The _mode_ of a set of observations is that value which occurs most often or with the greatest frequency.

The mode does not always exist. This is certainly true when all observations occur with the same frequency. For some sets of data there may be several values occurring with the greatest frequency in which case we have more than one mode.

Example 5. If the donations from the residents of Fairway Forest toward the Virginia Lung Association are recorded as 9, 10, 5, 9, 9, 7, 8, 6, 10, and 11 dollars, then 9 dollars, the value that occurs with the greatest frequency, is the mode.

Example 6. The number of movies attended last month by a random sample of 12 high school students were recorded as follows: 2, 0, 3, 1, 2, 4, 2, 5, 4, 0, 1, and 4. In this case, there are two modes, 2 and 4, since both 2 and 4 occur with the greatest frequency. The distribution is said to be **bimodal.**

Example 7. No mode exists for the sociology grades of Example 3, since each grade occurs only once.

In summary, let us consider the relative merits of the mean, median, and mode. The mean is the most commonly used measure of location in statistics. It is easy to calculate and it employs all available information. The distributions of means obtained in repeated sampling from a population are well known, and consequently the methods used in statistical inference for estimating μ are based on the sample mean. The only real disadvantage to the mean is that it may be affected adversely by extreme values. In Example 5 the mean contribution to the Virginia Lung Association was $8.40, which is fairly close to the mode or median, both of which are $9. However, if one of the contributions had been much larger, say $90 instead of $11, then the mean contribution is $16.30, a value considerably higher than the majority of gifts.

The median has the advantage of being easy to compute if the number of observations is relatively small. It is not influenced by extreme values and consequently gives a truer average, namely $9, if the highest contribution in

Example 5 is \$90 rather than \$11. In dealing with samples selected from populations, the sample means usually will not vary as much from sample to sample as will the medians. Therefore, if we are attempting to estimate the center of a population based on a sample value, the mean is more stable than the median. Hence a sample mean is likely to be closer than the sample median to the population mean.

The mode is the least used measure of the three. For small sets of data its value is almost useless if in fact it exists at all. Only in the case of a large mass of data does it have a significant meaning. Its two main advantages are that (1) it requires no calculation, and (2) it can be used for qualitative as well as quantitative data. Thus, if jogging is the preferred form of exercise expressed by most people, we say that jogging is the **modal choice.**

EXERCISES

1. The number of incorrect answers on a true–false competency test for a random sample of 15 students were recorded as follows: 2, 1, 3, 0, 1, 3, 6, 0, 3, 3, 5, 2, 1, 4, and 2. Find
   (a) the mean;
   (b) the median;
   (c) the mode.

2. The number of building permits issued last month to 12 construction firms in a small midwestern city were 4, 7, 0, 7, 11, 4, 1, 15, 3, 5, 8, and 7. Treating the data as a population, find
   (a) the mean;
   (b) the median;
   (c) the mode.

3. The reaction times for a random sample of 9 subjects to a stimulant were recorded as 2.5, 3.6, 3.1, 4.3, 2.9, 2.3, 2.6, 4.1, and 3.4 seconds. Calculate
   (a) the mean;
   (b) the median.

4. The employees of a local manufacturing plant pledged the following donations, in dollars, to the United Fund: 10, 40, 25, 5, 20, 10, 25, 50, 30, 10, 5, 15, 25, 50, 10, 30, 5, 25, 45, and 15. Treating the data as a population, calculate
   (a) the mean;
   (b) the mode.

5. According to ecology writer Jacqueline Killeen, phosphates contained in household detergents pass right through our sewer systems causing lakes to turn into swamps that eventually dry up into deserts. The following data show the amount of phosphates per load of laundry, in grams, for a random sample of various types of detergents used according to the prescribed directions:

Laundry Detergent	Phosphates per Load (gm)
A&P Blue Sail	48
Dash	47
Concentrated All	42
Cold Water All	42
Breeze	41
Oxydol	34
Ajax	31
Sears	30
Fab	29
Cold Power	29
Bold	29
Rinso	26

For the given phosphate data, find
(a) the mean;
(b) the median;
(c) the mode.

6. In studying the drying time of a new acrylic paint, the data, in hours, were coded by subtracting 5 from each observation. Find the sample mean for the drying times of 10 panels of wood using this paint if the coded measurements are 1.4, 0.8, 2.4, 0.5, 1.3, 2.8, 3.6, 3.2, 2.0, and 1.9 hours.

7. In estimating the mean breaking strength, in kilograms, of a new synthetic fishing line, the data were coded by subtracting 10 from each observation. Find the sample mean for 5 of these lines if the coded measurements are 0.4, -0.2, 1.5, 1.8, and -0.7.

8. Find the preferred measure of central location for the sample whose observations 18, 10, 11, 98, 22, 15, 11, 25, and 17 represent the number of automobiles sold during this past January by 9 different automobile agencies. Justify your choice.

9. A certain automobile averages 18 kilometers per liter on the highway. How many liters of gasoline are required to complete a 450-kilometer trip?

10. The average IQ of 10 students in a mathematics course is 114. If 9 of the students have IQs of 101, 125, 118, 128, 106, 115, 99, 118, and 109, what must be the other IQ?

11. At a recent convention in Daytona Beach, delegates could choose one of three optional recreational tours: Disneyworld, Marineland, and St. Augustine. Determine the modal choice of 15 delegates who made the following decisions: Disneyworld, St. Augustine, Disneyworld, Disneyworld, Marineland, St. Augustine, Marineland, Disneyworld, Marineland, Marineland, Disneyworld, St. Augustine, Disneyworld, Marineland, and Disneyworld.

12. **Midrange:** Another easily computed measure of location, the *midrange,* is defined as the mean of the largest and smallest values in a set of data. Find the midrange
    (a) for the random sample of Exercise 3;
    (b) for the population of Exercise 4.

13. **Weighted mean:** Often, we wish to average the k quantities x_1, x_2, ..., x_k by attaching more significance to some of the numbers than to others. We accomplish this by assigning weights w_1, w_2, ..., w_k to the k quantities, where the weights represent measures of their relative importance. The corresponding *weighted mean,* μ_w or $\bar{x}_w$, is given by $\sum_{i=1}^{k} w_i x_i \Big/ \sum_{i=1}^{k} w_i$. If all the weights are equal, we obtain the ordinary arithmetic mean of the k observations.
    (a) What is the average for a student who received grades of 85, 76, and 82 on 3 tests and a 79 on the final examination in a certain course if the final examination counts three times as much as each of the 3 tests?
    (b) On a vacation trip a family bought 21.3 liters of gasoline at 39.9 cents per liter, 18.7 liters at 42.9 cents per liter, and 23.5 liters at 40.9 cents per liter. Find the mean price paid per liter.
    (c) A savings and loan association makes one car loan of $5000 at 10.5% interest, a second car loan of $6300 at 10.8% interest, and a third car loan of $4500 at 11% interest. What is the average percentage return to the savings and loan association for these 3 loans?

14. **Combined mean:** Suppose that k finite populations having N_1, N_2, ..., N_k measurements, respectively, have means μ_1, μ_2, ..., μ_k. The *combined population mean,* μ_c, of all the populations is

$$\mu_c = \frac{\sum_{i=1}^{k} N_i \mu_i}{\sum_{i=1}^{k} N_i}.$$

If random samples of size n_1, n_2, ..., n_k, selected from these k populations, have the means $\bar{x}_1$, $\bar{x}_2$, ..., $\bar{x}_k$, respectively, the *combined sample mean,* $\bar{x}_c$, of all the sample data is

$$\bar{x}_c = \frac{\sum_{i=1}^{k} n_i \bar{x}_i}{\sum_{i=1}^{k} n_i}.$$

   (a) Three sections of a statistics class containing 28, 32, and 35 students averaged 83, 80, and 76, respectively, on the same final examination. What is the combined population mean for all 3 sections?
   (b) A survey of a random sample of people leaving an amusement park showed an average expenditure of $10.30 for the evening. The average expenditure

for the 20 girls in the sample was $9.70 and for the boys it was $11.10. How many boys are there in the random sample?

15. **Geometric mean:** The *geometric mean*, G, of k positive numbers $x_1, x_2, \ldots, x_k$ is the kth root of their product; that is,

$$G = \sqrt[k]{x_1 x_2 \cdots x_k}.$$

Note that the logarithm of the geometric mean of the k positive numbers equals the arithmetic mean of their logarithms. The geometric mean is used primarily to average data for which the ratio of consecutive terms remains approximately constant. This occurs, for example, with such data as rates of change, ratios, economic index numbers, population sizes over consecutive time periods, and the like.

(a) Find the geometric mean of 1, 4, and 128.

(b) On January 1 a savings account contains $1000. If no further deposits and no withdrawals are made during the year, and interest is earned at 5% compounded monthly, find the average amount of money in the savings account during the first 6 months. (Use logarithms.)

(c) Over a period of 4 consecutive years an employee has received 7.2, 8.6, 6.9, and 9.8% annual pay increases. The ratios, therefore, of each new salary to the previous year's salary are 1.072, 1.086, 1.069, and 1.098. Using logarithms, find the geometric mean for these four ratios and then determine the average percent increase for this employee over the 4-year period.

16. **Harmonic mean:** The *harmonic mean*, H, of k numbers $x_1, x_2, \ldots, x_k$ is the number k divided by the sum of the reciprocals of the k numbers; that is,

$$H = \frac{k}{\sum\limits_{i=1}^{k} \frac{1}{x_i}}.$$

In actual practice, the harmonic mean is most frequently used in averaging speeds for various distances covered where the distances remain constant, and also in finding the average cost of some commodity, such as mutual funds, when several different purchases are made by investing the same amount of money each time.

(a) On a vacation trip to Canada, a family travels 500 kilometers each day. If the trip lasts 3 days and the family averages 80 kilometers per hour the first day, 93 kilometers per hour the second day, and 87 kilometers per hour the third day, find the average speed for the entire trip.

(b) A college professor invests $100 a month in a mutual growth fund. For the last 4 months the prices per share in the fund have been $5.45, $5.76, $6.10, and $5.90. Calculate the average price per share paid by the professor over this period of time.

(c) If a carpenter spends $20 one month for nails costing $4 per carton and another $20 a month later for the same kind of nails that now cost $6 per carton, how much did he pay per carton on the average for the nails he purchased?

2.3

Measures of Variation

The three measures of central location discussed in Section 2.2 do not by themselves give an adequate description of our data. We need to know how the observations spread out from the average. It is quite possible to have two sets of observations with the same mean or median that differ considerably in the variability of their measurements about the average.

Consider the following measurements, in liters, for two samples of orange juice bottled by companies A and B:

Sample A	0.97	1.00	0.94	1.03	1.11
Sample B	1.06	1.01	0.88	0.91	1.14

Both samples have the same mean, 1.00 liters. It is quite obvious that company A bottles orange juice with a more uniform content than company B. We say that the variability or the dispersion of the observations from the average is less for sample A than for sample B. Therefore, in buying orange juice, we would feel more confident that the bottle we select will be closer to the advertised average if we buy from company A.

The most important statistics for measuring the variability of a set of data are the **range** and the **variance.** The simplest of these to compute is the range.

DEFINITION

Range. The *range* of a set of data is the difference between the largest and smallest number in the set.

Example 8. The IQs of 5 members of a family are 108, 112, 127, 118, and 113. Find the range.

Solution. The range of the 5 IQs is $127 - 108 = 19$.

In the case of the companies bottling orange juice, the range for company A is 0.17 liters compared to a range of 0.26 liters for company B, indicating a greater spread in the values for company B.

The range is a poor measure of variation, particularly if the size of the sample or population is large. It considers only the extreme values and tells us nothing about the distribution of numbers in between. Consider, for example, the following two sets of data, both with a range of 12:

Set A	3	4	5	6	8	9	10	12	15
Set B	3	7	7	7	8	8	8	9	15

In set A the mean and median are both 8, but the numbers vary over the entire interval from 3 to 15. In set B the mean and median are also 8, but most of the values are closer to the center of the data. Although the range fails to measure this variation between the upper and lower observations, it does have some useful applications. In industry the range for measurements on items coming off an assembly line might be specified in advance. As long as all measurements fall within the specified range, the process is said to be in control.

To overcome the disadvantage of the range, we shall consider a measure of variation, namely, the **variance,** that considers the position of each observation relative to the mean of the set. This is accomplished by examining the **deviations from the mean.** The deviation of an observation from the mean is found by subtracting the mean of our set of data from the given observation. For the finite population $x_1, x_2, \ldots, x_N$, the deviations are

$$x_1 - \mu, x_2 - \mu, \ldots, x_N - \mu.$$

Similarly, if our set of data is the random sample $x_1, x_2, \ldots, x_n$, the deviations are

$$x_1 - \bar{x}, x_2 - \bar{x}, \ldots, x_n - \bar{x}.$$

An observation greater than the mean will yield a positive deviation, whereas an observation smaller than the mean will produce a negative deviation. Comparing the deviations for the two sets of data above, we have the following:

Set A	-5	-4	-3	-2	0	1	2	4	7
Set B	-5	-1	-1	-1	0	0	0	1	7

Clearly, most of the deviations of set B are smaller in magnitude than those of set A, indicating less variation among the observations of set B. Our aim now is to obtain a single numerical measure of variation that incorporates all the deviations from the mean. The most obvious procedure would be to average the deviations, but, as the reader will be asked to show in Exercise 1 on page 42, the sum of the deviations from the mean is zero for any set of data, and consequently their mean is also zero. To circumvent this problem, we could find a measure of variation called the **mean deviation** (see Exercise 11 on page 43) whereby we compute the mean of the absolute values of the deviations. An absolute value of a number is the number without the associated algebraic sign. Thus the absolute value of -4 is simply 4.

In practice the mean of the absolute values of the deviations from the mean is seldom used. The use of absolute values makes its mathematical treatment awkward. Instead, we shall work with the squares of all the deviations in computing the variance. In the case of a finite population of size N, the variance, denoted by the symbol σ^2 (read: sigma squared), may be computed directly from the following summation formula:

DEFINITION

Population Variance. Given the finite population $x_1, x_2, \ldots, x_N$, the *population variance* is

$$\sigma^2 = \frac{\displaystyle\sum_{i=1}^{N} (x_i - \mu)^2}{N}.$$

Assuming that the two sets A and B are populations, we now use the deviations in the preceding table to calculate their variance. For set A

$$\sigma^2 = \frac{\displaystyle\sum_{i=1}^{9} (x_i - 8)^2}{9}$$

$$= \frac{(-5)^2 + (-4)^2 + \cdots + (4)^2 + (7)^2}{9}$$

$$= \frac{124}{9}$$

and for set B

$$\sigma^2 = \frac{\displaystyle\sum_{i=1}^{9} (x_i - 8)^2}{9}$$

$$= \frac{(-5)^2 + (-1)^2 + \cdots + (1)^2 + (7)^2}{9}$$

$$= \frac{78}{9}.$$

A comparison of the two variances shows that the data of set A are more variable than the data of set B.

By using the squares of the deviations to compute the variance, we obtain a number in squared units. That is, if the original measurements were in feet, the variance would be expressed in square feet. To get a measure of variation expressed in the same units as the raw data, as was the case for the range, we take the square root of the variance. Such a measure is called the **standard deviation.**

Example 9. The following scores were given by 6 judges for a gymnast's performance in the vault of an international meet: 7, 5, 9, 7, 8, and 6. Find the standard deviation of this population.

Solution. First, we compute

$$\mu = \frac{7 + 5 + 9 + 7 + 8 + 6}{6} = 7$$

and then

$$\sigma^2 = \frac{\displaystyle\sum_{i=1}^{6} (x_i - 7)^2}{6}$$

$$= \frac{(0)^2 + (-2)^2 + (2)^2 + (0)^2 + (1)^2 + (-1)^2}{6}$$

$$= \frac{5}{3}.$$

The standard deviation is then given by $\sigma = \sqrt{\tfrac{5}{3}} = 1.29$.

The variance of a sample, denoted by s^2, is a statistic. Therefore, different random samples of size n, selected from the same population, would generally yield different values for s^2. In most statistical applications the parameter σ^2 is unknown and is estimated by the value s^2. For our estimate to be good, it must be computed from a formula that on the average produces the true answer σ^2. That is, if we were to take all possible random samples of size n from a population and compute s^2 for each sample, the average of all the s^2 values should be equal to σ^2. A statistic that estimates the true parameter on the average is said to be **unbiased.**

Intuitively, we would expect the formula for s^2 to be the same summation formula as that used for σ^2, with the summation now extending over the sample observations and with μ replaced by $\bar{x}$. This is indeed done in many texts, but the values so computed for the sample variance tend to underestimate σ^2 on the average. To compensate for this bias, we replace n by $n - 1$ in the divisor.

DEFINITION

_____

Sample Variance. Given a random sample $x_1, x_2, \ldots, x_n$, the *sample variance* is

$$s^2 = \frac{\sum\limits_{i=1}^{n} (x_i - \bar{x})^2}{n - 1}.$$

_____

Example 10. A comparison of coffee prices at 4 randomly selected grocery stores in San Diego showed increases from the previous month of 12, 15, 17, and 20 cents for a 200-gram jar. Find the variance of this random sample of price increases.

Solution. Calculating the sample mean, we get

$$\bar{x} = \frac{12 + 15 + 17 + 20}{4} = 16 \text{ cents.}$$

Therefore,

$$s^2 = \frac{\sum\limits_{i=1}^{4} (x_i - 16)^2}{3}$$

$$= \frac{(12 - 16)^2 + (15 - 16)^2 + (17 - 16)^2 + (20 - 16)^2}{3}$$

$$= \frac{(-4)^2 + (-1)^2 + (1)^2 + (4)^2}{3}$$

$$= \frac{34}{3}.$$

If $\bar{x}$ is a decimal number that has been rounded off, we accumulate a large error using the sample-variance formula in the form given above. To avoid this, let us derive the more useful computational formula, as given in the following theorem.

THEOREM 2.1

Computing Formula for s^2. If s^2 is the variance of a random sample of size n, we may write

$$s^2 = \frac{n \sum_{i=1}^{n} x_i^2 - \left(\sum_{i=1}^{n} x_i \right)^2}{n(n-1)}.$$

Proof. By definition

$$s^2 = \frac{\sum_{i=1}^{n} (x_i - \bar{x})^2}{n-1} = \frac{\sum_{i=1}^{n} (x_i^2 - 2\bar{x}x_i + \bar{x}^2)}{n-1}.$$

Applying Theorems 1.1, through 1.3, we obtain

$$s^2 = \frac{\sum_{i=1}^{n} x_i^2 - 2\bar{x} \sum_{i=1}^{n} x_i + n\bar{x}^2)}{n-1}.$$

Replacing $\bar{x}$ by $\sum_{i=1}^{n} x_i / n$, and multiplying numerator and denominator by n, we obtain

$$s^2 = \frac{n \sum_{i=1}^{n} x_i^2 - \left(\sum_{i=1}^{n} x_i \right)^2}{n(n-1)}.$$

The **sample standard deviation,** denoted by s, is defined to be the positive square root of the sample variance.

Example 11. Find the variance of the data 3, 4, 5, 6, 6, and 7, representing the number of trout caught by a random sample of 6 fishermen on June 19, 1981, at Lake Muskoka.

Solution. In tabular form we write

x_i	x_i^2	
3	9	
4	16	
5	25	
6	36	
6	36	
7	49	
31	171	$n = 6$

Hence

$$s^2 = \frac{(6)(171) - (31)^2}{(6)(5)} = \frac{13}{6}.$$

Often, it is possible to simplify the computational procedure for calculating the variance of a set of data by using coding techniques. Recall that coding was used in Section 2.2 to compute the mean. The effects of coding on the variance by subtracting a constant from each observation or by dividing each observation by a constant will be of particular interest to us. We shall investigate these effects here only for random samples, but the results are equally valid for populations.

If we let $y_i = x_i + c$, it follows that $\bar{y} = \bar{x} + c$, and hence the variance of the y_i's is

$$s^2 = \frac{\displaystyle\sum_{i=1}^{n} (y_i - \bar{y})^2}{n - 1} = \frac{\displaystyle\sum_{i=1}^{n} [(x_i + c) - (\bar{x} + c)]^2}{n - 1}$$

$$= \frac{\displaystyle\sum_{i=1}^{n} (x_i - \bar{x})^2}{n - 1}.$$

Therefore, if each observation of a set of data is transformed to a new set by the addition (or subtraction) of a constant c, the variance of the original set of data is the same as the variance of the new set.

Now suppose we let $y_i = cx_i$, so that $\bar{y} = c\bar{x}$. It follows that the variance of the y_i's is

$$s^2 = \frac{\sum\limits_{i=1}^{n} (y_i - \bar{y})^2}{n - 1} = \frac{\sum\limits_{i=1}^{n} (cx_i - c\bar{x})^2}{n - 1}$$

$$= \frac{c^2 \sum\limits_{i=1}^{n} (x_i - \bar{x})^2}{n - 1}.$$

Therefore, if a set of data is transformed to a new set by multiplying (or dividing) each observation by a constant c, the variance of the original set is equal to the variance of the new set divided (or multiplied) by c^2.

Example 12. A random sample of 5 bank presidents indicated annual salaries of $63,000, $48,000, $52,000, $35,000, and $41,000. Find the variance of this set of data by using appropriate coding techniques.

Solution. If we divide all the salaries by 1000 and then subtract 50, we obtain the numbers 13, -2, 12, -15, and -9, for which

$$\sum_{i=1}^{5} x_i = -1 \quad \text{and} \quad \sum_{i=1}^{5} x_i^2 = 623.$$

Now, for the coded data

$$s^2 = \frac{(5)(623) - (-1)^2}{(5)(4)} = 155.7$$

and, after multiplying by 1000^2, the variance of the original set of salaries is $s^2 = 1.557 \times 10^8$.

The standard deviation seems to be the best measure of variation that we have. At this point, however, it has meaning only when comparing two or more sets of data having the same units of measurement and approximately the same mean. Therefore, we could compare the variances of the observations of two companies bottling orange juice, and the larger value would indicate the company whose product is more variable or less uniform provided that bottles of the same size were used. It would not be meaningful to compare the variance of a set of heights to the variance of a set of aptitude scores.

In Exercise 19 on page 45, and in Section 2.4, we show how the standard deviation can be used to describe a single set of observations.

2.4

Chebyshev's Theorem

In Sections 2.2 and 2.3 we described a set of observations—a population or a sample—by means of a center or average and the variability about this average. The two values most often used by statisticians are the mean and the standard deviation. If a distribution of measurements has a small standard deviation, we would expect most of the values to be grouped closely around the mean. However, a large value of the standard deviation indicates a greater variability, in which case we would expect the observations to be more spread out from the mean.

The Russian mathematician P. L. Chebyshev (1821–1894) discovered that the fraction of the measurements falling between any two values symmetric about the mean is related to the standard deviation. **Chebyshev's theorem** gives a conservative estimate of the fraction of measurements falling within k standard deviations of the mean for any fixed number k.

THEOREM 2.2

_____

Chebyshev's Theorem. At least the fraction $1 - 1/k^2$ of the measurements of any set of data must lie within k standard deviations of the mean.

_____

For $k = 2$ the theorem states that at least $1 - 1/2^2 = \frac{3}{4}$, or 75%, of the measurements must lie within 2 standard deviations on either side of the mean. That is, $\frac{3}{4}$ or more of the observations of a population must lie in the interval $\mu \pm 2\sigma$. Considering our set of data to be a sample, then for $k = 2$ the theorem states that at least $\frac{3}{4}$ of the measurements must lie in the interval $\bar{x} \pm 2s$. Similarly, the theorem says that at least $\frac{8}{9}$, or about 88.9%, of the measurements must lie in the interval $\mu \pm 3\sigma$ for a population or $\bar{x} \pm 3s$ for a sample. When $k = 1$ the theorem is not very helpful. It states that at least $1 - 1/1^2 = 0$ of the measurements must lie within 1 standard deviation on either side of the mean. In statistical inference we could use an interval computed from a sample to estimate the corresponding population interval. Hence the sample interval $\bar{x} \pm 2s$ could be used to estimate the interval $\mu \pm 2\sigma$ containing at least $\frac{3}{4}$ of the measurements of the population.

Example 13. If the IQs of a random sample of 1080 students at a large university have a mean score of 120 and a standard deviation of 8, use Chebyshev's theorem to determine the interval containing at least 810 of the IQs in the sample. From this interval draw a statistical inference concerning the IQs of all students at this university. In what range can we be sure that no more than 120 of the scores fall?

Solution. Solving the equation

$$1 - \frac{1}{k^2} = \frac{810}{1080} = \frac{3}{4},$$

we find that $k = 2$ and then

$$\bar{x} \pm 2s = 120 \pm (2)(8) = 120 \pm 16.$$

That is, the interval from 104 to 136 contains at least $\frac{3}{4}$ or at least 810 of the IQs of our sample. From this result one might make the inference that at least $\frac{3}{4}$ of the IQs for the entire university fall in the interval from 104 to 136. If

$$1 - \frac{1}{k^2} = \frac{960}{1080} = \frac{8}{9},$$

then $k = 3$ and at least $\frac{8}{9}$ or at least 960 of the IQs of our sample must lie in the interval $120 + (3)(8)$, or from 96 to 144. Therefore, no more than 120 of the IQs fall below 96 and above 144.

   Chebyshev's theorem holds for any distribution of observations and, for this reason, the results are usually weak. The value given by the theorem is a lower bound only. That is, we know that *no less* than $\frac{3}{4}$ of a set of measurements must lie within 2 standard deviations on either side of their mean, but we never know if there are considerably more than $\frac{3}{4}$ of the values in this interval unless we actually find the interval and count the measurements within it. In Chapter 3 we shall see that with large sets of data, stronger results are usually possible.

2.5
z Scores

In assessing the accomplishments of a student in chemistry and economics during the summer session at a certain college, we might compare her numerical grades for the two courses. If we assume that the student made a grade of 82 in chemistry and a grade of 89 in economics, can we conclude that she is a better student in economics than in chemistry? Perhaps we should consider how this student performed relative to the other students in each of her classes. Is it not possible that one examination was much more difficult than the other and she actually did better in chemistry relative to the other students enrolled in chemistry than she did in economics? After all, the mean grade in chemistry was 68 and the standard deviation was 8, whereas the distribution of economics grades had a mean of 80 and a standard deviation of 6.

The problem before us, then, is one of comparing two observations from two different populations in order to determine their relative rank. In our illustration the entire set of chemistry grades constitutes one of the populations, the entire set of economics grades represents the other population, and the student's grades are the two observations. One method for ranking these two observations is to convert the individual observations into **standard units** known as *z* **scores** or *z* **values.**

DEFINITION

z Score. An observation, *x*, from a population with mean μ and standard deviation σ, has a *z score* or *z value* defined by

$$z = \frac{x - \mu}{\sigma}.$$

A *z* score measures how many standard deviations an observation is above or below the mean. Since σ is never negative, a positive *z* score measures the number of standard deviations an observation is above the mean, and a negative *z* score gives the number of standard deviations an observation is below the mean. Note that the units of the denominator and the numerator of a *z* score cancel. Hence a *z* score is unitless, thereby permitting a comparison of two observations relative to their groups, measured in completely different units.

Let us now compute the *z* scores corresponding to our student's grades in chemistry and economics. For chemistry we obtain

$$z = \frac{82 - 68}{8} = 1.75$$

and for economics

$$z = \frac{89 - 80}{6} = 1.50.$$

We see that the student had a grade in chemistry that was 1.75 standard deviations above the mean of the chemistry grades, whereas in economics she was only 1.50 standard deviations above the mean of the economics grades. Comparing these two *z* scores, we can now say that the student's relative performance in chemistry was higher than her performance in economics.

Example 14. Different typing skills are required for secretaries depending on whether one is working in a law office, an accounting firm, or for a research mathematical group at a major university. In order to evaluate candidates for these positions, an employment agency administers three distinct standardized typing samples. A time penalty has been incorporated into the scoring of each sample based on the number of typing errors. The mean and standard deviation for each test, together with the score achieved by a recent applicant, are given in Table 2.1.

TABLE 2.1
Data for Standardized Typing Samples

Sample	Applicant's Score	Mean	Standard Deviation
Law	141 sec	180 sec	30 sec
Accounting	7 min	10 min	2 min
Scientific	33 min	26 min	5 min

For what type of position does this applicant seem to be best suited?

Solution. First we compute the z score for each sample.

$$\text{Law:} \qquad z = \frac{141 - 180}{30} = -1.3.$$

$$\text{Accounting: } z = \frac{7 - 10}{2} = -1.5.$$

$$\text{Scientific:} \quad z = \frac{33 - 26}{5} = 1.4.$$

Since speed is of primary importance, we are looking for the z score that represents the greatest number of standard deviations to the left of the mean, and in our case that would be -1.5. Therefore, this particular applicant ranks higher among typists in accounting firms than when compared to typists in the other two areas, and consequently should be placed with an accounting firm.

EXERCISES

1. Given the random sample $x_1, x_2, \ldots, x_n$, show that $\sum_{i=1}^{n} (x_i - \bar{x}) = 0$.

2. With reference to the population of building permits issued to the 12 construction firms in Exercise 2 on page 27, calculate
   (a) the range;
   (b) the variance.

3. With reference to the sample of reaction times for the 9 subjects receiving the stimulant in Exercise 3 on page 27, calculate
   (a) the range;
   (b) the variance using the definition on page 35.

4. The number of goals scored by a college lacrosse team for a given season are 4, 9, 0, 1, 3, 24, 12, 3, 30, 12, 7, 13, 18, 4, 5, and 15. Treating the data as a population, calculate the standard deviation.

5. With reference to the random sample of incorrect answers on a true–false competency test for the 15 students in Exercise 1 on page 27, calculate the variance using
   (a) the formula defining s^2 on page 35;
   (b) the computational formula for s^2 of Theorem 2.1.

6. The grade-point averages of 20 college seniors selected at random from the graduating class are as follows:

$$
\begin{array}{cccc}
3.2 & 1.9 & 2.7 & 2.4 \\
2.8 & 2.9 & 3.8 & 3.0 \\
2.5 & 3.3 & 1.8 & 2.5 \\
3.7 & 2.8 & 2.0 & 3.2 \\
2.3 & 2.1 & 2.5 & 1.9
\end{array}
$$

   Calculate the standard deviation.

7. Verify that the variance of the sample 4, 9, 3, 6, 4, and 7 is 5.1, and using this fact find
   (a) the variance of the sample 12, 27, 9, 18, 12, and 21;
   (b) the variance of the sample 9, 14, 8, 11, 9, and 12.

8. Find the variance for the population consisting of the measurements $\frac{1}{2}$, $\frac{1}{4}$, $\frac{1}{3}$, $\frac{1}{2}$, and $\frac{1}{6}$ by using the coded data 6, 3, 4, 6, and 2.

9. Faculty salaries for a random sample of teachers in the public school system of a certain town were coded by dividing each salary by 1000. Find the variance of these salaries if the coded observations are 18, 15, 21, 19, 13, 15, 14, 23, 18, and 16 dollars.

10. A taxi company tested a random sample of 10 steel-belted radial tires of a certain brand and recorded the following tread wear: 48,000, 53,000, 45,000, 61,000, 59,000, 56,000, 63,000, 49,000, 53,000, and 54,000 kilometers. Find the standard deviation of this set of data by first dividing each observation by 1000 and then subtracting 55.

11. **Mean deviation:** The *mean deviation* of a sample of n observations is defined to be $\sum_{i=1}^{n} |x_i - \bar{x}|/n$. Find the mean deviation of the sample 2, 3, 5, 7, and 8.

12. If the distribution of IQs of college students in Arizona has a mean $\mu = 123$ and a standard deviation $\sigma = 9$, use Chebyshev's theorem to determine the interval containing
    (a) at least $\frac{3}{4}$ of the IQs;
    (b) no more than $\frac{1}{9}$ of the IQs.

13. A coffee-maker is regulated so that it takes an average of 5.8 minutes to brew a cup of coffee with a standard deviation of 0.6 minute. According to Chebyshev's theorem, what percentage of the times that this coffee-maker is used will the brewing time take anywhere from
    (a) 4.6 minutes to 7.0 minutes?
    (b) 3.4 minutes to 8.2 minutes?
    (c) 4.3 minutes to 7.3 minutes?

14. A study of the nicotine contents of a certain brand of cigarette shows that on the average one cigarette contains 1.52 milligrams of nicotine with a standard deviation of 0.07 milligram. According to Chebyshev's theorem, between what values must the nicotine content be for
    (a) at least $\frac{24}{25}$ of all cigarettes of this brand?
    (b) at least $\frac{48}{49}$ of all cigarettes of this brand?

15. In Exercise 6 calculate the percentage of grade-point averages falling in the intervals $\bar{x} \pm 2s$ and $\bar{x} \pm 3s$. Do these results agree with Chebyshev's theorem?

16. An automobile salesman made a profit of $245 on a subcompact model for which the average profit has been $200 with a standard deviation of $50. Later on the same day he made a profit of $620 on a large luxury model for which the average profit has been $500 with a standard deviation of $150. For which of these two models is the salesman's profit relatively higher?

17. The minimum temperatures for three widely separated locations across the United States on January 1, 1980, were recorded as 15°F at location A, 62°F at location B, and 3°C at location C. A check with the weather stations at these three locations produced the following data for these locations over the last 20 years:

	Location		
	A	B	C
Mean temperature	20°F	71°F	5°C
Standard deviation	4.5°F	6.9°F	2.3°C

Relatively speaking, which location on January 1, 1980, experienced the coolest day?

18. Jane works for a New York theatrical agency whose employees had an average income this past year of $28,000 with a standard deviation of $3000. How much did Jane earn this past year if her z score is -0.8?

19. **Coefficient of variation:** The standard deviation does not by itself tell us much about the variability of a single set of data. Perhaps a more appropriate measure is the *coefficient of variation,* defined by

$$V = \frac{s}{\bar{x}} \cdot 100\% \qquad \text{or} \qquad V = \frac{\sigma}{\mu} \cdot 100\%,$$

which expresses the standard deviation as a percentage of the mean. Since V is a measure of *relative variation* expressed as a percent, the coefficient of variation can be used to compare the variability of two or more sets of data even when the observations are expressed in different units of measurement.

(a) Compute the coefficient of variation for the distribution of goals scored by the lacrosse team in Exercise 4.

(b) The weights of 10 boxes of a certain brand of cereal have a mean content of 278 grams with a standard deviation of 9.64 grams. If these boxes were purchased at 10 different stores and the average price per box is $1.29 with a standard deviation of $0.09, can you conclude that the weights are relatively more homogeneous than the prices?

3 Statistical Description of Data

Often, we are confronted with the problem of disseminating large masses of statistical data in compact form. Although numerical measures of location and variation are certainly useful compact descriptions of a set of observations, they do not by themselves identify all the important features of the data. Considerable information can be retrieved from large masses of data when they are summarized and displayed by means of appropriate tables, charts, and graphs.

Frequency Distributions

Important characteristics of a large mass of data can be readily assessed by grouping the data into different classes and then determining the number of observations that fall in each of the classes. Such an arrangement, in tabular form, is called a **frequency distribution.**

Data that are presented in the form of a frequency distribution are called **grouped data.** We often group the data of a sample into intervals to produce a better overall picture of the unknown population, but in so doing we lose the identity of the individual observations in the sample.

TABLE 3.1
Frequency Distribution for
the Weights of 50 Pieces
of Luggage

Weight (kilograms)	Number of Pieces
7– 9	2
10–12	8
13–15	14
16–18	19
19–21	7

Table 3.1 is a frequency distribution of the weights of 50 pieces of luggage, recorded to the nearest kilogram, belonging to the passengers on a commercial flight from Denver to Chicago. For these data we have used the 5 class intervals 7–9, 10–12, 13–15, 16–18, and 19–21. The smallest and largest values that can fall in a given class interval are referred to as its **class limits.** For the interval 10–12, the smaller number, 10, is the **lower class limit,** and the larger number, 12, is the **upper class limit.** The original data were recorded to the nearest kilogram, so the 8 observations in the interval 10–12 are the weights of all the pieces of luggage weighing more than 9.5 kilograms but less than 12.5 kilograms. The numbers 9.5 and 12.5 are called the **class boundaries** for the given interval. For the interval 10–12, the number 9.5 is called the **lower class boundary** and 12.5 is called the **upper class boundary.** However, 12.5 would also be the lower class boundary for the interval 13–15. Class boundaries are always carried out to one more decimal place than the recorded observations. This ensures that no observation can fall precisely on a class boundary and thereby avoids any confusion as to which class interval the observation belongs. The number of observations falling in a particular class is called the **class frequency** and is denoted by the letter f.

The numerical difference between the upper and lower class boundaries of a class interval is defined to be the **class width.** In practice it is desirable to have equal class widths whenever possible. We shall denote this common width by c. Hence, in Table 3.1, we have $c = 3$ kilograms.

The midpoint between the upper and lower class boundaries or class limits of a class interval is called the **class mark** or **class midpoint.** For Table 3.2, the class marks are 8, 11, 14, 17, and 20.

TABLE 3.2
Frequency Distribution for the Weights of
50 Pieces of Luggage

Class Interval	Class Boundaries	Class Mark, x	Frequency, f
7– 9	6.5– 9.5	8	2
10–12	9.5–12.5	11	8
13–15	12.5–15.5	14	14
16–18	15.5–18.5	17	19
19–21	18.5–21.5	20	7

Table 3.1 is the type of frequency distribution that we might see in a published report or newspaper. For statistical purposes it is usually more advantageous to give a more detailed distribution, as demonstrated in Table 3.2 for the same data. Table 3.2 contains information that we shall find useful in calculating other descriptive properties of our data.

To illustrate the construction of a frequency distribution, consider the data of Table 3.3, which represents the lives of 40 similar car batteries recorded to the nearest tenth of a year. The batteries were guaranteed to last 3 years.

TABLE 3.3
Car Battery Lives

2.2	4.1	3.5	4.5	3.2	3.7	3.0	2.6
3.4	1.6	3.1	3.3	3.8	3.1	4.7	3.7
2.5	4.3	3.4	3.6	2.9	3.3	3.9	3.1
3.3	3.1	3.7	4.4	3.2	4.1	1.9	3.4
4.7	3.8	3.2	2.6	3.9	3.0	4.2	3.5

We must first decide on the number of classes into which the data are to be grouped. This is done somewhat arbitrarily, although we are guided by the size of our sample. Certainly, we want fewer classes than observations, or we gain nothing by grouping. Since we lose the identity of the individual observations by grouping, too few classes would be giving away or destroying too much information. Usually, we choose between 5 and 20 class intervals. The smaller the number of data available, the smaller is our choice for the number of classes. For the data of Table 3.3, let us choose 7 class intervals. The class width must be large enough so that the 7 class intervals accommodate all the data. To determine the approximate class width, we divide the range by the number of intervals. Therefore, in our illustration the range is $4.7 - 1.6 = 3.1$, and the class width can be no less than $3.1/7 = 0.443$. Since the class width should have the same number of significant places as the observations, we choose $c = 0.5$.

We must now decide where to start the bottom interval. If we begin the lowest class interval at 1.5, the lower class boundary for this interval will be 1.45. To this we add the class width, 0.5, and find the upper class boundary to be 1.95. Therefore, the upper class limit for the first interval must be 1.9. Note that we have first found the lower class limit for the first interval and then, before writing down the upper class limit, we determined the class boundaries. Proceeding in this order for the first class interval avoids frequently made errors. The midpoint or mark for this class is the average of the upper and lower class limits, $(1.5 + 1.9)/2 = 1.7$.

The remaining intervals and class boundaries are now obtained by adding the class width 0.5 to each class limit and class boundary until we reach the seventh interval, which contains the highest observation in the set of data. Finally, we count the number of observations falling in each class and record the appropriate number in the frequency column. As a check, total the frequency column to see that all observations have been accounted for. The frequency distribution for the data of Table 3.3 is given in Table 3.4.

TABLE 3.4
Frequency Distribution of Battery Lives

Class Interval	Class Boundaries	Class Midpoint	Frequency, f
1.5–1.9	1.45–1.95	1.7	2
2.0–2.4	1.95–2.45	2.2	1
2.5–2.9	2.45–2.95	2.7	4
3.0–3.4	2.95–3.45	3.2	15
3.5–3.9	3.45–3.95	3.7	10
4.0–4.4	3.95–4.45	4.2	5
4.5–4.9	4.45–4.95	4.7	3

The steps in grouping a large set of data into a frequency distribution may be summarized as follows:

1. Decide on the number of class intervals required.
2. Determine the range.
3. Divide the range by the number of classes to estimate the approximate width of the interval.
4. List the lower class limit of the bottom interval and then the lower class boundary. Add the class width to the lower class boundary to obtain the upper class boundary. Write down the upper class limit.
5. List all the class limits and class boundaries by adding the class width to the limits and boundaries of the previous interval.
6. Determine the class marks of each interval by averaging the class limits or the class boundaries.
7. Tally the frequencies for each class.
8. Sum the frequency column and check against the total number of observations.

Variations of Table 3.4 are obtained by listing the **relative frequencies** or **percentages** for each interval. The relative frequency of each class can be obtained by dividing the class frequency by the total frequency. A table listing relative frequencies is called a **relative frequency distribution.** If each relative frequency is multiplied by 100%, we have a **percentage distribution.** The relative frequency distribution for the data in Table 3.2 is given in Table 3.5.

TABLE 3.5
Relative Frequency Distribution for 50 Pieces of Luggage

Class Interval	Class Boundaries	Class Mark, x	Relative Frequency
7– 9	6.5– 9.5	8	0.04
10–12	9.5–12.5	11	0.16
13–15	12.5–15.5	14	0.28
16–18	15.5–18.5	17	0.38
19–21	18.5–21.5	20	0.14

In many situations we are concerned not with the number of observations in a given class but in the number that fall above or below a specified value. For example, in Table 3.4 the number of batteries lasting less than 3 years is

7. The total frequency of all values less than the upper class boundary of a given class interval is called the **cumulative frequency** up to and including that class. A table such as Table 3.6, showing the cumulative frequencies, is called a **cumulative frequency distribution.**

TABLE 3.6
Cumulative Frequency Distribution
of Battery Lives

Class Boundaries	Cumulative Frequency
Less than 1.45	0
Less than 1.95	2
Less than 2.45	3
Less than 2.95	7
Less than 3.45	22
Less than 3.95	32
Less than 4.45	37
Less than 4.95	40

Two additional forms of Table 3.6 are possible using relative frequencies and percentages. Such distributions are called **relative cumulative frequency distributions** and **percentage cumulative distributions.** The percentage cumulative distribution enables one to read off the percentage of observations falling below certain specified values. For example, in Table 3.7 we can see that 80% of the batteries last less than 3.95 years.

TABLE 3.7
Percentage Cumulative Distribution
of Battery Lives

Class Boundaries	Cumulative Percent
Less than 1.45	0
Less than 1.95	5.0
Less than 2.45	7.5
Less than 2.95	17.5
Less than 3.45	55.0
Less than 3.95	80.0
Less than 4.45	92.5
Less than 4.95	100.0

The examples that we have discussed so far deal strictly with **numerical** or **quantitative data.** Of course, frequency tables can also be constructed for **categorical** or **qualitative data.** We illustrate this in Table 3.8 for the number of automobiles belonging to a rental agency at a large metropolitan airport.

TABLE 3.8
Automobiles Purchased by Rental
Agency

Model	Number
Buick Regal	12
Chevrolet Monte Carlo	15
Oldsmobile Cutlass	15
Pontiac Grand Prix	9
Ford Thunderbird	5
Ford Fairmont	10
Pontiac Phoenix	12
Chevrolet Citation	20
Buick Skylark	13
Toyota Corolla	5
Honda Accord	6
Cadillac Seville	2
Lincoln Versailles	3

As before, we must decide how many categories or classes to use. These categories must be chosen so as to accommodate all the data and so that no item is placed under more than one category. The concepts of class limits, class boundaries, and class marks are of no concern when constructing frequency distributions using categorical data.

3.2
Graphical
Representations

The information provided by a frequency distribution in tabular form is easier to grasp if presented graphically. Most people find a visual picture beneficial in comprehending the essential features of a frequency distribution. A widely used form of graphic presentation of numerical data is the **bar chart** shown in Figure 3.1 for the data of Table 3.4.

One can quickly observe from the bar chart that most of the batteries lasted from 3.0 to 3.4 years, only a very few batteries lasted less than 2.5 years and no battery lasted longer than 4.9 years. In a bar chart the base of each bar corresponds to a class interval of the frequency distribution and the heights of the bars represent the frequencies associated with each class.

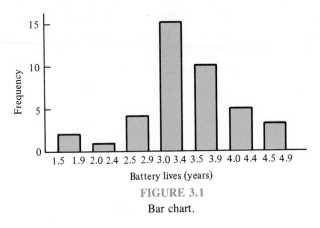

FIGURE 3.1

Bar chart.

Although the bar chart provides immediate information about a set of data in a condensed form, we are usually more interested in a related pictorial representation called a **histogram.** A histogram differs from a bar chart in that the bases of each bar are the class boundaries rather than the class limits. The use of class boundaries for the bases eliminates the spaces between the bars to give the solid appearance of Figure 3.2.

For some problems it will be more convenient to let the vertical axis represent relative frequencies or percentages. The graphs, called **relative frequency histograms** or **percentage histograms,** have exactly the same shape as the frequency histogram but a different vertical scale.

In viewing a histogram, the eye tends to compare the areas of the different rectangles rather than their heights. Although this is appropriate for class intervals of equal width, it can be very misleading if some of the class widths differ. Unscrupulous individuals have been known to deliberately misrepresent

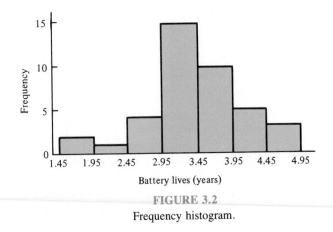

FIGURE 3.2

Frequency histogram.

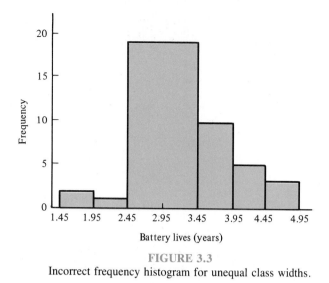

FIGURE 3.3
Incorrect frequency histogram for unequal class widths.

data by erroneously constructing histograms with unequal class widths. Suppose, for example, that we combine the two class intervals 2.5–2.9 and 3.0–3.4 of Table 3.4 into the single interval 2.5–3.4 containing the 19 observations of the combined frequencies 4 and 15.

In Figure 3.3, we get the mistaken impression that well over half of the observations fall in the longer class interval 2.5–3.4 when the actual number is just one less than half. To correct for this misconception, we must reduce the height of this new rectangle by the inverse of the factor that extends the class interval. Since we doubled the class width by combining the two intervals, we must therefore divide the height of this new rectangle by 2 to give the correct visual picture as shown in Figure 3.4. Of course, now that areas and not heights represent the frequencies, we have no further need for the vertical axis and it is therefore omitted.

A second useful way of presenting numerical data in graphic form is by

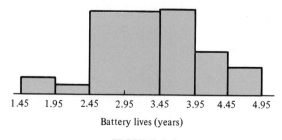

FIGURE 3.4
Correct frequency distribution for unequal class widths.

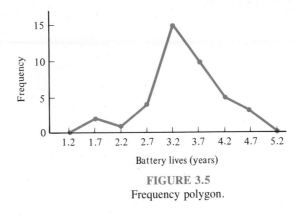

FIGURE 3.5
Frequency polygon.

means of a **frequency polygon.** Frequency polygons are constructed by plotting class frequencies against class marks and connecting the consecutive points by straight lines.

A polygon is a many-sided *closed* figure. To close the frequency polygon, an additional class interval is added to both ends of the distribution, each with zero frequency. For our example, the midpoints of these two additional classes will be 1.2 and 5.2. These two points enable us to connect both ends to the horizontal axis, resulting in a polygon. The frequency polygon for the data of Table 3.4 is shown in Figure 3.5. We can obtain the frequency polygon very quickly from the histogram by joining the midpoints of the tops of adjacent rectangles and then adding the two intervals at each end.

If we wish to compare two sets of data with unequal sample sizes by constructing two frequency polygons on the same graph, we must use relative frequencies or percentages. A graph similar to Figure 3.3, but using relative frequencies or percentages, is called a **relative frequency polygon** or a **percentage polygon.**

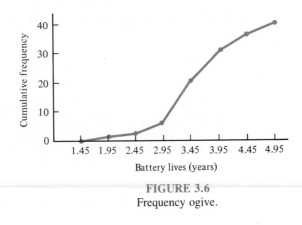

FIGURE 3.6
Frequency ogive.

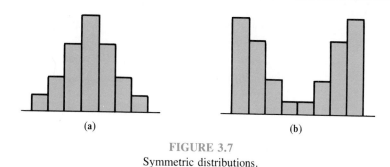

FIGURE 3.7
Symmetric distributions.

A second line graph, called a **cumulative frequency polygon,** or **ogive,** is obtained by plotting the cumulative frequency less than any upper class boundary against the upper class boundary and joining all the consecutive points by straight lines. The cumulative frequency polygon for the data of Table 3.6 is shown in Figure 3.6. If relative cumulative frequencies or percentages had been used, we would call the graph a **relative frequency ogive** or a **percentage ogive.**

3.3

Symmetry and Skewness

The shape or distribution of a set of measurements is best displayed by means of a histogram. Some of the many possible shapes that might arise are illustrated in Figures 3.7 and 3.8. A distribution is said to be **symmetric** if it can be folded along a vertical axis so that the two sides coincide. We see that the distributions in Figure 3.7 are indeed symmetric, although quite different in appearance. A distribution that lacks symmetry with respect to a vertical axis is said to be **skewed.**

The distribution illustrated in Figure 3.8*a* is said to be skewed to the right, or **positively skewed,** since it has a long right tail compared to a much

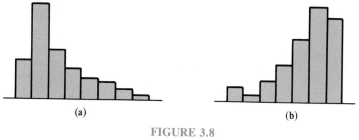

FIGURE 3.8
Skewed distributions.

shorter left tail. In Figure 3.8*b* the distribution is skewed to the left, or **negatively skewed.**

For a symmetric distribution of measurements, the mean and median are both located at the same position along the horizontal axis. However, if the data are skewed to the right as in Figure 3.8*a*, the large values in the right tail are not offset by correspondingly low values in the left tail and consequently the mean will be greater than the median. In Figure 3.8*b* the reverse is true, and the small values in the left tail will make the mean less than the median. We shall use this behavior between the mean and median relative to the standard deviation to define a numerical measure of skewness.

DEFINITION

Pearsonian Coefficient of Skewness. The *Pearsonian coefficient of skewness* is given by

$$ SK = \frac{3(\bar{x} - \tilde{x})}{s} \quad \text{or} \quad SK = \frac{3(\mu - \tilde{\mu})}{\sigma}. $$

For a perfectly symmetrical distribution the mean and median are identical and the value of SK is zero. When the distribution is skewed to the left, the mean is less than the median and the value of SK will be negative. However, if the distribution is skewed to the right, the mean is greater than the median and the value of SK will be positive. In general, the values of SK will fall between -3 and 3.

Example 1. Compute the Pearsonian coefficient of skewness for the distribution of battery lives in Table 3.4.

Solution. Assuming the data of Table 3.4 to be a sample, we find that $\bar{x} = 3.41$, $\tilde{x} = 3.4$, and $s = 0.70$. Therefore,

$$ SK = \frac{3(3.41 - 3.4)}{0.70} = 0.04, $$

indicating only a very slight amount of skewness to the right. With such a small value of SK, we could essentially say that the distribution is symmetrical.

Although histograms assume a wide variety of shapes, fortunately most distributions that we meet in practice can be represented approximately by

bell-shaped histograms similar to Figure 3.7*a* for which the Pearsonian coefficient of skewness will be very close to zero. This was certainly true for the distribution of battery lives depicted in Figure 3.2. In fact, it will be true for any set of data where the frequency of the observations falling in the various classes decreases at roughly the same rate as we get farther out in the tails of the distribution. For example, the distribution of all the grades recorded for the first semester at the local community college should have this bell-shaped histogram, since we would expect a majority of C's, somewhat fewer D's and B's, and still fewer F's and A's. It would seem reasonable to make the same assumption for the distribution of lives of a certain type of fuel valve for an automobile. If the value lasts 30 months on the average, we would not be surprised if some had to be replaced a little before 30 months and some a little beyond that time, whereas only a very few are likely to fail during the first 6 months or to last as long as 5 years.

These bell-shaped distributions play a major role in the field of statistical inference. Some are understandably more variable than others, as reflected by a flatter and wider histogram. In Figure 3.9*a* most of the observations are close to the mean, with very few falling in the extreme tails of the distribution. However, in Figure 3.9*b*, the measurements still cluster about the mean, but a greater number are recorded farther from the mean in both directions. The standard deviation is larger for the data represented by Figure 3.9*b* than for the data of Figure 3.9*a*, since large deviations of the observations from the mean occur more frequently and small deviations less frequently.

Chebyshev's theorem tells us that at least $\frac{3}{4}$ or $\frac{8}{9}$ of the observations of any distribution, bell-shaped or not, will be within 2 or 3 standard deviations of the mean, respectively. If the distribution happens to be somewhat bell-shaped, we can state a rule that gives even stronger results.

DEFINITION

_____

Empirical Rule. Given a bell-shaped distribution of measurements, then approximately

68% of the observations lie within 1 standard deviation of the mean.
95% of the observations lie within 2 standard deviations of the mean.
99.7% of the observations lie within 3 standard deviations of the mean.

_____

From Example 1, we find that $\bar{x} = 3.41$ and $s = 0.70$ for the data of Table 3.4. Now the Empirical Rule states that approximately 68% or 27 of the 40 observations should be contained in the interval $\bar{x} \pm s = 3.41 \pm 0.70$, or from 2.71 to 4.11. An actual count shows that 28 of the 40 observations

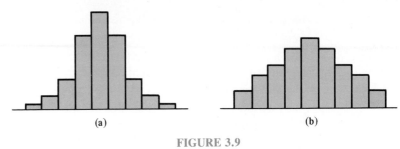

(a)   (b)

FIGURE 3.9

Variability of observations about the mean.

fall in the given interval. Similarly, 95% or 38 of the 40 observations should fall in the interval $\bar{x} \pm 2s = 3.41 \pm (2)(0.70)$, or from 2.01 to 4.81. The actual count this time shows that exactly 38 of the 40 observations fall in the specified interval. The interval $\bar{x} \pm 3s = 3.41 \pm (3)(0.70)$, or from 1.31 to 5.51, contains all the measurements. By Chebyshev's theorem we could only have concluded that at least 30 observations will fall in the interval from 2.01 to 4.81 and at least 36 observations will fall between 1.31 and 5.51.

Example 2. A company pays its 1000 employees an average hourly wage of $6.25 with a standard deviation of 60 cents. Assuming the population of wages to be approximately bell-shaped, use the Empirical Rule to describe the variability of the data.

Solution. To describe the data by the Empirical Rule, we first compute the intervals

$$\mu \pm \sigma = \$6.25 \pm \$0.60, \text{ or } \$5.65 \text{ to } \$6.85;$$
$$\mu \pm 2\sigma = \$6.25 \pm (2)(\$0.60), \text{ or } \$5.05 \text{ to } \$7.45;$$
$$\mu \pm 3\sigma = \$6.25 \pm (3)(\$0.60), \text{ or } \$4.05 \text{ to } \$8.05.$$

We may conclude that approximately 680 employees earn from $5.65 to $6.85 an hour, 950 earn from $5.05 to $7.45 an hour, and 997 earn from $4.05 to $8.05 an hour. We may also conclude from these figures additional information concerning the number of employees in various nonoverlapping intervals. For example, if 950 employees earn from $5.05 to $7.45 an hour and 680 earn from $5.65 to $6.85 an hour, then $950 - 680 = 270$ employees fall in the intervals $5.05 to $5.65 and $6.85 to $7.45 per hour. The assumption of a bell-shaped distribution allows us to conclude, then, that these 270 employees are divided approximately in half between these two intervals. That is, 135 earn from $5.05 to $5.65 an hour and another 135 earn from $6.85 to $7.45 an hour. Similar reasoning allows us to conclude that only 1 or 2 employees earn more than $8.05 an hour.

3.4

Percentiles, Deciles, and Quartiles

In Section 2.2 we discussed only measures of *central* location. There are several other measures of location that describe or locate the position of certain noncentral pieces of data relative to the entire set of data. These measures, often referred to as **fractiles** or **quantiles,** are values below which a specific fraction or percentage of the observations in a given set must fall. Of special interest are those fractiles commonly referred to as **percentiles, deciles,** and **quartiles.** Let us first define a percentile.

DEFINITION

_____

Percentile. Percentiles are values that divide a set of observations into 100 equal parts. These values, denoted by P_1, P_2, ..., P_{99}, are such that 1% of the data falls below P_1, 2% falls below P_2, ..., and 99% falls below P_{99}.

_____

To illustrate the procedure in calculating a percentile, let us find P_{85} for the distribution of battery lives in Table 3.3. First we must rank the given data in increasing order of magnitude as displayed in Table 3.9. Since the table contains 40 observations, we seek the value below which $(85/100) \times 40 = 34$ observations fall. As seen from Table 3.9, P_{85} could be any value between 4.1 years and 4.2 years. In order to give a unique value, we shall define P_{85} to be the value midway between these two observations. Therefore, $P_{85} = 4.15$ years. This procedure works very well whenever the number of observations below the given percentile is a whole number. However, when the required number of observations is fractional, it is customary to use the next highest whole number to find the required percentile. For example, in finding P_{48} we seek the value below which $(48/100) \times 40 = 19.2$ observations fall. Rounding up to the next integer, we use the 20th observation as our location point. Hence $P_{48} = 3.4$ years.

TABLE 3.9
Car Battery Lives by Rank

1.6	2.6	3.1	3.2	3.4	3.7	3.9	4.3
1.9	2.9	3.1	3.3	3.4	3.7	3.9	4.4
2.2	3.0	3.1	3.3	3.5	3.7	4.1	4.5
2.5	3.0	3.2	3.3	3.5	3.8	4.1	4.7
2.6	3.1	3.2	3.4	3.6	3.8	4.2	4.7

Although one can always determine a percentile from the original data, it may be advantageous and less time-consuming to calculate a percentile directly from the frequency distribution. In grouping the data, we have chosen to ignore the identity of the individual observations. The only information that remains, assuming the original raw data have been discarded, is the number of observations falling in each class interval. To evaluate a percentile from a frequency distribution, we assume the measurements within a given class interval to be uniformly distributed between the lower and upper class boundaries. This is equivalent to interpreting a percentile as a value below which a specific fraction or percentage of the *area* of a histogram falls. To illustrate the calculation of a percentile from a frequency distribution, we consider the following example.

Example 3. Find P_{48} for the distribution of battery lives in Table 3.4.

Solution. We are seeking the value below which $(48/100) \times 40 = 19.2$ of the observations fall. The fact that the observations are assumed uniformly distributed over the class interval permits us to use fractional observations, as is the case here. There are 7 observations falling below the class boundary 2.95. We still need 12.2 of the next 15 observations falling between 2.95 and 3.45. Therefore, we must go a distance $(12/15) \times 0.5 = 0.41$ beyond 2.95. Hence

$$P_{48} = 2.95 + 0.41$$
$$= 3.36 \text{ years,}$$

compared with 3.4 years obtained above from the ungrouped data. Therefore, we conclude that 48% of all batteries of this type will last less than 3.36 years.

DEFINITION

Decile. Deciles are values that divide a set of observations into 10 equal parts. These values, denoted by $D_1, D_2, ..., D_9$, are such that 10% of the data falls below D_1, 20% falls below D_2, ..., and 90% falls below D_9.

Deciles are found in exactly the same way that we found percentiles. To find D_7 for the distribution of battery lives, we need the value below which $(70/100) \times 40 = 28$ of the observations in Table 3.9 fall. Since this can be any value between 3.7 years and 3.8 years, we take their average and hence $D_7 = 3.75$ years. Therefore, we conclude that 70% of all batteries of this type will last less than 3.75 years.

Example 4. Use the frequency distribution of Table 3.4 to find D_7 for the distribution of battery lives.

Solution. We need the value below which $(70/100) \times 40 = 28$ observations fall. There are 22 observations falling below 3.45. We still need 6 of the next 10 observations and therefore must go a distance $(6/10) \times 0.5 = 0.3$ beyond 3.45. Hence

$$D_7 = 3.45 + 0.3$$
$$= 3.75 \text{ years,}$$

which is identical to the value obtained by using the ungrouped data.

DEFINITION

_____

Quartile. Quartiles are values that divide a set of observations into 4 equal parts. These values, denoted by Q_1, Q_2, and Q_3, are such that 25% of the data falls below Q_1, 50% falls below Q_2, and 75% falls below Q_3.

_____

To find Q_1 for the distribution of battery lives, we need the value below which $(25/100) \times 40 = 10$ of the observations in Table 3.9 fall. Since the 10th and 11th measurements are both equal to 3.1 years, their average will also be 3.1 years, and hence $Q_1 = 3.1$ years.

Example 5. Use the frequency distribution of Table 3.2 to find Q_3 for the distribution of weights of 50 pieces of luggage.

Solution. We need the value below which $(75/100) \times 50 = 37.5$ observations fall. There are 24 observations falling below 15.5 kilograms. We still need 13.5 of the next 19 observations and therefore must go a distance $(13.5/19) \times 3 = 2.1$ beyond 15.5. Hence

$$Q_3 = 15.5 + 2.1$$
$$= 17.6 \text{ kilograms.}$$

Therefore, we conclude that 75% of all 50 pieces of luggage weigh less than 17.6 kilograms.

The 50th percentile, fifth decile, and second quartile of a distribution are all equal to the same value, commonly referred to as the *median*. All the quartiles and deciles are percentiles. For example, the seventh decile is the 70th percentile and the first quartile is the 25th percentile. Any percentile, decile, or quartile can also be estimated from a percentage ogive.

1. Find the class boundaries, class marks, and class widths for the following intervals:
   (a) 7 − 13;   (b) (−5) − (−1);   (c) 10.4 − 18.7;
   (d) 0.346 − 0.418;   (e) (−2.75) − 1.35;   (f) 78.49 − 86.72.

2. If the class marks of a frequency distribution of weights of miniature poodles are
   6.5, 8.5, 10.5, 12.5, and 14.5 kilograms, find
   (a) the class width;
   (b) the class boundaries;
   (c) the class limits.

3. Find the appropriate class width for each of the following distributions with 10
   intervals:

	Low Score	High Score
(a)	7.5	18.65
(b)	53	149
(c)	0.392	0.514
(d)	−15	0

4. The following scores represent the final examination grade for an elementary
   statistics course:

23	60	79	32	57	74	52	70	82	36
80	77	81	95	41	65	92	85	55	76
52	10	64	75	78	25	80	98	81	67
41	71	83	54	64	72	88	62	74	43
60	78	89	76	84	48	84	90	15	79
34	67	17	82	69	74	63	80	85	61

   Using 9 intervals with the lowest starting at 10,
   (a) set up a frequency distribution;
   (b) construct a cumulative frequency distribution.

5. The following data represent the length of life in minutes, measured to the nearest
   tenth, of a random sample of 50 black flies subjected to a new spray in a controlled
   laboratory experiment:

2.4	0.7	3.9	2.8	1.3
1.6	2.9	2.6	3.7	2.1
3.2	3.5	1.8	3.1	0.3
4.6	0.9	3.4	2.3	2.5
0.4	2.1	2.3	1.5	4.3
1.8	2.4	1.3	2.6	1.8
2.7	0.4	2.8	3.5	1.4
1.7	3.9	1.1	5.9	2.0
5.3	6.3	0.2	2.0	1.9
1.2	2.5	2.1	1.2	1.7

Using 8 intervals with the lowest starting at 0.1,
(a) set up a percentage distribution;
(b) construct a percentage cumulative distribution.

6. The following data represent the charges billed by a plumber for his last 30 home service calls:

$39.12	$61.74	$37.29	$44.35	$57.29
64.10	48.25	67.25	58.95	39.95
38.42	55.80	44.35	38.75	63.91
51.50	40.15	60.29	41.26	49.32
36.07	46.01	41.13	67.29	45.68
63.55	62.12	36.85	45.97	42.89

Using 6 intervals with the lowest starting at $36.05,
(a) set up a relative frequency distribution;
(b) construct a cumulative relative frequency distribution.

7. Asked to categorize new automobiles as being subcompact, compact, intermediate, large, or luxury models, a large distributor reported the following data for the last 30 sales: subcompact, subcompact, intermediate, large, subcompact, compact, luxury, intermediate, subcompact, subcompact, subcompact, intermediate, compact, subcompact, intermediate, large, subcompact, subcompact, compact, luxury, intermediate, subcompact, compact, compact, luxury, subcompact, intermediate, intermediate, subcompact, and intermediate. Construct a frequency distribution for these categorical data.

8. The American Physics Association reported the following data for graduating physics majors by geographic region in 1979:

Geographic Region	Number of Seniors
New England	524
Middle Atlantic	818
E.N. Central	815
W.N. Central	367
S. Atlantic	679
E.S. Central	196
W.S. Central	436
Mountain	346
Pacific	783

Illustrate these categorical data by means of a bar chart. (Of course, the width of each bar is arbitrary for categorical data.)

9. For the grouped data of Exercise 4,
   (a) construct a frequency histogram;
   (b) construct a frequency polygon;
   (c) construct a frequency ogive.

10. For the grouped data of Exercise 5,
    (a) construct a percentage histogram;
    (b) construct a percentage polygon;
    (c) construct a percentage ogive.

11. For the grouped data of Exercise 6,
    (a) construct a relative frequency histogram;
    (b) construct a relative frequency polygon;
    (c) construct a relative frequency ogive.

12. Treating the distribution of statistics grades in Exercise 4 as a population, find
    (a) the mean;
    (b) the median;
    (c) the standard deviation;
    (d) the Pearsonian coefficient of skewness.

13. For the distribution of lives of black flies subjected to the new spray in Exercise 5, find
    (a) the mean;
    (b) the median;
    (c) the standard deviation;
    (d) the Pearsonian coefficient of skewness.

14. Treating the distribution of charges billed by the plumber in Exercise 6 as a sample, find
    (a) the mean;
    (b) the median;
    (c) the standard deviation;
    (d) the Pearsonian coefficient of skewness.
    Explain why the Empirical Rule does not apply to this set of data.

15. Mathematics achievement test scores for 1000 students were found to have a mean and standard deviation of 500 and 100, respectively. If the distribution of scores is bell-shaped, approximately how many of the scores
    (a) fall in the interval 400–600?
    (b) exceed 700?
    (c) fall below 200?

16. A pharmaceutical store delivers an average of 40 prescription-filled orders a day, with a standard deviation of 8 orders. Assuming a bell-shaped population for the daily number of prescriptions delivered, what percentage of the time will the driver make
    (a) between 24 and 56 deliveries?
    (b) fewer than 32 deliveries?

17. In preparation for a spring clearance sale at a women's apparel shop, 2000 brochures were placed in preaddressed envelopes, sealed, and stamped. This process took an average of 7 seconds per envelope with a standard deviation of 1.8 seconds. Assuming this population of times to be approximately bell-shaped, use the Empirical Rule to find the number of envelopes that took
    (a) from 5.2 to 8.8 seconds to prepare;
    (b) from 3.4 to 10.6 seconds to prepare;
    (c) from 1.6 to 12.4 seconds to prepare.

18. Using the results from Exercise 12(a) and (c), find the number of grades in the intervals $\mu \pm \sigma$ and $\mu \pm 2\sigma$ and compare with the expected number given by the Empirical Rule.

19. Using the results from Exercise 13(a) and (c), find the number of lives of the black flies that fall in the intervals $\bar{x} \pm s$ and $\bar{x} \pm 2s$ and compare with the expected number given by the Empirical Rule.

20. **Mean of grouped data:** To evaluate the *mean of grouped data*, we assume that all the observations within a given class interval are equal to the class midpoint or class mark. Since some of the observations in the interval are likely to fall below the class mark and others above it, this would seem to be a reasonable assumption. Now, if $x_1, x_2, \ldots, x_k$ are the class marks of a set of grouped data with corresponding class frequencies $f_1, f_2, \ldots, f_k$, then the mean of our grouped data, μ or $\bar{x}$, is given by

$$\sum_{i=1}^{k} f_i x_i \Big/ \sum_{i=1}^{k} f_i.$$

    (a) Find the mean for the grouped data of Exercise 4(a) and compare your answer with the result obtained in Exercise 12(a).
    (b) Find the mean for the grouped data of Exercise 5(a) and compare your answer with the result obtained in Exercise 13(a).

21. **Sample variance of grouped data:** If $x_1, x_2, \ldots, x_k$ are the class marks of a set of grouped sample data with corresponding class frequencies $f_1, f_2, \ldots, f_k$, then the *sample variance of the grouped data* is

$$s^2 = \frac{n \sum_{i=1}^{k} f_i x_i^2 - \left(\sum_{i=1}^{k} f_i x_i \right)^2}{n(n-1)},$$

    where $n = \sum_{i=1}^{k} f_i$. Find the variance of the grouped data in Table 3.4 on page 50.

22. Find P_{28}, Q_2, and D_6 for the distribution of statistics grades in Exercise 4 by using
    (a) the ungrouped data;
    (b) the grouped data.

23. Find P_{89}, Q_3, and D_4 for the distribution of lives of black flies in Exercise 5 by using
    (a) the ungrouped data;
    (b) the grouped data.

24. Using the ranked data in Table 3.9 on page 61, find P_{23}, Q_3, and D_8 for the distribution of car battery lives.

25. Referring to the frequency distribution of Table 3.2 on page 49 for the weights of 50 pieces of luggage, find P_{56}, Q_1, and D_2.

26. **Mean deviation of grouped data:** If x_1, x_2, ..., x_k are the class marks of a set of grouped sample data with corresponding class frequencies f_1, f_2, ..., f_k, then the *mean deviation of the grouped data is* $\sum_{i=1}^{k} f_i \, | \, x_i - \bar{x} \, | \, / \sum_{i=1}^{k} f_i$. Find the mean deviation for the grouped data of Table 3.4 on page 50.

27. **Interquartile range:** An alternative measure of variation based on quartiles is the difference $Q_3 - Q_1$, called the *interquartile range*. It measures the length of the interval that contains the middle 50% of the data.
    (a) Use the value of Q_1 on page 63 and the value of Q_3 from Exercise 24 to calculate the interquartile range.
    (b) Use the value of Q_1 from Exercise 25 and the value of Q_3 from Example 5 to calculate the interquartile range.

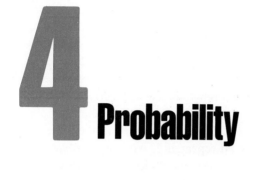

4 Probability

Perhaps it was man's unquenchable thirst for gambling that led to the early development of probability theory. In an effort to increase their winnings, gamblers called upon the mathematicians to provide optimum strategies for various games of chance. Some of the mathematicians providing these strategies were Pascal, Leibniz, Fermat, and James Bernoulli. As a result of this early development of probability theory, statistical inference, with all its predictions and generalizations, has branched out far beyond games of chance to encompass many other fields associated with chance occurrences such as politics, business, weather forecasting, and scientific research. To evaluate probabilities associated with chance outcomes, we have a number of elaborate counting techniques at our disposal. However, because probabilities obey certain mathematical laws, their computation can often be somewhat simplified.

4.1
Sample Space

In statistics we use the word *experiment* to describe any process that generates a set of data. An example of a statistical experiment might be the tossing of a coin. In this experiment there are only two possible outcomes, heads or tails. An experiment was carried out when the lives of 40 car batteries were recorded in Table 3.3 on page 49. Measuring the amount of rainfall at Lake Placid each year for the month of July can also be considered an experiment. We are particularly interested in the observations obtained by repeating the same experiment several times. In most cases the outcomes will depend on chance and, therefore, cannot be predicted with certainty. If a chemist runs an analysis several times under the same conditions, he will obtain different measurements, indicating an element of chance in the experimental procedure. Even when a coin is tossed repeatedly, we cannot be certain that a given toss will result in a head. However, we do know the entire set of possibilities for each toss.

DEFINITION

Sample Space. The set of all possible outcomes of a statistical experiment is called the *sample space* and is represented by the symbol S.

Each outcome in a sample space is called an **element** or a **member** of the sample space or simply a **sample point.** If the sample space has a finite number of elements, we may *list* the members separated by commas and enclosed in brackets. Thus the sample space S, of possible outcomes when a coin is tossed, may be written

$$S = \{H, T\},$$

where H and T correspond to "heads" and "tails," respectively.

Example 1. Consider the experiment of tossing a die. If we are interested in the number that shows on the top face, then the sample space would be

$$S_1 = \{1, 2, 3, 4, 5, 6\}.$$

On the other hand, if we are interested only in whether the number is even or odd, then the sample space is simply

$$S_2 = \{even, odd\}.$$

Example 1 illustrates the fact that more than one sample space can be used to describe the outcomes of an experiment. In this case S_1 provides more information than S_2. If we know which element in S_1 occurs, we can tell which outcome in S_2 occurs; however, a knowledge of what happens in S_2 in no way helps us to know which element in S_1 occurs. In general, it is desirable to use a sample space that gives the most information concerning the outcomes of the experiment.

In some experiments it will be helpful to list the elements of the sample space systematically by means of a **tree diagram.**

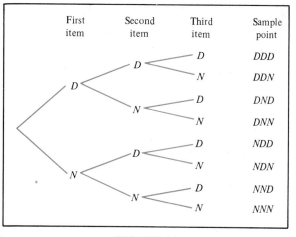

FIGURE 4.1

Tree diagram for Example 2.

Example 2. Suppose that 3 items are selected at random from a manufacturing process. Each item is inspected and classified defective, D, or nondefective, N. To list the elements of the sample space providing the most information, we construct the tree diagram of Figure 4.1. Now, the various paths of the tree give the distinct sample points. Starting at the top with the first path, we get the sample point DDD, indicating the possibility that all three items inspected are defective. Proceeding along the other paths, the sample space is

$$S = \{DDD, DDN, DND, DNN, NDD, NDN, NND, NNN\}.$$

Sample spaces with a large or infinite number of sample points are best described by a *statement* or *rule*. For example, if the possible outcomes of an experiment are the set of cities in the world with a population over 1 million, our sample space is written.

$$S = \{x|x \text{ is a city with a population over 1 million}\},$$

which reads "*S* is the set of all *x* such that *x* is a city with a population over 1 million." The vertical bar is read "such that." Similarly, if *S* is the set of all points (x, y) on the boundary of a circle of radius 2 centered at the origin, we write

$$S = \{(x, y)|x^2 + y^2 = 4\}.$$

Whether we describe the sample space by the rule method or by listing the elements will depend on the specific problem at hand. The rule method has practical advantages, particularly in the many experiments where a listing becomes a very tedious chore.

4.2
Events

In any given experiment we may be interested in the occurrence of certain **events,** rather than in the outcome of a specific element in the sample space. For instance, we might be interested in the event *A* that the outcome when a die is tossed is divisible by 3. This will occur if the outcome is an element of the subset $A = \{3, 6\}$ of the sample space S_1 in Example 1. As an additional illustration, we might be interested in the event *B* that the number of defectives is greater than 1 in Example 2. This will occur if the outcome is an element of the subset $B = \{DDN, DND, NDD, DDD\}$ of the sample space *S*.

To each event we assign a collection of sample points that constitutes a subset of the sample space. This subset represents all the elements for which the event is true.

DEFINITION

Event. An *event* is a subset of a sample space.

Example 3. Given the sample space $S = \{t|t \geq 0\}$, where *t* is the life in years of a certain electronic component, then the event *A* that the component fails before the end of the fifth year is the subset $A = \{t|0 \leq t < 5\}$.

DEFINITION

Simple and Compound Events. If an event is a set containing only one element of the sample space, then it is called a *simple event.* A *compound event* is one that can be expressed as the union of simple events.

Example 4. The event of drawing a heart from a deck of 52 playing cards is the subset A = {heart} of the sample space S = {heart, spade, club, diamond}. Therefore, A is a simple event. Now the event B of drawing a red card is a compound event, since B = {heart $\cup$ diamond} = {heart, diamond}.

Note that the union of simple events produces a compound event that is still a subset of the sample space. We should also note that if the 52 cards of the deck were the elements of the sample space rather than the 4 suits, then the event A of Example 4 would be a compound event.

DEFINITION

Null Space. The *null space* or *empty space* is a subset of the sample space that contains no elements. We denote this event by the symbol $\varnothing$.

If we let A be the event of detecting a microscopic organism by the naked eye in a biological experiment, then $A = \varnothing$. Also, if $B = \{x|x$ is a nonprime factor of 7\}, then B must be the null set, since the only possible factors of 7 are the prime numbers 1 and 7.

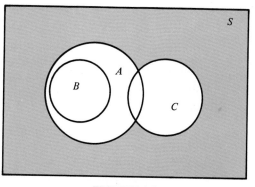

FIGURE 4.2
Events of the sample space.

The relationships between events and the corresponding sample space can be illustrated pictorially by means of a **Venn diagram.** In a Venn diagram we might let the sample space be a rectangle and represent events by circles drawn inside the rectangle. Thus, in Figure 4.2, we see that events A, B, and C are all subsets of the sample space S. It is also clear that event B is a subset of event A; events B and C have no sample points in common; and events A and C have at least one sample point in common. Figure 4.2 might, therefore, depict a situation in which one selects a card at random from an ordinary deck of 52 playing cards and observes whether the following events occur:

> *A*: the card is red,
> *B*: the card is the jack, queen, or king of diamonds,
> *C*: the card is an ace.

Clearly, the only sample points common to events *A* and *C* are the two red aces.

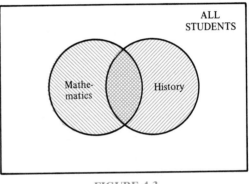

FIGURE 4.3
Events indicated by shading.

Sometimes, it is convenient to shade various areas of the Venn diagram as in Figure 4.3. In this case we take all the students of a certain college to be our sample space. The event representing those students taking mathematics has been shaded by drawing straight lines in one direction and the event representing those students studying history has been shaded by drawing lines in a different direction. The doubly shaded, or crosshatched area represents the event that a student enrolled in both mathematics and history, and the unshaded part of the diagram corresponds to those students who are studying subjects other than mathematics or history.

4.3
Operations with Events

We now consider certain operations with events that will result in the formation of new events. These new events will be subsets of the same sample space as the given events.

DEFINITION

Intersection of Events. The *intersection* of two events A and B, denoted by the symbol $A \cap B$, is the event containing all elements that are common to A and B.

The elements in the set $A \cap B$ represent the simultaneous occurrence of both A and B and therefore must be those elements and only those that belong to both A and B. These elements may either be listed or defined by the rule $A \cap B = \{ x \mid x \in A \text{ and } x \in B \}$, where the symbol $\in$ means "is an element of" or "belongs to." In the Venn diagram in Figure 4.4 the shaded area corresponds to the event $A \cap B$.

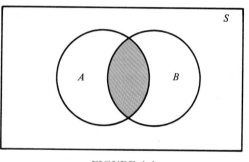

FIGURE 4.4
Intersection of A and B.

Example 5. Let $A = \{1, 2, 3, 4, 5\}$ and $B = \{2, 4, 6, 8\}$; then $A \cap B = \{2, 4\}$.

Example 6. If R is the set of all taxpayers and S is the set of all people over 65 years of age, then $R \cap S$ is the set of all taxpayers who are over 65 years of age.

Example 7. Let $P = \{a, e, i, o, u\}$ and $Q = \{r, s, t\}$; then $P \cap Q = \varnothing$. That is, P and Q have no elements in common.

In certain statistical experiments it is by no means unusual to define two events A and B that cannot both occur simultaneously. The events A and B are then said to be **mutually exclusive.** Stated more formally, we have the following definition:

DEFINITION

> *Mutually Exclusive Events.* Two events A and B are *mutually exclusive* if $A \cap B = \varnothing$; that is, A and B have no elements in common.

Two mutually exclusive events A and B are illustrated in the Venn diagram in Figure 4.5. By shading the areas corresponding to the events A and B, we find no overlapping shaded area representing the event $A \cap B$. Hence $A \cap B$ is empty.

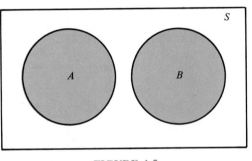

FIGURE 4.5
Mutually exclusive events.

Example 8. Suppose that a die is tossed. Let A be the event that an even number turns up and B the event that an odd number shows. The events $A = \{2, 4, 6\}$ and $B = \{1, 3, 5\}$ have no points in common, since an even and an odd number cannot occur simultaneously on a single toss of a die. Therefore, $A \cap B = \varnothing$, and consequently the events A and B are mutually exclusive.

Often, one is interested in the occurrence of at least one of two events associated with an experiment. Thus, in the die-tossing experiment, if $A = \{2, 4, 6\}$ and $B = \{4, 5, 6\}$, we might be interested in either A or B occurring, or both A and B occurring. Such an event, called the **union** of A and B, will occur if the outcome is an element of the subset $\{2, 4, 5, 6\}$.

DEFINITION

> *Union of Events.* The *union* of two events A and B, denoted by the symbol $A \cup B$, is the event containing all the elements that belong to A or to B or to both.

The elements of $A \cup B$ may be listed or defined by the rule $A \cup B = \{x \mid x \in A \text{ or } x \in B\}$. In the Venn diagram in Figure 4.6 the area representing the elements of the event $A \cup B$ has been shaded.

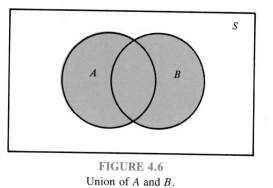

FIGURE 4.6
Union of A and B.

Example 9. Let $A = \{2, 3, 5, 8\}$ and $B = \{3, 6, 8\}$; then $A \cup B = \{2, 3, 5, 6, 8\}$.

Example 10. If $M = \{x \mid 3 < x < 9\}$ and $N = \{y \mid 5 < y < 12\}$, then $M \cup N = \{z \mid 3 < z < 12\}$.

Suppose that we consider the smoking habits of the employees of some manufacturing firm as our sample space. Let the subset of smokers correspond to some event. Then all the nonsmokers correspond to some event, also a subset of S, which is called the **complement** of the set of smokers.

DEFINITION

Complement of an Event. The _complement_ of an event A with respect to S is the set of all elements of S that are not in A. We denote the complement of A by the symbol A'.

The elements of A' may be listed or defined by the rule $A' = \{x \mid x \in S \text{ and } x \notin A\}$. In the Venn diagram in Figure 4.7 the area representing the elements of the event A' has been shaded.

Example 11. Let R be the event that a red card is selected from an ordinary deck of 52 playing cards and let S be the entire deck. Then R' is the event that the card selected from the deck is not red but a black card.

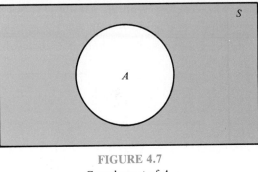

FIGURE 4.7
Complement of A.

Example 12. Consider the sample space $S = \{$book, dog, cigarette, coin, map, war$\}$. Let $A = \{$dog, war, book, cigarette$\}$. Then $A' = \{$coin, map$\}$.

Several results that follow from the foregoing definitions which may easily be verified by means of Venn diagrams are

1. $A \cap \varnothing = \varnothing$.   **2.** $A \cup \varnothing = A$.   **3.** $A \cap A' = \varnothing$.
4. $A \cup A' = S$.   **5.** $S' = \varnothing$.   **6.** $\varnothing' = S$.
7. $(A')' = A$.

EXERCISES

1. List the elements of each of the following sample spaces:
   (a) The set of integers between 1 and 50 divisible by 8.
   (b) The set $S = \{x \mid x^2 + 4x - 5 = 0\}$.
   (c) The set of outcomes when a coin is tossed until a tail or three heads appear.
   (d) The set $S = \{x \mid x \text{ is a continent}\}$.
   (e) The set $S = \{x \mid 2x - 4 \geq 0 \text{ and } x < 1\}$.

2. A newly married couple is planning to have 3 children. List the elements of the sample space S_1 using the letter M for "male" and F for "female." Define a second sample space S_2 where the elements represent the number of females.

3. An experiment involves tossing a pair of dice, 1 green and 1 red, and recording the numbers that come up. If x equals the outcome on the green die and y the outcome on the red die, describe the sample space S
   (a) by listing the elements (x, y);
   (b) by using the rule method.

4. An experiment consists of tossing a die and then flipping a coin once if the number on the die is even. If the number on the die is odd, the coin is flipped twice. Using the notation $4H$, for example, to denote the simple event that the

die comes up 4 and then the coin comes up heads, and 3HT to denote the simple event that the die comes up 3 followed by a head and then a tail on the coin, list the 18 elements of the sample space S.

5. Two jurors are selected from 4 alternates to serve at a murder trial. Using the notation A_1A_3, for example, to denote the simple event that alternates 1 and 3 are selected, list the 6 elements of the sample space S.

6. Use the rule method to describe the sample space S that consists of all points (x, y) on the boundary or in the interior of a circle of radius 3 centered at the origin.

7. For the sample space of Exercise 3,
   (a) list the elements corresponding to the event A that the sum is greater than 8;
   (b) list the elements corresponding to the event B that a 2 occurs on either die;
   (c) list the elements corresponding to the event C that a number greater than 4 comes up on the green die;
   (d) list the elements corresponding to the event $A \cap C$;
   (e) list the elements corresponding to the event $A \cap B$;
   (f) list the elements corresponding to the event $B \cap C$;
   (g) construct a Venn diagram to illustrate the intersections and unions of the events A, B, and C.

8. For the sample space of Exercise 4,
   (a) list the elements corresponding to the event A that a number less than 3 occurs on the die;
   (b) list the elements corresponding to the event B that 2 tails occur;
   (c) list the elements corresponding to the event A';
   (d) list the elements corresponding to the event $A' \cap B$;
   (e) list the elements corresponding to the event $A \cup B$.

9. An experiment consists of asking 3 women at random if they wash their dishes with brand X detergent.
   (a) List the elements of a sample space S using the letter Y for "yes" and N for "no."
   (b) List the elements of S corresponding to event E that at least 2 of the women use brand X.
   (c) Define an event that has as its elements the points {YYY, NYY, YYN, NYN}.

10. The résumés of 2 male applicants for a college teaching position in psychology are placed in the same file as the résumés of 2 female applicants. Two positions become available and the first, at the rank of assistant professor, is filled by selecting one of the 4 applicants at random. The second position, at the rank of instructor, is then filled by selecting at random one of the remaining 3 applicants. Using the notation M_2F_1, for example, to denote the simple event that the first position is filled by the second male applicant and the second position is then filled by the first female applicant,

(a) list the elements of the sample space S;

(b) list the elements of S corresponding to event A that the position of assistant professor is filled by a male applicant;

(c) list the elements of S corresponding to event B that exactly one of the 2 positions was filled by a male applicant;

(d) list the elements of S corresponding to event C that neither position was filled by a male applicant.

(e) list the elements of S corresponding to the event $A \cap B$;

(f) list the elements of S corresponding to the event $A \cup C$;

(g) construct a Venn diagram to illustrate the intersections and unions of the events A, B, and C.

11. An experiment consists of rolling a die until a 3 appears.

(a) Construct a sample space for this experiment.

(b) List the elements in E, the event that the 3 appears before the fifth roll.

12. Construct a Venn diagram to illustrate the possible intersections and unions for the following events relative to the sample space S consisting of all students at Roanoke College:

J: a student is a junior,

M: a student is a mathematics major,

W: a student is a woman.

13. If $S = \{1, 2, 3, 4, 5, 6, 7, 8, 9\}$, $A = \{2, 4, 7, 9\}$, $B = \{1, 3, 5, 7, 9\}$, $C = \{2, 3, 4, 5\}$, and $D = \{1, 6, 7\}$, list the elements of the sets corresponding to the following events:

(a) $A' \cup C$; (b) $B \cap C'$; (c) $(S \cap B')'$;

(d) $(C' \cap D) \cup B$; (e) $(B \cap C') \cup A$; (f) $A \cap C \cap D'$.

14. If $S = \{x \mid 0 < x < 12\}$, $M = \{x \mid 1 < x < 9)$, and $N = \{x \mid 0 < x < 5\}$, find

(a) $M \cup N$; (b) $M \cap N$; (c) $M' \cap N'$.

15. Consider the sample space

$$S = \{\text{automobile, bus, train, bicycle, boat, motorcycle, airplane}\}$$

representing various modes of travel, and the events

$$A = \{\text{bus, train, airplane}\},$$
$$B = \{\text{train, automobile, boat}\},$$
$$C = \{\text{bicycle}\}.$$

List the elements of the sets corresponding to the following events:

(a) $A' \cup B$; (b) $B \cap C' \cap A$; (c) $(A' \cup B) \cap (A' \cap C)$.

16. Construct a Venn diagram to illustrate the intersections and unions of the events *A*, *B*, and *C* in Exercise 15.

17. Which of the following pairs of events are mutually exclusive?
    (a) A golfer scoring the lowest round in a 72-hole tournament and losing the tournament.
    (b) A poker player getting two pairs and three of a kind on the same 5-card hand.
    (c) A mother giving birth to a baby girl and a set of twin daughters on the same day.
    (d) A tennis player losing the last game and winning the match.

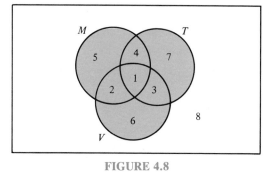

FIGURE 4.8
Venn diagram for Exercise 18.

18. Suppose that a family is leaving on a summer vacation in their camper and that *M* is the event that they will experience mechanical problems, *T* is the event that they will receive a ticket for committing a traffic violation, and *V* is the event that they will arrive at a campsite with no vacancies. Referring to the Venn diagram of Figure 4.8, state in words the events represented by the following regions:
    (a) Region 5.
    (b) Region 3.
    (c) Regions 1 and 2 together.
    (d) Regions 4 and 7 together.
    (e) Regions 3, 6, 7, and 8 together.

19. Referring to Exercise 18 and the Venn diagram of Figure 4.8, list the numbers of the regions that represent the following events:
    (a) The family will experience no mechanical problems and commit no traffic violation, but will find a campsite with no vacancies.
    (b) The family will experience both mechanical problems and trouble in locating a campsite with a vacancy, but will not receive a ticket for a traffic violation.
    (c) The family will either have mechanical trouble or find a campsite with no vacancies, but will not receive a ticket for committing a traffic violation.
    (d) The family will not arrive at a campsite with no vacancies.

20. By comparing appropriate regions of Venn diagrams, verify that
(a) $(A \cap B) \cup (A \cap B') = A$;
(b) $A' \cup (B' \cup C) = (A \cap B') \cup (A' \cup C)$.

4.4

Counting Sample Points

One of the problems that the statistician must consider and attempt to evaluate is the element of chance associated with the occurrence of certain events when an experiment is performed. These problems belong in the field of probability, a subject to be introduced in Section 4.5. In many cases we shall be able to solve a probability problem by counting the number of points in the sample space without actually listing each element. The fundamental principle of counting, often referred to as the **multiplication rule**, is stated in the following theorem.

THEOREM 4.1

Multiplication Rule. If an operation can be performed in n_1 ways, and if for each of these a second operation can be performed in n_2 ways, then the two operations can be performed together in $n_1 n_2$ ways.

> **Example 13.** How many sample points are in the sample space when a pair of dice is thrown once?
>
> **Solution.** The first die can land in any of 6 ways. For each of these 6 ways the second die can also land in 6 ways. Therefore, the pair of dice can land in $(6)(6) = 36$ ways. The student is asked to list these 36 elements in Exercise 3 on page 78.

Theorem 4.1 may be extended to cover any number of operations. In Example 2, for instance, the first item can be classified in 2 ways, defective or nondefective, and likewise for the second and third items, resulting in the $(2)(2)(2) = 8$ possibilities displayed in the tree diagram of Figure 4.1. The general multiplication rule covering k operations is stated in the following theorem.

THEOREM 4.2

Generalized Multiplication Rule. If an operation can be performed in n_1 ways, if for each of these a second operation can be performed in n_2 ways, if for each of the first two a third operation can be performed in n_3 ways, and so on, then the sequence of k operations can be performed in $n_1 n_2 \cdots n_k$ ways.

Example 14. How many lunches are possible consisting of soup, a sandwich, dessert, and a drink if one can select from 4 soups, 3 kinds of sandwiches, 5 desserts, and 4 drinks?

Solution. The total number of lunches would be $(4)(3)(5)(4) = 240$.

Example 15. How many even three-digit numbers can be formed from the digits 1, 2, 5, 6, and 9 if each digit can be used only once?

Solution. Since the number must be even, we have only 2 choices for the units position. For each of these we have 4 choices for the hundreds position and 3 choices for the tens position. Therefore, we can form a total of $(2)(4)(3) = 24$ even three-digit numbers.

Frequently, we are interested in a sample space that contains as elements all possible orders or arrangements of a group of objects. For example, we might want to know how many arrangements are possible for sitting 6 people around a table, or we might ask how many different orders are possible to draw 2 lottery tickets from a total of 20. The different arrangements are called **permutations.**

DEFINITION

Permutation. A *permutation* is an arrangement of all or part of a set of objects.

Consider the three letters *a, b,* and *c.* The possible permutations are *abc, acb, bac, bca, cab,* and *cba.* Thus we see that there are 6 distinct arrangements. Using Theorem 4.2, we could have arrived at the answer without actually listing the different orders. There are 3 positions to be filled from the letters *a, b,* and *c.* Therefore, we have 3 choices for the first position, and 2 for the second, leaving only 1 choice for the last position, giving a total of $(3)(2)(1) = 6$ permutations. In general, n distinct objects can be arranged in $n(n - 1)(n - 2) \cdots (3)(2)(1)$ ways. We represent this product by the symbol $n!$, which is read "n factorial." Three objects can be arranged in $3! = (3)(2)(1) = 6$ ways. By definition $1! = 1$ and $0! = 1$.

THEOREM 4.3

The number of permutations of n distinct objects is n!.

The number of permutations of the four letters *a, b, c,* and *d* will be $4! = 24$. Let us now consider the number of permutations that are possible by taking the 4 letters 2 at a time. These would be *ab, ac, ad, ba, ca, da, bc, cb, bd, db, cd,* and *dc.* Using Theorem 4.1, we have 2 positions to fill with 4 choices for the first and 3 choices for the second, a total of $(4)(3) = 12$ permutations. In general, n distinct objects taken r at a time can be arranged in $n(n - 1)(n - 2) \cdots (n - r + 1)$ ways. We represent this product by the symbol $_nP_r = n!/(n - r)!$.

THEOREM 4.4

_____

The number of permutations of n distinct objects taken r at a time is

$$_nP_r = \frac{n!}{(n - r)!}.$$

_____

Example 16. Two lottery tickets are drawn from 20 for first and second prizes. Find the number of sample points in the space S.

Solution. The total number of sample points is

$$_{20}P_2 = \frac{20!}{18!} = (20)(19) = 380.$$

Example 17. How many ways can a basketball team schedule 3 exhibition games with 3 teams if they are all available on any of 5 possible dates?

Solution. The total number of possible schedules is

$$_5P_3 = \frac{5!}{2!} = (5)(4)(3) = 60.$$

Permutations that occur by arranging objects in a circle are called **circular permutations.** Two circular permutations are not considered different, unless corresponding objects in the two arrangements are preceded or followed by a different object as we proceed in a clockwise direction. For example, if 4 people are playing bridge, we do not have a new permutation if they all move one position in a clockwise direction. By considering 1 person in a fixed position and arranging the other 3 in 3! ways, we find that there are 6 distinct arrangements for the bridge game.

THEOREM 4.5

_____

The number of permutations of n distinct objects arranged in a circle is $(n - 1)!$.

_____

So far we have considered permutations of distinct objects. That is, all the objects were completely different or distinguishable. Obviously, if the letters b and c are both equal to x, then the 6 permutations of the letters a, b, and c become axx, axx, xax, xax, xxa, and xxa, of which only 3 are distinct. Therefore, with 3 letters, 2 being the same, we have $3!/2! = 3$ distinct permutations. With the 4 letters a, b, c, and d we had 24 distinct permutations. If we let $a = b = x$ and $c = d = y$, we can only list the following: $xxyy$, $xyxy$, $yxxy$, $yyxx$, $xyyx$, and $yxyx$. Thus we have $4!/2!2! = 6$ distinct permutations.

THEOREM 4.6

The number of distinct permutations of n things of which n_1 are of one kind, n_2 of a second kind, ..., n_k of a kth kind is

$$\frac{n!}{n_1! n_2! \cdots n_k!}.$$

Example 18. How many different ways can 3 red, 4 yellow, and 2 blue bulbs be arranged in a string of Christmas tree lights with 9 sockets?

Solution. The total number of distinct arrangements is

$$\frac{9!}{3! 4! 2!} = 1260.$$

Often, we are concerned with the number of ways of partitioning a set of n objects into r subsets, called **cells**. A partition has been achieved if the intersection of every possible pair of the r subsets is the empty set $\varnothing$ and if the union of all subsets gives the original set. The order of the elements within a cell is of no importance. Consider the set $\{a, e, i, o, u\}$. The possible partitions into 2 cells, in which the first cell contains 4 elements and the second cell 1 element, are $\{(a, e, i, o), (u)\}$, $\{(a, i, o, u), (e)\}$, $\{(e, i, o, u), (a)\}$, $\{(a, e, o, u), (i)\}$, and $\{(a, e, i, u), (o)\}$. We see that there are 5 such ways to partition a set of 5 elements into 2 subsets or cells containing 4 elements in the first cell and 1 element in the second.

The number of partitions for this illustration is denoted by

$$\binom{5}{4,\ 1} = \frac{5!}{4! 1!} = 5,$$

where the top number represents the total number of elements and the bottom numbers represent the number of elements going into each cell. We state this more generally in the following theorem.

THEOREM 4.7

The number of ways of partitioning a set of n objects into r cells with n_1 elements in the first cell, n_2 elements in the second, and so on, is

$$\binom{n}{n_1, n_2, \ldots, n_r} = \frac{n!}{n_1! n_2! \cdots n_r!},$$

where $n_1 + n_2 + \cdots + n_r = n$.

Example 19. How many ways can 7 people be assigned to 1 triple and 2 double rooms?

Solution. The total number of possible partitions would be

$$\binom{7}{3, 2, 2} = \frac{7!}{3! 2! 2!} = 210.$$

In several problems we are interested in the number of ways of *selecting* r objects from n without regard to order. These selections are called **combinations**. A combination creates a partition with 2 cells, one cell containing the r objects selected and the other cell containing the $n - r$ objects that are left.

The number of such combinations, denoted by $\binom{n}{r, n - r}$, is usually shortened to $\binom{n}{r}$, since the number of elements in the second cell must be $n - r$.

THEOREM 4.8

The number of combinations of n distinct objects taken r at a time is

$$\binom{n}{r} = \frac{n!}{r!(n - r)!}.$$

Example 20. From 4 Republicans and 3 Democrats find the number of committees of 3 that can be formed with 2 Republicans and 1 Democrat.

Solution. The number of ways of selecting 2 Republicans from 4 is

$$\binom{4}{2} = \frac{4!}{2!2!} = 6.$$

The number of ways of selecting 1 Democrat from 3 is

$$\binom{3}{1} = \frac{3!}{1!2!} = 3.$$

Using Theorem 4.1, we find the number of committees that can be formed with 2 Republicans and 1 Democrat to be $(6)(3) = 18$.

It is of interest to note that the number of permutations of the r objects making up each of the $\binom{n}{r}$ combinations in Theorem 4.8 is $r!$. Consequently, the number of permutations of n distinct objects taken r at a time is related to the number of combinations by the formula

$$_nP_r = \binom{n}{r} r!.$$

EXERCISES

1. Registrants at a large convention are offered 6 sightseeing tours on each of 3 days. In how many ways can a person arrange to go on one of these tours?

2. In a medical study patients are classified in 8 ways according to whether they have blood type AB^+, AB^-, A^+, A^-, B^+, B^-, O^+, O^-, and also according to whether their blood pressure is low, normal, or high. Find the number of ways in which a patient can be classified.

3. If an experiment consists of throwing a die and then drawing a letter at random from the English alphabet, how many points are in the sample space?

4. A college freshman must take a science course, a humanities course, and a mathematics course. If she may select any of 6 science courses, any of 4 humanities, and any of 4 mathematics courses, how many ways can she arrange her program?

5. A developer of a new subdivision offers a prospective home buyer a choice of 4 designs, 3 different heating systems, a garage or carport, and a patio or screened porch. How many different plans are available to this buyer?

6. In how many different ways can a true–false test consisting of 9 questions be answered?

7. If a multiple-choice test consists of 5 questions each with 4 possible answers of which only one is correct,
    (a) how many different ways can a student check off one answer to each question?
    (b) how many ways can a student check off one answer to each question and get all the questions wrong?

8. (a) How many distinct permutations can be made from the letters of the word "columns"?
    (b) How many of these permutations start with the letter "m"?

9. How many distinct permutations can be made from the letters of the word "infinity"?

10. (a) How many ways can 6 people be lined up to get on a bus?
    (b) If a certain 3 persons insist on following each other, how many ways are possible?
    (c) If a certain 2 persons refuse to follow each other, how many ways are possible?

11. A contractor wishes to build 9 houses, each different in design. In how many ways can he place these homes on a street if 6 lots are on one side of the street and 3 lots are on the opposite side?

12. (a) How many three-digit numbers can be formed from the digits 0, 1, 2, 3, 4, 5, and 6 if each digit can be used only once?
    (b) How many of these are odd numbers?
    (c) How many are greater than 330?

13. In how many ways can 4 boys and 5 girls sit in a row if the boys and girls must alternate?

14. Four married couples have bought 8 seats in a row for a concert. In how many different ways can they be seated
    (a) with no restrictions?
    (b) if each couple is to sit together?
    (c) if all the men sit together to the right of all the women?

15. How many ways can the 5 starting positions on a basketball team be filled with 8 men who can play any of the positions?

16. Find the number of ways in which 6 teachers can be assigned to 4 sections of an introductory psychology course if no teacher is assigned to more than one section.

17. Three lottery tickets are drawn from 40 for first, second, and third prizes. Find the number of sample points in S for awarding the three prizes if no contestant can win more than one prize.

18. In how many ways can 5 different trees be planted in a circle?

19. In how many ways can a caravan of 8 covered wagons from Arizona be arranged in a circle?

20. In how many ways can 3 oaks, 4 pines, and 2 maples be arranged along a property line if one does not distinguish between trees of the same kind?

21. A college plays 12 football games during a season. In how many ways can the team end the season with 7 wins, 3 losses, and 2 ties?

22. Nine people are going on a skiing trip in 3 cars that will hold 2, 4, and 5 passengers, respectively. How many ways is it possible to transport the 9 people to the ski lodge using all cars?

23. How many ways are there to select 3 candidates from 8 equally qualified recent graduates for openings in an accounting firm?

24. In a California study, Dean Lester Breslow and Dr. James Enstrom of the University of California at Los Angeles' School of Public Health concluded that by following 7 simple health rules a man's life can be extended by eleven years on the average and a woman's life by seven. These 7 rules are: no smoking, regular exercise, use alcohol moderately, get seven to eight hours of sleep, maintain proper weight, eat breakfast, and do not eat between meals. In how many ways can a person adopt 5 of these rules to follow
    (a) if the person presently violates all 7 rules?
    (b) if the person never drinks and always eats breakfast?

25. From a group of 4 men and 5 women, how many committees of size 3 are possible
    (a) with no restrictions?
    (b) with 1 man and 2 women?
    (c) with 2 men and 1 woman if a certain man must be on the committee?

26. How many bridge hands are possible containing 4 spades, 6 diamonds, 1 club, and 2 hearts?

27. From 4 red, 5 green, and 6 yellow apples, how many selections of 9 apples are possible if 3 of each color are to be selected?

28. A shipment of 12 television sets contains 3 defective sets. In how many ways can a hotel purchase 5 of these sets and receive at least 2 of the defective sets?

4.5
Probability of an Event

The statistician is basically concerned with drawing conclusions or inferences from experiments involving uncertainties. For these conclusions and inferences to be accurately interpreted, an understanding of probability theory is essential.

What do we mean when we make the statements "John will probably win the tennis match," "I have a 50:50 chance of getting an even number when a die is tossed," "I am not likely to win at bingo tonight," or "Most of our graduating class will probably be married within 3 years"? In each case we are expressing an outcome of which we are not certain, but because of past information or from an understanding of the structure of the experiment, we have some degree of confidence in the validity of the statement.

The mathematical theory of probability for finite sample spaces provides a set of real numbers called **weights** or **probabilities,** ranging from 0 to 1, which allow us to evaluate the likelihood of occurrence of events. To every point in the sample space we assign a probability such that the sum of all probabilities is 1. If we have reason to believe that a certain sample point is quite likely to occur when the experiment is conducted, the probability assigned should be close to 1. On the other hand, a probability closer to zero is assigned to a sample point that is not likely to occur. In many experiments, such as tossing a coin or a die, all the sample points have the same chance of occurring and are assigned equal probabilities. For points outside the sample space, that is, for simple events that cannot possibly occur, we assign a probability of zero.

To find the probability of an event A, we sum all the probabilities assigned to the sample points in A. This sum is called the **probability of** A and is denoted by $P(A)$. Thus the probability of the set $\varnothing$ is zero and the probability of S is 1.

DEFINITION

Probability of an Event. The *probability* of an event A is the sum of the probabilities of all sample points in A. Therefore,

$$0 \leq P(A) \leq 1, \qquad P(\varnothing) = 0, \qquad P(S) = 1.$$

Example 21. A coin is tossed twice. What is the probability that at least 1 head occurs?

Solution. The sample space for this experiment is

$$S = \{HH,\ HT,\ TH,\ TT\}.$$

If the coin is balanced, each of these outcomes would be equally likely to occur. Therefore, we assign a probability of w to each sample point. Then $4w = 1$ or $w = \frac{1}{4}$. If A represents the event of at least 1 head occurring, then $P(A) = \frac{3}{4}$.

Example 22. A die is loaded in such a way that an even number is twice as likely to occur as an odd number. If E is the event that a number less than 4 occurs on a single toss of the die, find $P(E)$.

Solution. The sample space is $S = \{1, 2, 3, 4, 5, 6\}$. We assign a probability of w to each odd number and a probability of $2w$ to each even number. Since the sum of the probabilities must be 1, we have $9w = 1$ or $w = \frac{1}{9}$. Hence probabilities of $\frac{1}{9}$ and $\frac{2}{9}$ are assigned to each odd and even number, respectively. Therefore,

$$P(E) = \tfrac{1}{9} + \tfrac{2}{9} + \tfrac{1}{9} = \tfrac{4}{9}.$$

If the sample space for an experiment contains N elements all of which are equally likely to occur, we assign probabilities equal to $1/N$ to each of the N points. The probability of any event A containing n of these N sample points is then the ratio of the number of elements in A to the number of elements in S.

THEOREM 4.9

———

If an experiment can result in any one of N different equally likely outcomes, and if exactly n of these outcomes correspond to event A, then the probability of event A is

$$P(A) = \frac{n}{N}.$$

———

Example 23. If a card is drawn from an ordinary deck, find the probability that it is a heart.

Solution. The number of possible outcomes is 52, of which 13 are hearts. Therefore, the probability of event A of getting a heart is $P(A) = \frac{13}{52} = \frac{1}{4}$.

Example 24. In a poker hand consisting of 5 cards, find the probability of holding 2 aces and 3 jacks.

Solution. The number of ways of being dealt 2 aces from 4 is

$$\binom{4}{2} = \frac{4!}{2!2!} = 6$$

and the number of ways of being dealt 3 jacks from 4 is

$$\binom{4}{3} = \frac{4!}{3!1!} = 4.$$

By the multiplication rule of Theorem 4.1, there are $n = (6)(4) = 24$ hands with 2 aces and 3 jacks. The total number of 5-card poker hands, all of which are equally likely, is

$$N = \binom{52}{5} = \frac{52!}{5!47!} = 2,598,960.$$

Therefore, the probability of event C of getting 2 aces and 3 jacks in a 5-card poker hand is

$$P(C) = \frac{24}{2,598,960} = 0.9 \times 10^{-5}.$$

If the probabilities cannot be assumed equal, they must be assigned on the basis of prior knowledge or experimental evidence. For example, if a coin is not balanced, we could estimate the two probabilities by tossing the coin a large number of times and recording the outcomes. The true probabilities would be the fractions of heads and tails that occur in the long run. This method of arriving at probabilities is known as the **relative frequencey** definition of probability.

To find a numerical value that represents adequately the probability of winning at tennis, we must depend on our past performance at the game as well as that of our opponent and to some extent in our belief in being able to win. Similarly, to find the probability that a horse will win a race, we must arrive at a probability based on the previous records of all the horses entered in the race. Intuition would undoubtedly also play a part in determining the size of the bet that one might be willing to wager. The use of intuition, personal beliefs, and other indirect information in arriving at probabilities is referred to as the **subjective** definition of probability.

4.6

Additive Rules   Often, it is easier to calculate the probability of an event from known probabilities of other events. This may well be true if the event in question can be represented as the union of two or more other events or as the complement of some event. Several important laws that frequently simplify the computation of probabilities follow. The first, called the **additive rule,** applies to unions of events.

THEOREM 4.10

Additive Rule. If A and B are any two events, then

$$P(A \cup B) = P(A) + P(B) - P(A \cap B).$$

Consider the Venn diagram in Figure 4.9. The $P(A \cup B)$ is the sum of the probabilities of the sample points in $A \cup B$. Now $P(A) + P(B)$ is the sum of all the probabilities in A plus the sum of all the probabilities in B. Therefore, we have added the probabilities in $A \cap B$ twice. Since these probabilities add up to give $P(A \cap B)$, we must subtract this probability once to obtain the sum of the probabilities in $A \cup B$, which is $P(A \cup B)$.

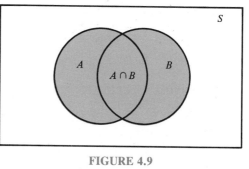

FIGURE 4.9
Venn diagram for Theorem 4.10.

COROLLARY 1

If A and B are mutually exclusive, then

$$P(A \cup B) = P(A) + P(B).$$

The corollary is an immediate result of Theorem 4.10, since if A and B are mutually exclusive, $A \cap B = \emptyset$ and then $P(A \cap B) = P(\emptyset) = 0$. In general, we write

COROLLARY 2

If A_1, A_2, A_3, ..., A_n are mutually exclusive, then

$$P(A_1 \cup A_2 \cup \cdots \cup A_n) = P(A_1) + P(A_2) + \cdots + P(A_n).$$

Note that if A_1, A_2, ..., A_n is a partition of a sample space S, then

$$P(A_1 \cup A_2 \cup \cdots \cup A_n) = P(A_1) + P(A_2) + \cdots + P(A_n)$$
$$= P(S)$$
$$= 1.$$

Example 25. The probability that a student passes mathematics is $\frac{2}{3}$, and the probability that he passes English is $\frac{4}{9}$. If the probability of passing at least one course is $\frac{4}{5}$, what is the probability that he will pass both courses?

Solution. If M is the event "passing mathematics" and E the event "passing English," then by transposing the terms in Theorem 4.10 we have

$$P(M \cap E) = P(M) + P(E) - P(M \cup E)$$
$$= \tfrac{2}{3} + \tfrac{4}{9} - \tfrac{4}{5}$$
$$= \tfrac{14}{45}.$$

Example 26. What is the probability of getting a total of 7 or 11 when a pair of dice is tossed?

Solution. Let A be the event that 7 occurs and B the event that 11 comes up. Now a total of 7 occurs for 6 of the 36 sample points and a total of 11 occurs for only 2 of the sample points. Since all sample points are equally likely, we have $P(A) = \frac{1}{6}$ and $P(B) = \frac{1}{18}$. The events A and B are mutually exclusive, since a total of 7 and 11 cannot both occur on the same toss. Therefore,

$$P(A \cup B) = P(A) + P(B)$$
$$= \tfrac{1}{6} + \tfrac{1}{18}$$
$$= \tfrac{2}{9}.$$

Often it is more difficult to calculate the probability that an event occurs than it is to calculate the probability that the event does not occur. Should this be the case for some event A, we simply find $P(A')$ first and then using Theorem 4.11, find $P(A)$ by subtraction.

THEOREM 4.11

If A and A' are complementary events, then

$$P(A) + P(A') = 1.$$

Proof. Since $A \cup A' = S$ and the events A and A' are mutually exclusive, then

$$
\begin{aligned}
1 &= P(S) \\
&= P(A \cup A') \\
&= P(A) + P(A').
\end{aligned}
$$

Example 27. A coin is tossed 6 times in succession. What is the probability that at least 1 head occurs?

Solution. Let E be the event that at least 1 head occurs. The sample space S consists of $2^6 = 64$ sample points, since each toss can result in 2 outcomes. Now, $P(E) = 1 - P(E')$, where E' is the event that no head occurs. This can happen in only one way—when all tosses result in a tail. Therefore, $P(E') = \frac{1}{64}$ and $P(E) = 1 - \frac{1}{64} = \frac{63}{64}$.

EXERCISES

1. Find the errors in each of the following statements:
   (a) The probabilities that an automobile salesperson will sell 0, 1, 2, or 3 cars on any given day in February are, respectively, 0.19, 0.38, 0.29, and 0.15.
   (b) The probability that it will rain tomorrow is 0.40 and the probability that it will not rain tomorrow is 0.52.
   (c) The probabilities that a printer will make 0, 1, 2, 3, or 4 or more mistakes in printing a document are, respectively, 0.19, 0.34, -0.25, 0.43, and 0.29.
   (d) On a single draw from a deck of playing cards the probability of selecting a heart is $\frac{1}{4}$, the probability of selecting a black card is $\frac{1}{2}$, and the probability of selecting both a heart and a black card is $\frac{1}{8}$.

2. Referring to Exercise 7 on page 79, find
   (a) the probability of event A;
   (b) the probability of event C;
   (c) the probability of event $A \cap C$.

3. Three men are seeking public office. Candidates A and B are given about the same chance of winning, but candidate C is given twice the chance of either A or B.
   (a) What is the probability that C wins?
   (b) What is the probability that A does not win?

4. A box contains 500 envelopes of which 75 contain $100 in cash, 150 contain $25, and 275 contain $10. An envelope may be purchased for $25. What is the sample space for the different amounts of money? Assign probabilities to the sample points and then find the probability that the first envelope purchased contains less than $100.

5. A 5-sided die with sides numbered 1, 2, 3, 4, and 5 is constructed so that the 1 and 5 occur twice as often as the 2 and 4, which occur three times as often as the 3. What is the probability that a perfect square occurs when this die is tossed once?

6. If A and B are mutually exclusive events and $P(A) = 0.3$ and $P(B) = 0.5$, find
   (a) $P(A \cup B)$;   (b) $P(A')$;   (c) $P(A' \cap B)$.
   Hint: Construct Venn diagrams and fill in the probabilities associated with the various regions.

7. If A, B, and C are mutually exclusive events and $P(A) = 0.2$, $P(B) = 0.3$, and $P(C) = 0.2$, find
   (a) $P(A \cup B \cup C)$;   (b) $P[A' \cap (B \cup C)]$;   (c) $P(B \cup C')'$.
   Hint: Construct Venn diagrams as in Exercise 6.

8. If a letter is chosen at random from the English alphabet, find the probability that the letter
   (a) is a vowel exclusive of y;
   (b) precedes the letter j;
   (c) follows the letter g.

9. If a permutation of the word "white" is selected at random, find the probability that the permutation
   (a) begins with a consonant;
   (b) ends with a vowel;
   (c) has the consonants and vowels alternating.

10. If each coded item in a catalog begins with 3 distinct letters followed by 4 distinct nonzero digits, find the probability of randomly selecting one of these coded items with the first letter a vowel and the last digit even.

11. A pair of dice is tossed. Find the probability of getting
    (a) a total of 8;
    (b) at most a total of 5.

12. Two cards are drawn in succession from a deck without replacement. What is the probability that both cards are greater than 2 and less than 8?

13. If 3 books are picked at random from a shelf containing 5 novels, 3 books of poems, and a dictionary, what is the probability that
    (a) the dictionary is selected?
    (b) 2 novels and 1 book of poems are selected?

14. In a poker hand consisting of 5 cards, find the probability of holding
    (a) 3 aces;
    (b) 4 hearts and 1 club.

15. In a game of *Yahtzee,* where 5 dice are tossed simultaneously, find the probability of getting
    (a) 2 pairs;
    (b) 4 of a kind.

16. In a college graduating class of 100 students, 54 studied mathematics, 69 studied history, and 35 studied both mathematics and history. If one of these students is selected at random, find the probability that
    (a) the student takes mathematics or history;
    (b) the student does not take either of these subjects;
    (c) the student takes history but not mathematics.

17. Referring to the important health practices advocated by the California study in Exercise 24 on page 89, suppose that in a senior college class of 500 students it is found that 210 smoke, 258 drink alcoholic beverages, 216 eat between meals, 122 smoke and drink alcoholic beverages, 83 eat between meals and drink alcoholic beverages, 97 smoke and eat between meals, and 52 engage in all three of these bad health practices. If a member of this senior class is selected at random, find the probability that the student
    (a) smokes but does not drink alcoholic beverages;
    (b) eats between meals and drinks alcoholic beverages but does not smoke;
    (c) neither smokes nor eats between meals.

18. The probability that an American industry will locate in Munich is 0.7, the probability that it will locate in Brussels is 0.4, and the probability that it will locate in either Munich or Brussels or both is 0.8. What is the probability that the industry will locate
    (a) in both cities?
    (b) in neither city?

19. From past experiences a stockbroker believes that under present economic conditions a customer will invest in tax-free bonds with a probability of 0.6, will invest in mutual funds with a probability of 0.3, and will invest in both tax-free bonds and mutual funds with a probability of 0.15. At this time, find the probability that a customer will invest
    (a) in either tax-free bonds or mutual funds;
    (b) in neither tax-free bonds nor mutual funds.

20. In a certain federal prison it is known that $\frac{2}{3}$ of the inmates are under 25 years of age. It is also known that $\frac{3}{5}$ of the inmates are male and the $\frac{5}{8}$ of the inmates are female or 25 years of age or older. What is the probability that a prisoner selected at random from this prison is female and at least 25 years old?

21. The probabilities that a service station will pump gas into 0, 1, 2, 3, 4, or 5 or more cars during a certain 30-minute period are 0.03, 0.18, 0.24, 0.28, 0.10, and 0.17. Find the probability that in this 30-minute period
    (a) more than 2 cars receive gas;
    (b) at most 4 cars receive gas;
    (c) 4 or more cars receive gas.

22. **Odds:** The *odds* in favor of an event E are quoted as a to b if and only if $P(E)$ $= a/(a + b)$.
    (a) If an insurance company quotes odds of 3 to 1 for the event that an individual 70 years of age will survive another 10 years, what is the probability assigned to this event?
    (b) If the probability of a successful transplant operation is $\frac{1}{8}$, what are the odds against this type of surgery?

4.7
Conditional Probability

The probability of an event B occurring when it is known that some event A has occurred is called a **conditional probability** and is denoted by $P(B|A)$. The symbol $P(B|A)$ is usually read "the probability that B occurs given that A occurs" or simply "the probability of B, given A."

Consider the event B of getting a perfect square when a die is tossed. The die is constructed so that the even numbers are twice as likely to occur as the odd numbers. Based on the sample space $S = \{1, 2, 3, 4, 5, 6\}$, with probabilities of $\frac{1}{9}$ and $\frac{2}{9}$ assigned, respectively, to the odd and even numbers, the probability of B occurring is $\frac{1}{3}$. Now suppose that it is known that the toss of the die resulted in a number greater than 3. We are now dealing with a reduced sample space $A = \{4, 5, 6\}$, which is a subset of S. To find the probability that B occurs, relative to the space A, we must first assign new probabilities to the elements of A proportional to their original probabilities such that their sum is 1. Assigning a probability of w to the odd number in A and a probability of $2w$ to the two even numbers, we have $5w = 1$ or $w = \frac{1}{5}$. Relative to the space A, we find that B contains the single element 4. Denoting this event by the symbol $B|A$, we write $B|A = \{4\}$, and hence

$$P(B|A) = \tfrac{2}{5}.$$

This example illustrates that events may have different probabilities when considered relative to different sample spaces.

We can also write

$$P(B|A) = \frac{2}{5} = \frac{\frac{2}{9}}{\frac{5}{9}} = \frac{P(A \cap B)}{P(A)},$$

where $P(A \cap B)$ and $P(A)$ are found from the original sample space S. In other words, a conditional probability relative to a subspace A of S may be calculated directly from the probabilities assigned to the elements in the original sample space S.

DEFINITION

_____

Conditional Probability. The *conditional probability* of B, given A, denoted by $P(B|A)$, is defined by the equation

$$P(B|A) = \frac{P(A \cap B)}{P(A)} \qquad \text{if } P(A) > 0.$$

_____

As an additional illustration suppose that our sample space S is the population of adults in a small town who have completed the requirements for a college degree. We shall categorize them according to sex and employment status.

	Employed	Unemployed
Male	460	40
Female	140	260

One of these individuals is to be selected at random for a tour throughout the country to publicize the advantages of establishing new industries in the town. We shall be concerned with the following events:

> M: a man is chosen,
> E: the chosen one is employed.

Using the reduced sample space E, we find that

$$P(M|E) = \frac{460}{600} = \frac{23}{30}.$$

Let $n(A)$ denote the number of elements in any set A. Using this notation, we can write

$$P(M|E) = \frac{n(E \cap M)}{n(E)} = \frac{n(E \cap M)/n(S)}{n(E)/n(S)} = \frac{P(E \cap M)}{P(E)},$$

where $P(E \cap M)$ and $P(E)$ are found from the original sample space S. To verify this result, note that

$$P(E) = \frac{600}{900} = \frac{2}{3}$$

and

$$P(E \cap M) = \frac{460}{900} = \frac{23}{45}.$$

Hence

$$P(M|E) = \frac{\frac{23}{45}}{\frac{2}{3}} = \frac{23}{30},$$

as before.

Example 28. The probability that a regularly scheduled flight departs on time is $P(D) = 0.83$, the probability that it arrives on time is $P(A) = 0.92$, and the probability that it departs and arrives on time is $P(D \cap A) = 0.78$. Find the probability that a plane (a) arrives on time given that it departed on time, and (b) departed on time given that it has arrived on time.

Solution.
(a) The probability that a plane arrives on time given that it departed on time is

$$P(A|D) = \frac{P(D \cap A)}{P(D)}$$

$$= \frac{0.78}{0.83} = 0.94.$$

(b) The probability that a plane departed on time given that it has arrived on time is

$$P(D|A) = \frac{P(D \cap A)}{P(A)}$$

$$= \frac{0.78}{0.92} = 0.85.$$

In the die-tossing experiment discussed on page 98, we noted that $P(B|A) = \frac{2}{5}$ while $P(B) = \frac{1}{3}$. That is, $P(B|A) \neq P(B)$, indicating that B *depends* on A. Now consider an experiment in which 2 cards are drawn in succession from an ordinary deck, with replacement. The events are defined as

> A: the first card is an ace,
> B: the second card is a spade.

Since the first card is replaced, our sample space for both the first and second draws consists of 52 cards, containing 4 aces and 13 spades. Hence

$$P(B|A) = \tfrac{13}{52} = \tfrac{1}{4}$$

and

$$P(B) = \tfrac{13}{52} = \tfrac{1}{4}.$$

That is, $P(B|A) = P(B)$. When this is true, the events A and B are said to be **independent**.

DEFINITION

──

Independent Events. Two events A and B are *independent* if either

$$P(B|A) = P(B)$$

or

$$P(A|B) = P(A).$$

Otherwise, A and B are *dependent*.

──

The condition $P(B|A) = P(B)$ implies that $P(A|B) = P(A)$, and conversely. For the card-drawing experiment where we showed that $P(B|A) = P(B) = \frac{1}{4}$, we also can see that $P(A|B) = P(A) = \frac{1}{13}$.

4.8
Multiplicative Rules

Multiplying the formula defined in Section 4.7 for conditional probability on both sides by $P(A)$, we obtain the following important **multiplicative rule,** which enables us to calculate the probability that two events will both occur:

THEOREM 4.12

_____

Multiplicative Rule. If in an experiment the events A and B can both occur, then

$$P(A \cap B) = P(A)P(B|A).$$

_____

Thus the probability that both A and B occur is equal to the probability that A occurs multiplied by the probability that B occurs, given that A occurs. Since the events $A \cap B$ and $B \cap A$ are equivalent, it follows from Theorem 4.12 that we can also write

$$P(A \cap B) = P(B \cap A) = P(B)P(A|B).$$

In other words, it does not matter which event is referred to as A and which event is referred to as B.

Example 29. Suppose that we have a fuse box containing 20 fuses, of which 5 are defective. If 2 fuses are selected at random and removed from the box in succession without replacing the first, what is the probability that both fuses are defective?

Solution. We shall let A be the event that the first fuse is defective and B the event that the second fuse is defective; then we interpret $A \cap B$ as the event that A occurs, and then B occurs after A has occurred. The probability of first removing a defective fuse is $\frac{1}{4}$ and then the probability of removing a second defective fuse from the remaining 4 is $\frac{4}{19}$. Hence

$$P(A \cap B) = \left(\frac{1}{4}\right)\left(\frac{4}{19}\right) = \frac{1}{19}.$$

If in Example 29 the first fuse is replaced and the fuses thoroughly rearranged before the second is removed, then the probability of a defective fuse on the second selection is still $\frac{1}{4}$, that is, $P(B|A) = P(B)$ and the events A and B are independent. When this is true, we can substitute $P(B)$ for $P(B|A)$ in Theorem 4.12 to obtain the following **special multiplicative rule:**

THEOREM 4.13

Special Multiplicative Rule. If two events A and B are independent, then

$$P(A \cap B) = P(A)P(B).$$

Therefore, to obtain the probability that two independent events will both occur, we simply find the product of their individual probabilities.

Example 30. A small town has one fire engine and one ambulance available for emergencies. The probability that the fire engine is available when needed is 0.98, and the probability that the ambulance is available when called is 0.92. In the event of an injury resulting from a burning building, find the probability that both the ambulance and the fire engine will be available.

Solution. Let A and B represent the respective events that the fire engine and the ambulance are available. Then

$$
\begin{aligned}
P(A \cap B) &= P(A)P(B) \\
&= (0.98)(0.92) \\
&= 0.9016.
\end{aligned}
$$

Example 31. One bag contains 4 white balls and 3 black balls, and a second bag contains 3 white balls and 5 black balls. One ball is drawn from the first bag and placed unseen in the second bag. What is the probability that a ball now drawn from the second bag is black?

Solution. Let B_1, B_2, and W_1 represent, respectively, the drawing of a black ball from bag 1, a black ball from bag 2, and a white ball from bag 1. We are interested in the union of the mutually exclusive events $(B_1 \cap B_2)$ and $W_1 \cap B_2)$. Now

$$
\begin{aligned}
P[(B_1 \cap B_2) &\cup (W_1 \cap B_2)] \\
&= P(B_1 \cap B_2) + P(W_1 \cap B_2) \\
&= P(B_1)P(B_2|B_1) + P(W_1)P(B_2|W_1) \\
&= (\tfrac{3}{7})(\tfrac{6}{9}) + (\tfrac{4}{7})(\tfrac{5}{9}) = \tfrac{38}{63}.
\end{aligned}
$$

Theorems 4.12 and 4.13 may be generalized to cover any number of events, as stated in the following theorem.

THEOREM 4.14

Generalized Multiplicative Rule. If in an experiment the events A_1, A_2, A_3, ..., A_k can occur, then

$$P(A_1 \cap A_2 \cap A_3 \cap \cdots \cap A_k)$$
$$= P(A_1)P(A_2|A_1)P(A_3|A_1 \cap A_2) \cdots P(A_k|A_1 \cap A_2 \cap \cdots \cap A_{k-1}).$$

If the events A_1, A_2, A_3, ..., A_k are independent, then

$$P(A_1 \cap A_2 \cap A_3 \cap \cdots A_k) = P(A_1)P(A_2)\,P(A_3) \cdots P(A_k).$$

Example 32. Three cards are drawn in succession, without replacement, from an ordinary deck of playing cards. Find the probability that the first card is a red ace, the second card is a ten or jack, and the third card is greater than 3 but less than 7.

Solution. First we define the events

A_1: the first card is a red ace,
A_2: the second card is a ten or jack,
A_3: the third card is greater than 3 but less than 7.

Now

$$P(A_1) = \tfrac{2}{52},$$

$$P(A_2|A_1) = \tfrac{8}{51},$$

$$P(A_3|A_1 \cap A_2) = \tfrac{12}{50},$$

and hence by Theorem 4.14,

$$P(A_1 \cap A_2 \cap A_3) = P(A_1)P(A_2|A_1)P(A_3|A_1 \cap A_2)$$

$$= (\tfrac{2}{52})(\tfrac{8}{51})(\tfrac{12}{50})$$

$$= \frac{8}{5525}.$$

Example 33. A coin is biased so that a head is twice as likely to occur as a tail. If the coin is tossed 3 times, what is the probability of getting 2 tails and 1 head?

Solution. The sample space for the experiment consists of the 8 elements

$$S = \{HHH, HHT, HTH, THH, HTT, THT, TTH, TTT\}.$$

However, with an unbalanced coin it is no longer possible to assign equal probabilities to each sample point. To find the probabilities, first consider the sample space $S_1 = \{H, T\}$, which represents the outcomes when the coin is tossed once. Assigning probabilities of w and $2w$ for getting a tail and a head, respectively, we have $3w = 1$ or $w = \frac{1}{3}$. Hence $P(H) = \frac{2}{3}$ and $P(T) = \frac{1}{3}$. Now let A be the event of getting 2 tails and 1 head in the 3 tosses of the coin. Then

$$A = \{TTH, THT, HTT\},$$

and since the outcomes on each of the 3 tosses are independent, it follows from Theorem 4.14 that

$$P(TTH) = P(T \cap T \cap H) = P(T)P(T)P(H)$$
$$= (\tfrac{1}{3})(\tfrac{1}{3})(\tfrac{2}{3}) = \tfrac{2}{27}.$$

Similarly,

$$P(THT) = P(HTT) = \tfrac{2}{27}$$

and hence

$$P(A) = \tfrac{2}{27} + \tfrac{2}{27} + \tfrac{2}{27} = \tfrac{2}{9}.$$

EXERCISES

1. If R is the event that a convict committed armed robbery and D is the event that the convict pushed dope, state in words what probabilities are expressed by
   (a) $P(R|D)$;   (b) $P(D'|R)$;   (c) $P(R'|D')$.

2. A class in advanced physics is comprised of 10 juniors, 30 seniors, and 10 graduate students. The final grades showed that 3 of the juniors, 10 of the seniors, and 5 of the graduate students received an A for the course. If a student is chosen at random from this class and is found to have earned an A, what is the probability that he or she is a senior?

3. A random sample of 200 adults are classified below according to sex and the level of education attained.

	Male	Female
Elementary	38	45
Secondary	28	50
College	22	17

If a person is picked at random from this group, find the probability that
(a) the person is a male, given that the person has a secondary education;
(b) the person does not have a college degree, given that the person is a female.

4. In the senior year of a high school graduating class of 100 students, 42 studied mathematics, 68 studied psychology, 54 studied history, 22 studied both mathematics and history, 25 studied both mathematics and psychology, 7 studied history but neither mathematics nor psychology, 10 studied all three subjects, and 8 did not take any of the three. If a student is selected at random, find the probability that
(a) a person enrolled in psychology takes all three subjects;
(b) a person not taking psychology is taking both history and mathematics.

5. A pair of dice is thrown. If it is known that one die shows a 4, what is the probability that
(a) the other die shows a 5?
(b) the total of both dice is greater than 7?

6. A card is drawn from an ordinary deck and we are told that it is red. What is the probability that the card is greater than 2 but less than 9?

7. The probability that an automobile being filled with gasoline will also need an oil change is 0.25; the probability that it needs a new oil filter is 0.40; and the probability that both the oil and filter need changing is 0.14.
(a) If the oil had to be changed, what is the probability that a new oil filter is needed?
(b) If a new oil filter is needed, what is the probability that the oil has to be changed?

8. The probability that a married man watches a certain television show is 0.4 and the probability that a married woman watches the show is 0.5. The probability that a man watches the show, given that his wife does, is 0.7. Find the probability that
(a) a married couple watches the show;
(b) a wife watches the show given that her husband does;
(c) at least 1 person of a married couple will watch the show.

9. The probability that a vehicle entering the Luray Caverns has Canadian license plates is 0.12; the probability that it is a camper is 0.28; and the probability that it is a camper with Canadian license plates is 0.09. What is the probability that
   (a) a camper entering the Luray Caverns has Canadian license plates?
   (b) a vehicle with Canadian license plates entering the Luray Caverns is a camper?
   (c) a vehicle entering the Luray Caverns does not have Canadian plates or is not a camper?

10. The probability that the lady of the house is home when the Avon representative calls is 0.6. Given that the lady of the house is home, the probability that she makes a purchase is 0.4. Find the probability that the lady of the house is home and makes a purchase when the Avon representative calls.

11. The probability that a doctor correctly diagnoses a particular illness is 0.7. Given that the doctor makes an incorrect diagnosis, the probability that the patient enters a law suit is 0.9. What is the probability that the doctor makes an incorrect diagnosis and the patient sues?

12. One bag contains 4 white balls and 3 black balls, and a second bag contains 3 white balls and 5 black balls. One ball is drawn at random from the second bag and is placed unseen in the first bag. What is the probability that a ball now drawn from the first bag is white?

13. A real estate agent has 8 master keys to open several new homes. Only 1 master key will open any given house. If 40% of these homes are usually left unlocked, what is the probability that the real estate agent can get into a specific home if the agent selects 3 master keys at random before leaving the office?

14. A town has 2 fire engines operating independently. The probability that a specific fire engine is available when needed is 0.96.
    (a) What is the probability that neither is available when needed?
    (b) What is the probability that a fire engine is available when needed?

15. If the probability that Tom will be alive in 20 years is 0.7 and the probability that Nancy will be alive in 20 years is 0.9, what is the probability that neither will be alive in 20 years?

16. The probability that a person visiting his dentist will have an X-ray is 0.6; the probability that a person who has an X-ray will also have a cavity filled is 0.3; and the probability that a person who has had an X-ray and a cavity filled will also have a tooth extracted is 0.1. What is the probability that a person visiting his dentist will have an X-ray, a cavity filled, and a tooth extracted?

17. Find the probability of randomly selecting 4 good quarts of milk in succession from a cooler containing 20 quarts of which 5 have spoiled, by using
    (a) the first formula of Theorem 4.14 on page 104;
    (b) the formulas of Theorems 4.8 and 4.9 on pages 86 and 91, respectively.

18. From a box containing 6 black balls and 4 green balls, 3 balls are drawn in succession, each ball being replaced in the box before the next draw is made. What is the probability that
    (a) all 3 are the same color?
    (b) each color is represented?

19. An allergist claims that 50% of the patients she tests are allergic to some type of weed. What is the probability that
    (a) exactly 3 of her next 4 patients are allergic to weeds?
    (b) none of her next 4 patients are allergic to weeds?

20. A coin is biased so that a head is twice as likely to occur as a tail. If the coin is tossed 4 times, what is the probability of getting
    (a) exactly 2 tails?
    (b) at least 3 heads?

4.9

Bayes' Rule

Let us now return to the illustration of Section 4.7, where an individual is being selected at random from the adults of a small town to tour the country and publicize the advantages of establishing new industries in the town. Suppose we are now given the additional information that 36 of those employed and 12 of those unemployed are members of the Rotary Club. We wish to find the probability of the event A that the individual selected is a member of the Rotary Club. Referring to Figure 4.10, we can write A as the union of the two mutually exclusive events $E \cap A$ and $E' \cap A$. Hence

$$A = (E \cap A) \cup (E' \cap A),$$

and by Corollary 4.1 of Theorem 4.10, and then Theorem 4.12, we can write

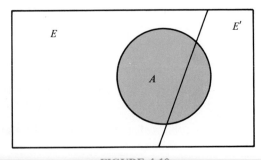

FIGURE 4.10
Venn diagram showing the events A, E, and E'.

$$P(A) = P[(E \cap A) + (E' \cap A)]$$
$$= P(E \cap A) + P(E' \cap A)$$
$$= P(E)P(A|E) + P(E')P(A|E').$$

The data of Section 4.7, together with the additional data given above for the set A, enables us to compute

$$P(E) = \frac{600}{900} = \frac{2}{3}, \qquad P(E') = \frac{1}{3},$$

$$P(A|E) = \frac{36}{600} = \frac{3}{50},$$

$$P(A|E') = \frac{12}{300} = \frac{1}{25}.$$

Displaying these probabilities by means of the tree diagram of Figure 4.11, in which the first branch yields the probability $P(E)P(A|E)$ and the second branch yields the probability $P(E')P(A|E')$, it follows that

$$P(A) = (\tfrac{2}{3})(\tfrac{3}{50}) + (\tfrac{1}{3})(\tfrac{1}{25})$$

$$= \tfrac{4}{75}.$$

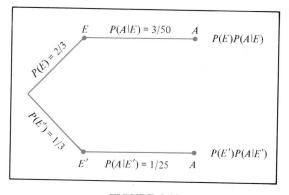

FIGURE 4.11
Tree diagram for data on page 99.

A generalization of the foregoing illustration to the case where the sample space is partitioned into k subsets is covered by the following theorem, sometimes called the **theorem of total probability** or the **rule of elimination***:

*The subsets $A_1, A_2, \ldots, A_k$ constitute a partition of the sample space S if $A_1 \cup A_2 \cup \cdots \cup A_k = S$ and $A_i \cap A_j = \varnothing$ for all $i \neq j$.

Theorem of Total Probability. If the events $B_1, B_2, \ldots, B_k$ constitute a partition of the sample space S such that $P(B_i) \neq 0$ for $i = 1, 2, \ldots, k$, then for any event A of S

$$P(A) = P(B_1)P(A|B_1) + P(B_2)P(A|B_2) + \cdots + P(B_k)P(A|B_k).$$

Proof. Consider the Venn diagram of Figure 4.12. The event A is seen to be the union of the mutually exclusive events $B_1 \cap A, B_2 \cap A, \ldots, B_k \cap A$; that is,

$$A = (B_1 \cap A) \cup (B_2 \cap A) \cup \cdots \cup (B_k \cap A).$$

Using Corollary 4.2 of Theorem 4.10, and then Theorem 4.12, we have

$$\begin{aligned}
P(A) &= P[(B_1 \cap A) \cup (B_2 \cap A) \cup \cdots \cup (B_k \cap A)] \\
&= P[B_1 \cap A) + P(B_2 \cap A) + \cdots + P(B_k \cap A) \\
&= P(B_1)P(A|B_1) + P(B_2)P(A|B_2) + \cdots + P(B_k)P(A|B_k).
\end{aligned}$$

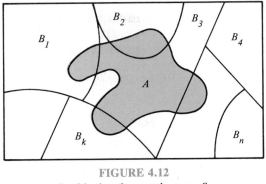

FIGURE 4.12
Partitioning the sample space S.

Example 34. Three members of a private country club have been nominated for the office of president. The probability that Mr. Adams will be elected is 0.3, the probability that Mr. Brown will be elected is 0.5, and the probability that Ms. Cooper will be elected is 0.2. Should Mr. Adams be elected, the probability for an increase in membership fees is 0.8. Should Mr. Brown or Ms. Cooper be elected, the corresponding probabilities for an increase in fees are 0.1 and 0.4. What is the probability that there will be an increase in membership fees?

Solution. Consider the following events:

$$A: \text{ membership fees are increased,}$$
$$B_1: \text{ Mr. Adams is elected,}$$
$$B_2: \text{ Mr. Brown is elected,}$$
$$B_3: \text{ Ms. Cooper is elected.}$$

Applying the rule of elimination, we can write

$$P(A) = P(B_1)P(A|B_1) + P(B_2)P(A|B_2) + P(B_3)P(A|B_3).$$

Referring to the tree diagram of Figure 4.13, we find that the three branches give the probabilities

$$P(B_1)P(A|B_1) = (0.3)(0.8) = 0.24,$$
$$P(B_2)P(A|B_2) = (0.5)(0.1) = 0.05,$$
$$P(B_3)P(A|B_3) = (0.2)(0.4) = 0.08,$$

and hence

$$P(A) = 0.24 + 0.05 + 0.08$$
$$= 0.37.$$

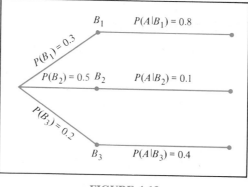

FIGURE 4.13
Tree diagram for Example 34.

Instead of asking for $P(A)$ by the rule of elimination, suppose that we now consider the problem of finding the conditional probability $P(B_3|A)$ in Example 34. In other words, if it is known that membership fees have increased, what is the probability that Ms. Cooper was elected president of the club? Questions of this type can be answered by using the following theorem, called **Bayes' rule:**

THEOREM 4.16

Bayes' Rule. If the events B_1, B_2, ..., B_k constitute a partition of the sample space S where $P(B_i) \neq 0$ for $i = 1, 2, ..., k$, then for any event A in S such that $P(A) \neq 0$,

$$P(B_r|A) = \frac{P(B_r)P(A|B_r)}{P(B_1)P(A|B_1) + P(B_2)P(A|B_2) + \cdots + P(B_k)P(A|B_k)},$$

for $r = 1, 2, ..., k$.

Proof. By the definition of conditional probability,

$$P(B_r|A) = \frac{P(B_r \cap A)}{P(A)}$$

$$= \frac{P(B_r)P(A|B_r)}{P(A)}.$$

Now, substituting the expression for $P(A)$ from Theorem 4.15, we obtain the desired result and the proof is complete.

Example 35. With reference to Example 34, if someone is considering joining the club but delays his decision for several weeks only to find out that the fees have been increased, what is the probability that Ms. Cooper was elected president of the club?

Solution. Using Bayes' rule to write

$$P(B_3|A) = \frac{P(B_3)P(A|B_3)}{P(B_1)P(A|B_1) + P(B_2)P(A|B_2) + P(B_3)P(A|B_3)},$$

and then substituting the probabilities calculated in Example 34, we have

$$P(B_3|A) = \frac{0.08}{0.24 + 0.05 + 0.08} = \frac{8}{37}.$$

In view of the fact that fees have increased, this result suggests that Ms. Cooper is probably not the president of the club.

EXERCISES

1. In a certain region of the country it is known from past experience that the probability of selecting an adult over 40 years of age with cancer is 0.02. If the probability of a doctor correctly diagnosing a person with cancer as having the disease is 0.78 and the probability of incorrectly diagnosing a person without cancer as having the disease is 0.06, what is the probability that a person is diagnosed as having cancer?

2. Police plan to enforce speed limits by using radar traps at 4 different locations within the city limits. The radar traps at each of the locations L_1, L_2, L_3, and L_4 are operated 40%, 30%, 20%, and 30% of the time, and if a person who is speeding on his way to work has probabilities of 0.2, 0.1, 0.5, and 0.2, respectively, of passing through these locations, what is the probability that he will receive a speeding ticket?

3. Referring to Exercise 1, what is the probability that a person diagnosed as having cancer actually has the disease?

4. If in Exercise 2 the person received a speeding ticket on his way to work, what is the probability that he passed through the radar trap located at L_2?

5. Suppose that colored balls are distributed in three indistinguishable boxes as follows:

	Box		
	1	*2*	*3*
Red	2	4	3
White	3	1	4
Blue	5	3	3

A box is selected at random from which a ball is drawn at random.
(a) Find the probability that the ball is red.
(b) Given that the ball is red, what is the probability that box 3 was selected?

6. A large industrial firm uses 3 local motels to provide overnight accommodations for its clients. From past experience it is known that 20% of the clients are assigned rooms at the Ramada Inn, 50% at the Sheraton, and 30% at the Lakeview Motor Lodge. If the plumbing is faulty in 5% of the rooms at the Ramada Inn, in 4% of the rooms at the Sheraton, and in 8% of the rooms at the Lakeview Motor Lodge, what is the probability that
(a) a client will be assigned a room with faulty plumbing?
(b) a person with a room having faulty plumbing was assigned accommodations at the Lakeview Motor Lodge?

7. A commuter owns 2 cars, 1 a compact and 1 a standard model. About $\frac{3}{4}$ of the time he uses the compact to travel to work, and about $\frac{1}{4}$ of the time the larger car

is used. When he uses the compact car, he usually gets home by 5:30 P.M. about 75% of the time; if he uses the standard-sized car, he gets home by 5:30 P.M. about 60% of the time (but he enjoys the air conditioner in the larger car). If he gets home at 5:35 P.M., what is the probability that he used the compact car?

8. A truth serum given to a suspect is known to be 90% reliable when the person is guilty and 99% reliable when the person is innocent. In other words, 10% of the guilty are judged innocent by the serum and 1% of the innocent are judged guilty. If the suspect was selected from a group of suspects of which only 5% have ever committed a crime, and the serum indicates that he is guilty, what is the probability that he is innocent?

5 Distributions of Random Variables

The generalizations associated with statistical inferences are subject to uncertainties, since we are dealing with only partial information obtained from a subset of the data of interest. To cope with these uncertainties, an understanding of probability theory is essential in order to provide a mathematical model that theoretically describes the behavior of the population associated with the statistical experiment. These theoretical models, which are very similar to relative frequency distributions, are called probability distributions.

5.1

Concept of a Random Variable

The term *statistical experiment* has been used to describe any process by which one or more chance measurements are obtained. Often, we are not interested in the details associated with each sample point but only in some numerical description of the outcome. For example, the sample space giving a detailed description of each possible outcome when one tosses a coin 3 times may be written

$$S = \{HHH, HHT, HTH, THH, HTT, THT, TTH, TTT\}.$$

If one is concerned only with the number of heads that fall, then a numerical value of 0, 1, 2, or 3 will be assigned to each sample point.

The numbers 0, 1, 2, and 3 are random quantities determined by the outcome of an experiment. They may be thought of as the values assumed by some **random variable** X, which in this case represents the number of heads when a coin is tossed 3 times.

DEFINITION

Random Variable. A function whose value is a real number determined by each element in the sample space is called a *random variable*.

We shall use a capital letter, say X, to denote a random variable and its corresponding small letter, x in this case, for one of its values. In the coin-tossing illustration above, we notice that the random variable X assumes the value 2 for all elements in the subset

$$E = \{HHT, HTH, THH\}$$

of the sample space S. That is, each possible value of X represents an event that is a subset of the sample space for the given experiment.

Example 1. Two balls are drawn in succession without replacement from an urn containing 4 red balls and 3 black balls. The possible outcomes and the values y of the random variable Y, where Y is the number of red balls, are

Sample Space	y
RR	2
RB	1
BR	1
BB	0

Example 2. A hatcheck girl returns 3 hats at random to 3 customers who had previously checked them. If Smith, Jones, and Brown, in that order, receive one of the 3 hats, list the sample points for the possible orders of returning the hats and find the values m of the random variable M that represents the number of correct matches.

Solution. If S, J, and B stand for Smith's, Jones', and Brown's hats, respectively, then the possible arrangements in which the hats may be returned and the number of correct matches are

Sample Space	m
SJB	3
SBJ	1
JSB	1
JBS	0
BSJ	0
BJS	1

In each of the two preceding examples the sample space contains a finite number of elements. On the other hand, when a die is thrown until a 5 occurs, we obtain a sample space with an unending sequence of elements, namely,

$$S = \{F, NF, NNF, NNNF, \ldots\},$$

where F and N represent, respectively, the occurrence and nonoccurrence of a 5. But even in this experiment, the number of elements can be equated to the number of whole numbers and in this sense can be counted.

DEFINITION _____

Discrete Sample Space. If a sample space contains a finite number of possibilities or an unending sequence with as many elements as there are whole numbers, it is called a *discrete sample space*.

_____

The outcomes of some statistical experiments may be neither finite nor countable. Such is the case, for example, when one conducts an investigation measuring the distances that a certain make of automobile will travel over a prescribed test course on 5 liters of gasoline. Assuming distance to be a variable measured to any degree of accuracy, then clearly we have an infinite number of possible distances in the sample space that cannot be equated to the number of whole numbers. Also, if one were to record the length of time for a chemical reaction to take place, once again the possible time intervals making up our sample space are infinite in number and uncountable. We see now that all sample spaces need not be discrete.

DEFINITION

Continuous Sample Space. If a sample space contains an infinite number of possibilities equal to the number of points on a line segment, it is called a *continuous sample space.*

Random variables defined over discrete and continuous sample spaces are called, respectively, **discrete random variables** and **continuous random variables.** In most practical problems continuous random variables represent *measured* data, such as all possible heights, weights, temperatures, distances, or life periods, whereas discrete random variables represent *count* data, such as the number of defectives in a sample of k items or the number of highway fatalities per year in a given state. Note that the random variables Y and M of Examples 1 and 2 both represent count data, Y the number of red balls and M the number of correct hat matches.

5.2
Discrete Probability Distributions

A discrete random variable assumes each of its values with a certain probability. In the case of tossing a coin 3 times, the variable X, representing the number of heads, assumes the value 2 with probability $\frac{3}{8}$, since 3 of the 8 equally likely sample points result in 2 heads and 1 tail. By assuming equal probabilities for the simple events in Example 2, the probability that no man gets back his right hat, that is, the probability that M assumes the value 0, is $\frac{1}{3}$. The possible values m of M and their probabilities are given by

m	0	1	3
$P(M = m)$	$\frac{1}{3}$	$\frac{1}{2}$	$\frac{1}{6}$

Note that the values of m exhaust all possible cases, and hence the probabilities add to 1.

Frequently, it is convenient to represent all the probabilities of a random variable X by a formula. Such a formula would necessarily be a function of the numerical values x that we shall denote by $f(x)$, $g(x)$, $r(x)$, and so forth. Hence we write $f(x) = P(X = x)$; that is, $f(3) = P(X = 3)$. The set of ordered pairs $(x, f(x))$ is called the **probability function** or **probability distribution** of the discrete random variable X.

DEFINITION

Discrete Probability Distribution. A table or a formula listing all possible values that a discrete random variable can take on, along with the associated probabilities, is called a *discrete probability distribution.*

Example 3. Find the probability distribution of the sum of the numbers when a pair of dice is tossed.

Solution. Let X be a random variable whose values x are the possible totals. Then x can be any integer from 2 to 12. Two dice can fall in $(6)(6) = 36$ ways, each with probability $\frac{1}{36}$. The $P(X = 3) = \frac{2}{36}$, since a total of 3 can occur in only 2 ways. Consideration of the other cases leads to the following probability distribution:

x	2	3	4	5	6	7	8	9	10	11	12
$P(X = x)$	$\frac{1}{36}$	$\frac{2}{36}$	$\frac{3}{36}$	$\frac{4}{36}$	$\frac{5}{36}$	$\frac{6}{36}$	$\frac{5}{36}$	$\frac{4}{36}$	$\frac{3}{36}$	$\frac{2}{36}$	$\frac{1}{36}$

Example 4. Find a formula for the probability distribution of the number of heads when a coin is tossed 4 times.

Solution. Since there are $2^4 = 16$ points in the sample space representing equally likely outcomes, the denominator for all probabilities, and therefore for our function, will be 16. To obtain the number of ways of getting, say 3 heads, we need to consider the number of ways of partitioning 4 outcomes into 2 cells, with 3 heads assigned to one cell and a tail assigned to the other. This can be done in $\binom{4}{3} = 4$ ways. In general x heads and $4 - x$ tails can occur in $\binom{4}{x}$ ways, where x can be 0, 1, 2, 3, or 4. Thus the probability distribution $f(x) = P(X = x)$ is

$$f(x) = \frac{\binom{4}{x}}{16}, \quad \text{for } x = 0, 1, 2, 3, 4.$$

It is often helpful to look at a probability distribution graphically in the form of a **probability histogram**, as demonstrated in Figure 5.1 for Example 4. The rectangles are constructed so that their bases of equal width are centered at each value of x and their heights are equal to the corresponding probabilities given by $f(x)$.

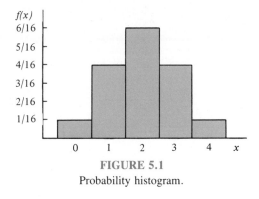

FIGURE 5.1
Probability histogram.

Since each rectangle in Figure 5.1 has a base of unit width, then $P(X = 0)$, $P(X = 1)$, $P(X = 2)$, $P(X = 3)$, and $P(X = 4)$ are equal to the *areas* of the rectangles centered at $x = 0$, $x = 1$, $x = 2$, $x = 3$, and $x = 4$, respectively. Even if the bases were not of unit width, we could adjust the heights of the rectangles to give areas that would still equal the probabilities of X assuming any of its values x. This concept of using areas to represent probabilities is necessary for our consideration of the probability distributions of continuous random variables.

Certain probability distributions are applicable to more than one physical situation. The probability distribution of Example 4, for example, also applies to the random variable Y, where Y is the number of red cards that occurs when 4 cards are drawn at random from a deck in succession with each card replaced and the deck shuffled before the next drawing. Special discrete probability distributions that can be applied to many different experimental situations will be considered in Chapter 6.

5.3
Continuous Probability Distributions

A continuous random variable has a probability of zero of assuming *exactly* any of its values. Consequently, its probability distribution cannot be given in tabular form. At first this may seem startling, but it becomes more plausible when we consider a particular example. Let us discuss a random variable whose values are the heights of all people over 21 years of age. Between any two values, say 163.5 and 164.5 centimeters, or even 163.99 and 164.01 centimeters, there are an infinite number of heights, of which only one is 164 centimeters. The probability of selecting a person at random who is exactly 164 centimeters tall, and not one of the infinitely large set of heights so close to 164 centimeters that you cannot humanly measure the difference, is extremely remote. Thus we assign a probability of zero to the event. This is not the case, however, if we talk about the probability of selecting a person who is at least 163 centimeters but not more than 165 centimeters tall. Now we are dealing with an interval rather than a point value of our random variable.

We shall concern ourselves with computing probabilities for various intervals of continuous random variables such as $P(a < X < b)$, $P(W > c)$, and so forth. Note that when X is continuous

$$P(a < X \le b) = P(a < X < b) + P(X = b)$$
$$= P(a < X < b).$$

That is, it does not matter whether we include an end point of the interval or not. This is not true when X is discrete.

Although the probability distribution of a continuous random variable cannot be presented in tabular form, it does have a formula. Such a formula would necessarily be a function of the numerical values of the continuous variable X and as such could be graphed as a continuous curve. The probability function portrayed by this curve is usually called a **probability density function,** or simply a **density function.** Most density functions that have practical applications in the analysis of statistical data are continuous for all values of X and their graphs may take any of several forms, some of which are shown in Figure 5.2. Because areas will be used to represent probabilities and probabilities are positive numerical values, the density function must lie entirely above the x axis.

A probability density function is constructed so that the area under its curve bounded by the x axis is equal to 1. If the density function is represented

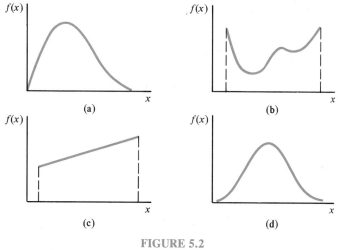

FIGURE 5.2
Typical density functions.

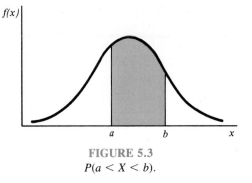

FIGURE 5.3
$P(a < X < b)$.

by the curve in Figure 5.3, then the probability that X assumes a value between a and b is equal to the shaded area under the density function between the ordinates at $x = a$ and $x = b$.

DEFINITION

Probability Density Function. The function with values $f(x)$ is called a *probability density function* for the continuous random variable X if the total area under its curve and above the x axis is equal to 1 and if the area under the curve between any two ordinates $x = a$ and $x = b$ gives the probability that X lies between a and b.

Example 5. A continuous random variable X that can assume values between $x = 2$ and $x = 4$ has a density function given by

$$f(x) = \frac{x + 1}{8}.$$

(a) Show that $P(2 < X < 4) = 1$.
(b) Find $P(X < 3.5)$.
(c) Find $P(2.4 < X < 3.5)$.

Solution
(a) Since the shaded region in Figure 5.4 is a trapezoid, the area is found by summing the heights of the parallel sides, multiplying by the length of the base, and dividing by 2. That is,

$$\text{area} = \frac{(\text{sum of the parallel sides}) \times \text{base}}{2}$$

$$= \frac{[f(2) + f(4)](2)}{2}.$$

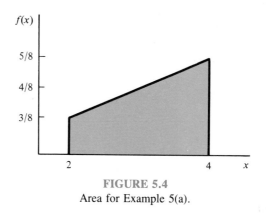

FIGURE 5.4
Area for Example 5(a).

Now, since $f(2) = \frac{3}{8}$ and $f(4) = \frac{5}{8}$, it follows that

$$P(2 < X < 4) = \frac{(\frac{3}{8} + \frac{5}{8})(2)}{2} = 1.$$

(b) As before, $f(2) = \frac{3}{8}$, and we find that $f(3.5) = 4.5/8$. Hence the shaded area of Figure 5.5 gives

$$P(X < 3.5) = \frac{(\frac{3}{8} + \frac{4.5}{8})(1.5)}{2}$$

$$= 0.70.$$

(c) We find that $f(2.4) = 3.4/8$, and together with $f(3.5) = 4.5/8$, we see from Figure 5.6 that

$$P(2.4 < X < 3.5) = \frac{(\frac{3.4}{8} + \frac{4.5}{8})(1.1)}{2}$$

$$= 0.54.$$

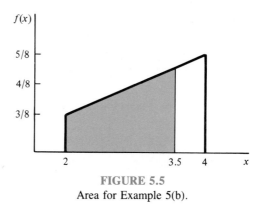

FIGURE 5.5
Area for Example 5(b).

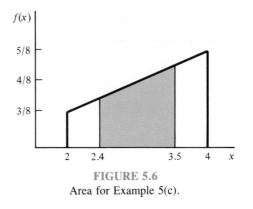

FIGURE 5.6
Area for Example 5(c).

In general, we do not have simple straight-line density functions as in Example 5. More often, the curve has a complicated formula and the computations for determining areas under such a curve are rather involved. Such is the case for the many continuous distributions that arise in experimental work that can be represented by the characteristic bell-shaped curve of Figure 5.3. Fortunately, areas have been computed and put in tabular form for all the continuous probability functions discussed in this text.

EXERCISES

1. Classify the following random variables as discrete or continuous.

   X: the number of automobile accidents each year in Virginia.
   Y: the length of time to play 18 holes of golf.
   M: the amount of milk produced yearly by a particular cow.
   N: the number of eggs laid each month by 1 hen.
   P: the number of building permits issued each month in a certain city.
   Q: the weight of grain in pounds produced per acre.

2. An overseas shipment of 5 foreign automobiles contains 2 that have slight paint blemishes. If an agency receives 3 of these automobiles at random, list the elements of the sample space S using the letters B and N for "blemished" and "nonblemished," respectively, and then to each sample point assign a value x of the random variable X representing the number of automobiles purchased by the agency with paint blemishes.

3. Let W be a random variable giving the number of heads minus the number of tails in three tosses of a coin. List the elements of the sample space S for the three tosses of the coin and to each sample point assign a value w of W.

4. A coin is flipped until 3 heads in succession occur. List only those elements of the sample space that require 6 or less tosses. Is this a discrete sample space? Explain.

5. From a box containing 4 dimes and 2 nickels, 3 coins are selected at random without replacement. Find the probability distribution for the total T of the 3 coins. Express the probability distribution graphically as a probability histogram.

6. From a box containing 4 black balls and 2 green balls, 3 balls are drawn in succession, each ball being replaced in the box before the next draw is made. Find the probability distribution for the number of green balls.

7. Find the probability distribution of the random variable W in Exercise 3, assuming the coin is biased so that a head is twice as likely to occur as a tail.

8. Find the probability distribution for the number of jazz records when 4 records are selected at random from a collection consisting of 5 jazz records, 2 classical records, and 3 polka records. Express your results by means of a formula.

9. Find a formula for the probability distribution of the random variable X representing the outcome when a single die is rolled once.

10. A shipment of 7 television sets contains 2 defective sets. A hotel makes a random purchase of 3 of the sets. If X is the number of defective sets purchased by the hotel, find the probability distribution of X. Express the results graphically as a probability histogram.

11. Three cards are drawn in succession from a deck without replacement. Find the probability distribution for the number of spades.

12. A continuous random variable X that can assume values between $x = 1$ and $x = 4$ has a density function given by $f(x) = \frac{1}{3}$.
    (a) Show that the area under the curve is equal to 1.
    (b) Find $P(1.5 < X < 3)$.
    (c) Find $P(X \geq 2.2)$.

13. A continuous random variable X that can assume values between $x = 2$ and $x = 5$ has a density function given by $f(x) = 2(1 + x)/27$.
    (a) Find $P(X < 4)$.
    (b) Find $P(3 \leq X < 4)$.

14. A continuous random variable X has the density function

$$f(x) = \begin{cases} x & \text{for } 0 < x < 1 \\ 2 - x & \text{for } 1 \leq x < 2 \\ 0 & \text{elsewhere.} \end{cases}$$

    (a) Show that $P(0 < X < 2) = 1$.
    (b) Find $P(X < 1.2)$.

5.4

Joint Probability Distributions

Our study of random variables and their probability distributions in the preceding sections was restricted to one-dimensional sample spaces, in that we recorded outcomes of an experiment as values assumed by a single random variable. There will be many situations, however, where we may find it desirable to record the simultaneous outcomes of several random variables. For example, we might measure the amount of precipitate P and the volume V of gas released from a controlled chemical experiment, giving rise to a two-dimensional sample space consisting of the outcomes (p, v); or one might be interested in the hardness H and tensile strength T of cold-drawn copper resulting in the outcomes (h, t). In a study to determine the likelihood of success in college, based upon high school data, one might use a three-dimensional sample space and record for each individual his aptitude test score, high school rank in class, and grade-point average at the end of the freshman year in college.

If X and Y are two discrete random variables, the probability distribution for their simultaneous occurrence can be represented by a function with values $f(x, y)$ for any pair of values (x, y) within the range of the random variables X and Y. It is customary to refer to this function as the **joint probability distribution** of X and Y. Hence, in the discrete case,

$$f(x, y) = P(X = x, Y = y);$$

that is, the values $f(x, y)$ give the probability that outcomes x and y occur at the same time. For example, if a television set is to be serviced and X represents the age to the nearest year of the set and Y represents the number of defective tubes in the set, then $f(5, 3)$ is the probability that the television set is 5 years old and needs 3 new tubes. When X and Y are continuous random variables, the **joint density function** $f(x, y)$ is a surface lying above the xy plane and $P[(X, Y) \in A]$, where A is any region in the xy plane, is equal to the volume of the right cylinder bounded by the base A and the surface.

In keeping with the mathematical prerequisites of this text, we shall consider for the most part only joint probability functions of discrete random variables.

DEFINITION

Joint Probability Distribution. A table or formula listing all possible values x and y of the discrete random variables X and Y, together with the associated probabilities $f(x, y)$, is called a *joint probability distribution.*

Example 6. Two refills for a ballpoint pen are selected at random from a box that contains 3 blue refills, 2 red refills, and 3 green refills. If X is the number of blue refills and Y is the number of red refills selected, find (a) the joint probability function $f(x, y)$ and (b) $P[(X, Y) \in A]$, where A is the region $\{(x, y)|x + y \leq 1\}$.

Solution

(a) The possible pairs of values (x, y) are $(0, 0)$, $(0, 1)$, $(1, 0)$, $(1, 1)$, $(0, 2)$, and $(2, 0)$. Now, $f(0, 1)$, for example, represents the probability that a red and a green refill are selected. The total number of equally likely ways of selecting any 2 refills from the 8 is $\binom{8}{2} = 28$. The number of ways of selecting 1 red from 3 red refills and 1 green from 3 green refills is $\binom{2}{1}\binom{3}{1} = 6$. Hence $f(0, 1) = \frac{6}{28} = \frac{3}{14}$. Similar calculations yield the probabilities for the other cases, which are presented in Table 5.1. Note that the probabilities sum to 1.

TABLE 5.1
Joint Probability Distribution
for Example 6

$f(x, y)$	x 0	1	2	Row Totals
0	$\frac{3}{28}$	$\frac{9}{28}$	$\frac{3}{28}$	$\frac{15}{28}$
y 1	$\frac{3}{14}$	$\frac{3}{14}$		$\frac{3}{7}$
2	$\frac{1}{28}$			$\frac{1}{28}$
Column Totals	$\frac{5}{14}$	$\frac{15}{28}$	$\frac{3}{28}$	1

In Chapter 6 it will become clear that the joint probability distribution of Table 5.1 can be represented by the formula

$$f(x, y) = \frac{\binom{3}{x}\binom{2}{y}\binom{3}{2 - x - y}}{\binom{8}{2}},$$

for $x = 0, 1, 2$; $y = 0, 1, 2$; $0 \leq x + y \leq 2$.

(b) $P[(X, Y) \in A] = P(X + Y \leq 1)$
$$= f(0, 0) + f(0, 1) + f(1, 0)$$
$$= \tfrac{3}{28} + \tfrac{3}{14} + \tfrac{9}{28}$$
$$= \tfrac{9}{14}.$$

Given the values $f(x, y)$ of the joint probability distribution of the discrete random variables X and Y, the one-dimensional probability distributions of X alone, with values $g(x)$, and Y alone, with values $h(y)$, are given by the column and row totals in Table 5.1. We define these two functions to be the **marginal distributions** of X and Y, respectively. Hence

$$g(0) = P(X = 0) = P(X = 0, Y = 0) + P(X = 0, Y = 1)$$
$$+ P(X = 0, Y = 2)$$
$$= f(0, 0) + f(0, 1) + f(0, 2)$$
$$= \tfrac{3}{28} + \tfrac{3}{14} + \tfrac{1}{28}$$
$$= \tfrac{5}{14},$$
$$g(1) = P(X = 1) = \tfrac{15}{28},$$
$$g(2) = P(X = 2) = \tfrac{3}{28}.$$

Similarly,

$$h(0) = P(Y = 0) = P(X = 0, Y = 0) + P(X = 1, Y = 0)$$
$$+ P(X = 2, Y = 0)$$
$$= f(0, 0) + f(1, 0) + f(2, 0)$$
$$= \tfrac{3}{28} + \tfrac{9}{28} + \tfrac{3}{28}$$
$$= \tfrac{15}{28},$$
$$h(1) = P(Y = 1) = \tfrac{3}{7},$$
$$h(2) = P(Y = 2) = \tfrac{1}{28}.$$

In tabular form the marginal distributions may be written as follows:

x	0	1	2		y	0	1	2
$g(x)$	$\tfrac{5}{14}$	$\tfrac{15}{28}$	$\tfrac{3}{28}$		$h(y)$	$\tfrac{15}{28}$	$\tfrac{3}{7}$	$\tfrac{1}{28}$

In Section 5.1 we stated that the value x of the random variable X represents an event that is a subset of the sample space. By using the definition of conditional probability as given in Chapter 4,

$$P(B|A) = \frac{P(A \cap B)}{P(A)}, \qquad P(A) \neq 0,$$

where A and B are now the events defined by $X = x$ and $Y = y$, respectively, then

$$P(Y = y|X = x) = \frac{P(X = x, Y = y)}{P(X = x)}$$

$$= \frac{f(x, y)}{g(x)}, \qquad g(x) > 0,$$

when X and Y are discrete random variables.

It is not difficult to show that the values $f(x, y)/g(x)$, which depend strictly on y with x fixed, satisfy all the conditions of a probability distribution. Expressing these values by the symbol $f(y|x)$, we have the following definition:

DEFINITION

_____

Conditional Distributions. The *conditional distribution* of the discrete random variable Y, given that $X = x$ is given by

$$f(y|x) = \frac{f(x, y)}{g(x)}, \qquad g(x) > 0,$$

Similarly, the *conditional distribution* of the discrete random variable X, given that $Y = y$, is given by

$$f(x|y) = \frac{f(x, y)}{h(y)}, \qquad h(y) > 0.$$

_____

Example 7. Referring to Example 6, find $f(x|1)$ for all values of x and determine $P(X = 0|Y = 1)$.

Solution. First we find

$$h(1) = f(0, 1) + f(1, 1) + f(2, 1) = \tfrac{3}{14} + \tfrac{3}{14} + 0 = \tfrac{3}{7}.$$

Now,

$$f(x|1) = \frac{f(x, 1)}{h(1)} = \frac{7}{3}f(x, 1), \qquad x = 0, 1, 2.$$

Therefore,

$$f(0|1) = \tfrac{7}{3}f(0, 1) = (\tfrac{7}{3})(\tfrac{3}{14}) = \tfrac{1}{2},$$
$$f(1|1) = \tfrac{7}{3}f(1, 1) = (\tfrac{7}{3})(\tfrac{3}{14}) = \tfrac{1}{2},$$
$$f(2|1) = \tfrac{7}{3}f(2, 1) = (\tfrac{7}{3})(0) = 0,$$

and the conditional distribution of X, given that $Y = 1$, is

x	0	1	2
$f(x\|1)$	$\tfrac{1}{2}$	$\tfrac{1}{2}$	0

Finally,

$$P(X = 0|Y = 1) = f(0|1) = \tfrac{1}{2}.$$

Suppose that the random variable X assumes the values 1, 2, 3, 4, 5, and 6 for the outcomes when a die is tossed, and Y assumes the values 0 and 1 for heads and tails, respectively, when a coin is tossed, then $f(3, 1)$ is the probability that both a 3 occurs on the die and a tail occurs on the coin if the die and coin are tossed together. In this experiment $f(3, 1) = P(X = 3, Y = 1) = P(X = 3)P(Y = 1) = g(3)h(1)$, since obtaining a 3 on a die and a tail on a coin are independent events. Because this is true for all possible pairs (x, y), the random variables X and Y are said to be **statistically independent** and we write $f(x, y) = g(x)h(y)$.

DEFINITION

Independent Random Variables. The random variables X and Y are said to be *statistically independent* if and only if

$$f(x, y) = g(x)h(y)$$

for all possible values of X and Y.

Checking for statistical independence of discrete random variables requires a very thorough investigation, since it is possible to have the product of the marginal distributions equal to the joint probability distribution for some but not all combinations of (x, y). If you can find any point (x, y) for which $f(x, y) \neq g(x)h(y)$, the discrete variables X and Y are not statistically independent.

Example 8. Show that the random variables of Example 6 are not statistically independent.

Solution. Let us consider the point $(0, 1)$. From Table 5.1 we find the three probabilities $f(0, 1)$, $g(0)$, and $h(1)$ to be

$$f(0, 1) = \tfrac{3}{14},$$

$$g(0) = \sum_{y=0}^{2} f(0, y) = \tfrac{3}{28} + \tfrac{3}{14} + \tfrac{1}{28} = \tfrac{5}{14},$$

$$h(1) = \sum_{x=0}^{2} f(x, 1) = \tfrac{3}{14} + \tfrac{3}{14} + 0 = \tfrac{3}{7}.$$

Clearly,

$$f(0, 1) \neq g(0)h(1),$$

and therefore X and Y are not statistically independent.

EXERCISES

1. Determine the value of c so that the following functions represent joint probability distributions of the random variables X and Y:
   (a) $f(x, y) = cxy$, for $x = 1, 2, 3; y = 1, 2, 3$.
   (b) $f(x, y) = c|x - y|$, for $x = -2, 0, 2; y = -2, 3$.

2. If the joint probability distribution of X and Y is given by

$$f(x, y) = \frac{(x + y)}{30}, \qquad \text{for } x = 0, 1, 2, 3; y = 0, 1, 2$$

   find
   (a) $P(X \leq 2, Y = 1)$;   (b) $P(X > 2, Y \leq 1)$;
   (c) $P(X > Y)$;   (d) $P(X + Y = 4)$.

3. From a sack of fruit containing 3 oranges, 2 apples, and 3 bananas a random sample of 4 pieces of fruit is selected. If X is the number of oranges and Y is the number of apples in the sample, find
   (a) the joint probability distribution of X and Y;
   (b) $P[(X, Y) \in A]$, where A is the region $\{(x, y)|x + y \leq 2\}$.

4. Consider an experiment that consists of 2 rolls of a balanced die. If X is the number of 4's and Y is the number of 5's obtained in the 2 rolls of the die, find
   (a) the joint probability distribution of X and Y;
   (b) $P[(X, Y) \in A]$, where A is the region $\{(x, y)|2x + y < 3\}$.

5. Let X denote the number of heads and Y the number of heads minus the number of tails when 3 coins are tossed. Find the joint probability distribution of X and Y.

6. Three cards are drawn without replacement from the 12 face cards (jacks, queens, and kings) of an ordinary deck of 52 playing cards. Let X be the number of kings selected and Y the number of jacks. Find
   (a) the joint probability distribution of X and Y;
   (b) $P[(X, Y) \in A]$, where A is the region $\{(x, y)|x + y \geq 2\}$.

7. Referring to Exercise 2, find
   (a) the marginal distribution of X;
   (b) the marginal distribution of Y.

8. A coin is tossed twice. Let Z denote the number of heads on the first toss and W the total number of heads on the 2 tosses. If the coin is unbalanced and a head has a 40% chance of occurring, find
   (a) the joint probability distribution of W and Z;
   (b) the marginal distribution of W;
   (c) the marginal distribution of Z;
   (d) the probability that at least 1 head occurs.

9. Referring to Exercise 3, find
   (a) $f(y|2)$ for all values of y;   (b) $P(Y = 0|X = 2)$.

10. Suppose that X and Y have the following joint probability distribution:

		x		
y		1	2	3
1		0	$\frac{1}{6}$	$\frac{1}{12}$
2		$\frac{1}{5}$	$\frac{1}{9}$	0
3		$\frac{2}{15}$	$\frac{1}{4}$	$\frac{1}{18}$

   (a) Evaluate the marginal distribution of X.
   (b) Evaluate the marginal distribution of Y.
   (c) Find $P(Y = 3|X = 2)$.

11. Suppose that X and Y have the following joint probability function:

		x	
y		2	4
1		0.10	0.15
3		0.20	0.30
5		0.10	0.15

(a) Find the marginal distribution of the random variable X.
(b) Find the marginal distribution of the random variable Y.

12. Determine whether the two random variables of Exercise 10 are dependent or independent.

13. Determine whether the two random variables of Exercise 11 are dependent or independent.

5.5

Mean of a Random Variable

If two coins are tossed 16 times and X is the number of heads that occur per toss, then the values of X can be 0, 1, and 2. Suppose that the experiment yields no heads, 1 head, and 2 heads, a total of 4, 7, and 5 times, respectively. The average number of heads per toss of the 2 coins is then

$$\bar{x} = \frac{(0)(4) + (1)(7) + (2)(5)}{16} = 1.06.$$

This is an average value and is not necessarily a possible outcome for the experiment. For instance, a salesman's average monthly income is not likely to be equal to any of his monthly paychecks.

Let us now restructure our computation of $\bar{x}$ so as to have the following equivalent form:

$$\bar{x} = (0)(\tfrac{4}{16}) + (1)(\tfrac{7}{16}) + (2)(\tfrac{5}{16}) = 1.06.$$

The numbers $\tfrac{4}{16}$, $\tfrac{7}{16}$, and $\tfrac{5}{16}$ are the fractions of the total tosses resulting in 0, 1, and 2 heads, respectively. These fractions are also the relative frequencies for the different values of X in our sample. In effect, then, one can calculate the mean of a set of data by knowing the distinct values that occur and their relative frequencies, without any knowledge of the total number of measurements or observations in our set of data. Therefore, if $\tfrac{4}{16}$ or $\tfrac{1}{4}$ of the tosses result in no heads, $\tfrac{7}{16}$ of the tosses result in 1 head, and $\tfrac{5}{16}$ of the tosses result in 2 heads, the mean number of heads per toss would be 1.06 no matter whether the total number of tosses was 16, 1000, or even 10,000.

Let us now use this method of relative frequencies in calculating the population mean for the number of heads per toss of two coins that we might expect in the long run. We shall refer to this population mean as the **mean of the random variable** X or the **mean of its distribution** and write it as μ_x or simply μ. It is also common among statisticians to refer to this mean as the **mathematical expectation** or the **expected value** of the random variable X

and denote it as $E(X)$. In other words, the mean or expected value of a random variable can be interpreted as the mean of the population or distribution whose observations are all the values of X that are generated by repeating the experiment over and over again indefinitely.

Assuming that fair coins were tossed, the sample space for our experiment is given by

$$S = \{HH, HT, TH, TT\}.$$

Since the four sample points are all equally likely, it follows that

$$P(X = 0) = P(TT) = \tfrac{1}{4},$$
$$P(X = 1) = P(TH) + P(HT) = \tfrac{1}{2},$$

and

$$P(X = 2) = P(HH) = \tfrac{1}{4},$$

where a typical element, say TH, indicates that the first toss resulted in a tail followed by a head on the second toss. Now, these probabilities are just the relative frequencies for the given events in the long run. Therefore,

$$\mu = E(X) = (0)(\tfrac{1}{4}) + (1)(\tfrac{1}{2}) + (2)(\tfrac{1}{4})$$
$$= 1.$$

This means that a person who tosses 2 coins over and over again will, on the average, get 1 head per toss.

The method described above for calculating the expected number of heads per toss of 2 coins suggests that the mean or expected value of any discrete random variable may be obtained by multiplying each of the values x_1, x_2, ..., x_n of the random variable X by its corresponding probability $f(x_1)$, $f(x_2)$, ..., $f(x_n)$ and summing the products. This is true, however, only if the random variable is discrete. Calculating the mean of a continuous random variable requires a knowledge of calculus and is therefore beyond the scope of this text.

DEFINITION

_____

Mean of a Random Variable. Let X be a discrete random variable with the probability distribution

x	x_1	x_2	$\cdots$	x_n
$P(X = x)$	$f(x_1)$	$f(x_2)$	$\cdots$	$f(x_n)$

The *mean* or *expected value* of X is

$$\mu = E(X) = \sum_{i=1}^{n} x_i f(x_i).$$

Example 9. Find the expected value of X, where X represents the outcome when a die is tossed.

Solution. Each of the numbers 1, 2, 3, 4, 5, and 6 occurs with probability $\frac{1}{6}$. Therefore,

$$\mu = E(X) = (1)(\tfrac{1}{6}) + (2)(\tfrac{1}{6}) + \cdots + (6)(\tfrac{1}{6}) = 3.5.$$

This means that a person will, on the average, roll 3.5.

Example 10. Find the expected number of boys on a committee of 3 selected at random from 4 boys and 3 girls.

Solution. Let X represent the number of boys on the committee. The probability distribution of X is given by

$$f(x) = \frac{\binom{4}{x}\binom{3}{3-x}}{\binom{7}{3}}, \quad \text{for } x = 0, 1, 2, 3.$$

A few simple calculations yield $f(0) = \frac{1}{35}$, $f(1) = \frac{12}{35}$, $f(2) = \frac{18}{35}$, and $f(3) = \frac{4}{35}$. Therefore,

$$\mu = E(X) = (0)(\tfrac{1}{35}) + (1)(\tfrac{12}{35}) + (2)(\tfrac{18}{35}) + (3)(\tfrac{4}{35}) = 1.7.$$

Thus if a committee of 3 is selected at random over and over again from 4 boys and 3 girls, it would contain on the average 1.7 boys.

Example 11. In a gambling game a man is paid $5 if he gets all heads or all tails when 3 coins are tossed, and he pays out $3 if either 1 or 2 heads show. What is his expected gain?

Solution. The sample space for the possible outcomes when 3 coins are tossed simultaneously, or equivalently if 1 coin is tossed three times, is

$$S = \{HHH, HHT, HTH, THH, HTT, THT, TTH, TTT\}.$$

One can argue that each of these possibilities is equally likely and occurs with probability equal to $\frac{1}{8}$. An alternative approach would be to apply the multiplicative rule of probability for independent events to each element of S. For example,

$$P(HHT) = P(H)P(H)P(T)$$
$$= (\tfrac{1}{2})(\tfrac{1}{2})(\tfrac{1}{2}) = \tfrac{1}{8}.$$

The random variable of interest is Y, the amount the gambler can win, and the possible values of Y are \$5 if event $E_1 = \{HHH, TTT\}$ occurs and $-\$3$ if event $E_2 = \{HHT, HTH, THH, HTT, THT, TTH\}$ occurs. Since E_1 and E_2 occur with probabilities $\frac{1}{4}$ and $\frac{3}{4}$, respectively, it follows that

$$\mu = E(Y) = (5)(\tfrac{1}{4}) + (-3)(\tfrac{3}{4}) = -1.$$

In this game the gambler will, on the average, lose \$1 per toss of the 3 coins.

A game is considered "fair" if the gambler will, on the average, come out even. Therefore, an expected gain of zero defines a fair game.

Now let us consider a new random variable $g(X)$, which depends on X; that is, each value of $g(X)$ is determined by knowing the values of X. For instance, $g(X)$ might be X^2 or $3X - 1$, so that whenever X assumes the value 2, $g(X)$ assumes the value $g(2)$. In particular, if X is a discrete random variable with probability distribution $f(x)$, $x = -1, 0, 1, 2$, and $g(X) = X^2$, then

$$P[g(X) = 0] = P(X = 0) = f(0),$$
$$P[g(X) = 1] = P(X = -1) + P(X = 1) = f(-1) + f(1),$$
$$P[g(X) = 4] = P(X = 2) = f(2),$$

so that the probability distribution of $g(X)$ may be written

$g(x)$	0	1	4
$P[g(X) = g(x)]$	$f(0)$	$f(-1) + f(1)$	$f(2)$

By the definition of an expected value of a random variable, we obtain

$$\mu_{g(X)} = E[g(X)]$$
$$= 0f(0) + 1[f(-1) + f(1)] + 4f(2)$$

$$= (-1)^2 f(-1) + (0)^2 f(0) + (1)^2 f(1) + (2)^2 f(2)$$

$$= \sum_{i=1}^{4} g(x_i) f(x_i),$$

where $x_1 = -1$, $x_2 = 0$, $x_3 = 1$, and $x_4 = 2$. This result is generalized in Theorem 5.1.

THEOREM 5.1 _____

Mean of a Function of One Variable. Let X be a discrete random variable with the probability distribution

x	x_1	x_2	$\cdots$	x_n
$P(X = x)$	$f(x_1)$	$f(x_2)$	$\cdots$	$f(x_n)$

The mean or expected value of the random variable g(X) is

$$\mu_{g(X)} = E[g(X)] = \sum_{i=1}^{n} g(x_i) f(x_i).$$

_____

Example 12. Suppose the number of cars, X, that pass through a car wash between 4:00 p.m. and 5:00 p.m. on any sunny Friday has the following probability distribution:

x	4	5	6	7	8	9
$P(X = x)$	$\frac{1}{12}$	$\frac{1}{12}$	$\frac{1}{4}$	$\frac{1}{4}$	$\frac{1}{6}$	$\frac{1}{6}$

Let $g(X) = 2X - 1$ represent the amount of money in dollars, paid to the attendant by the manager. Find the attendant's expected earnings for this particular time period.

Solution. By Theorem 5.1, the attendant can expect to receive

$$E[g(X)] = E(2X - 1) = \sum_{x=4}^{9} (2x - 1)f(x)$$

$$= (7)(\tfrac{1}{12}) + (9)(\tfrac{1}{12}) + (11)(\tfrac{1}{4}) + (13)(\tfrac{1}{4})$$

$$+ (15)(\tfrac{1}{6}) + (17)(\tfrac{1}{6})$$

$$= \$12.67.$$

Example 13. Let X be a random variable with probability distribution as follows:

x	0	1	2	3
$P(X = x)$	$\frac{1}{3}$	$\frac{1}{2}$	0	$\frac{1}{6}$

Find $E[(X - \mu)^2]$.

Solution. First we must find μ. Therefore,

$$\mu = E(X) = (0)(\tfrac{1}{3}) + (1)(\tfrac{1}{2}) + (2)(0) + (3)(\tfrac{1}{6}) = 1.$$

The problem now reduces to finding $E[(X - 1)^2]$. We have

$$
\begin{aligned}
E[(X - 1)^2] &= \sum_{x=0}^{3} (x - 1)^2 f(x) \\
&= (-1)^2 f(0) + (0)^2 f(1) + (1)^2 f(2) + (2)^2 f(3) \\
&= (1)(\tfrac{1}{3}) + (0)(\tfrac{1}{2}) + (1)(0) + 4(\tfrac{1}{6}) \\
&= 1.
\end{aligned}
$$

We shall now extend our concept of mathematical expectation to the case of two random variables X and Y with joint probabilities given by $f(x, y)$.

DEFINITION

Mean of a Function of Two Variables. Let X and Y be discrete random variables with joint probabilities given by $f(x, y)$, where $x = x_1, x_2, \ldots, x_m$ and $y = y_1, y_2, \ldots, y_n$. The mean or expected value of the random variable $g(X, Y)$ is

$$\mu_{g(X,Y)} = E[g(X, Y)] = \sum_{i=1}^{m} \sum_{j=1}^{n} g(x_i, y_j) f(x_i, y_j).$$

Example 14. Let X and Y be random variables with joint probability distribution given by Table 5.1. Find the expected value of $g(X, Y) = XY$.

Solution. By the preceding definition, we write

$$E(XY) = \sum_{x=0}^{2} \sum_{y=0}^{2} xy\, f(x, y)$$

$$= (0)(0)f(0, 0) + (0)(1)f(0, 1) + (0)(2)f(0, 2)$$
$$+ (1)(0)f(1, 0) + (1)(1)f(1, 1) + (2)(0)f(2, 0)$$
$$= f(1, 1) = \tfrac{3}{14}.$$

Note that if $g(X, Y) = X$ in the preceding definition, we have

$$\mu_X = E(X) = \sum_{i=1}^{m} \sum_{j=1}^{n} x_i f(x_i, y_j)$$

$$= \sum_{i=1}^{m} x_i g(x_i),$$

where $g(x_i)$ are the values of the marginal distribution of X. Therefore, in calculating μ_X over a two-dimensional space, one may use either the joint probability distribution of X and Y or the marginal distribution of X. Similarly, we define

$$\mu_Y = E(Y) = \sum_{i=1}^{m} \sum_{j=1}^{n} y_j f(x_i, y_j)$$

$$= \sum_{j=1}^{n} y_j h(y_j),$$

where $h(y_j)$ are the values of the marginal distribution of the random variable Y.

Example 15. Referring to the joint probability distribution of Table 5.1 on page 127, find μ_X and μ_Y.

Solution. By definition,

$$\mu_X = E(X) = \sum_{x=0}^{2} \sum_{y=0}^{2} xf(x, y) = \sum_{x=0}^{2} xg(x)$$
$$= (0)(\tfrac{10}{28}) + (1)(\tfrac{15}{28}) + (2)(\tfrac{3}{28})$$
$$= \tfrac{3}{4}$$

and

$$\mu_Y = E(Y) = \sum_{x=0}^{2} \sum_{y=0}^{2} yf(x, y) = \sum_{y=0}^{2} yh(y)$$
$$= (0)(\tfrac{15}{28}) + (1)(\tfrac{3}{7}) + (2)(\tfrac{1}{28})$$
$$= \tfrac{1}{2}.$$

5.6

**Variance of a
Random Variable**

A population whose observations are the values assumed by a random variable X, when an experiment is repeated over and over again indefinitely, not only has a mean μ as defined in Section 5.5, but also a variance that we write as σ_X^2 or simply σ^2. We shall refer to this population variance as the **variance of the random variable** X or the **variance of its distribution.**

In Chapter 2 we found σ^2 for a finite population by averaging the squares of the deviations from the mean. Thus, if the population consists of the 6 measurements 2, 5, 5, 8, 8, and 8, we note that $\mu = 6$ and the variance is

$$\sigma^2 = \frac{(2-6)^2 + (5-6)^2 + (5-6)^2 + (8-6)^2 + (8-6)^2 + (8-6)^2}{6}$$

$$= 5.$$

By observing that several of the deviations from the mean are repeated, we can also write

$$\sigma^2 = \frac{(2-6)^2(1) + (5-6)^2(2) + (8-6)^2(3)}{6}$$

$$= (2-6)^2(\tfrac{1}{6}) + (5-6)^2(\tfrac{1}{3}) + (8-6)^2(\tfrac{1}{2})$$

$$= 5.$$

In this last form one should easily recognize the numbers $\tfrac{1}{6}$, $\tfrac{1}{3}$, and $\tfrac{1}{2}$ to be the relative frequencies for the distinct values 2, 5, and 8 in our population. We see, therefore, that it is possible to compute μ and then σ^2 without knowing the actual size of the population, provided, of course, that we are given the distinct measurements that occur and their relative frequencies.

We shall now employ this method of using relative frequencies in calculating the variance of a discrete random variable X. Since relative frequencies in the long run are just the probabilities associated with each value of the discrete random variable, we can find the variance of the random variable by squaring the deviations from the mean, multiplying each squared deviation by its corresponding probability, and summing the products. In other words, $\sigma^2 = E[(X - \mu)^2]$.

DEFINITION

Variance of a Random Variable. Let X be a discrete random variable with the probability distribution

x	x_1	x_2	$\cdots$	x_n
$f(x)$	$f(x_1)$	$f(x_2)$	$\cdots$	$f(x_n)$

The *variance* of X is

$$\sigma^2 = E[(X - \mu)^2] = \sum_{i=1}^{n} (x_i - \mu)^2 f(x_i).$$

Calculating the variance of a continuous random variable requires advanced mathematics beyond the scope of this text and, therefore, will not be considered here.

Example 16. Calculate the variance of the random variable X in Example 10, where X is the number of boys on a committee of 3 selected at random from 4 boys and 3 girls.

Solution. In Exercise 10 we showed that $\mu = \frac{12}{7}$. Therefore,

$$\sigma^2 = \sum_{x=0}^{3} (x - \mu)^2 f(x)$$

$$= (0 - \tfrac{12}{7})^2(\tfrac{1}{35}) + (1 - \tfrac{12}{7})^2(\tfrac{12}{35}) + (2 - \tfrac{12}{7})^2(\tfrac{18}{35})$$

$$+ (3 - \tfrac{12}{7})^2(\tfrac{4}{35})$$

$$= \tfrac{24}{49}.$$

Example 17. Calculate the interval $\mu \pm 2\sigma$ for the random variable of Example 16. According to Chebyshev's theorem, how does one interpret this interval?

Solution. Based upon our calculations in Example 16, we have

$$\mu \pm 2\sigma = \tfrac{12}{7} \pm 2 \sqrt{\tfrac{24}{49}}$$

$$= 1.714 \pm 1.400.$$

Therefore, if one selects a committee of 3 at random from 4 boys and 3 girls, and repeats this process over and over again indefinitely, then according to Chebyshev's theorem, the number of boys on the committee will fall in the interval 0.314 to 3.114 at least $\frac{3}{4}$ of the time. That is, we could expect the committee to consist of 1, 2, or 3 boys at least $\frac{3}{4}$ of the time.

An alternative and preferred formula for finding σ^2, which often simplifies the calculations, is given in the following theorem.

THEOREM 5.2

Computing Formula for σ^2. *The variance of the random variable X is given by*

$$\sigma^2 = E(X^2) - \mu^2.$$

Proof. Using the rules of summation given by Theorems 1 and 2 in Sections 1.4, we can write

$$\sigma^2 = \sum_{i=1}^{n} (x_i - \mu)^2 f(x_i)$$

$$= \sum_{i=1}^{n} x_i^2 f(x_i) - 2\mu \sum_{i=1}^{n} x_i f(x_i) + \mu^2 \sum_{i=1}^{n} f(x_i).$$

Since $\mu = \sum_{i=1}^{n} x_i f(x_i)$ by definition and $\sum_{i=1}^{n} f(x_i) = 1$ for any discrete probability distribution, it follows that

$$\sigma^2 = \sum_{i=1}^{n} x_i^2 f(x_i) - \mu^2$$

$$= E(X^2) - \mu^2.$$

Example 18. Use the formula of Theorem 5.2 to recalculate the variance of the random variable in Exercise 16.

Solution. Once again referring to Exercise 10, we find that

$$\sum_{x=0}^{3} x^2 f(x) = (0)(\tfrac{1}{35}) + (1)(\tfrac{12}{35}) + (4)(\tfrac{18}{35}) + (9)(\tfrac{4}{35})$$

$$= \tfrac{24}{7}.$$

Now, since we had $\mu = \tfrac{12}{7}$, it follows that

$$\sigma^2 = \tfrac{24}{7} - (\tfrac{12}{7})^2 = \tfrac{24}{49},$$

which agrees with the result obtained in Example 16.

Example 19. The random variable X, representing the number of defective missiles when 3 missiles are fired, has the following probability distribution:

x	0	1	2	3
$P(X = x)$	0.51	0.38	0.10	0.01

Using Theorem 5.2, calculate σ^2.

Solution. First, we compute

$$\mu = (0)(0.51) + (1)(0.38) + (2)(0.10) + (3)(0.01)$$
$$= 0.61.$$

Now,

$$E(X)^2 = (0)(0.51) + (1)(0.38) + (4)(0.10) + (9)(0.01)$$
$$= 0.87.$$

Therefore,

$$\sigma^2 = 0.87 - (0.61)^2 = 0.4979.$$

We shall now extend our concept of the variance of a discrete random variable X to also include random variables related to X. For the random variable $g(X)$, the variance will be denoted by $\sigma^2_{g(X)}$ and is calculated by means of the following theorem:

THEOREM 5.3 _____

Variance of a Function of One Variable. Let X be a discrete random variable with the probability distribution

x	x_1	x_2	$\cdots$	x_n
$f(x)$	$f(x_1)$	$f(x_2)$	$\cdots$	$f(x_n)$

The variance of the random variable g(X) is

$$\sigma^2_{g(X)} = E\{[g(X) - \mu_{g(X)}]^2\} = \sum_{i=1}^{n} [g(x_i) - \mu_{g(X)}]^2 f(x_i).$$

_____

Proof. Since $g(X)$ is itself a random variable with mean $\mu_{g(X)}$ as defined in Theorem 5.1, it follows from the definition of the variance of a random variable on page 140 that

$$\sigma^2 = E\{[g(X) - \mu_{g(X)}]^2\}.$$

Now applying Theorem 5.1 again, we have

$$E\{[g(X) - \mu_{g(X)}]^2\} = \sum_{i=1}^{n} [g(x_i) - \mu_{g(X)}]^2 f(x_i).$$

Example 20. Calculate the variance of $g(X) = 2X + 3$, where X is a random variable with probability distribution

x	0	1	2	3
$f(x)$	$\frac{1}{4}$	$\frac{1}{8}$	$\frac{1}{2}$	$\frac{1}{8}$

Solution. First let us find the mean of the random variable $2X + 3$. According to Theorem 5.1,

$$\mu_{2X+3} = E(2X + 3) = \sum_{x=0}^{3} (2x + 3)f(x) = 6.$$

Now, using Theorem 5.3, we have

$$\sigma^2_{2X+3} = E\{[(2X + 3) - \mu_{2X+3}]^2\}$$

$$= E\{[(2X + 3) - 6]^2\}$$

$$= E(4X^2 + 12X + 3)$$

$$= \sum_{x=0}^{3} (4x^2 + 12x + 3)f(x)$$

$$= \tfrac{137}{8}.$$

5.7

Properties of the Mean and Variance

We shall now develop some useful properties that will simplify the calculations of means and variances of random variables that appear in later chapters. These properties will permit us to calculate expectations in terms of other expectations that are either known or are easily computed. All the results that we present here are valid for both discrete and continuous random variables. Proofs are given only for the discrete case.

THEOREM 5.4

If a and b are constants, then

$$\mu_{aX+b} = a\mu_X + b = a\mu + b.$$

Proof. By the definition of the mean of a random variable,

$$\mu_{aX+b} = E(aX + b) = \sum_{i=1}^{n} (ax_i + b)f(x_i)$$

$$= (ax_1 + b)f(x_1) + (ax_2 + b)f(x_2) + \cdots$$
$$+ (ax_n + b)f(x_n)$$
$$= a[x_1 f(x_1) + x_2 f(x_2) + \cdots + x_n f(x_n)]$$
$$+ b[f(x_1) + f(x_2) + \cdots + f(x_n)]$$
$$= a \sum_{i=1}^{n} x_i f(x_i) + b \sum_{i=1}^{n} f(x_i).$$

The first sum on the right is our definition of μ and the second sum equals 1. Therefore, we have

$$\mu_{aX+b} = a\mu + b.$$

COROLLARY 1

Setting $a = 0$, we see that $\mu_b = b$.

COROLLARY 2

Setting $b = 0$, we see that $\mu_{aX} = a\mu$.

Example 21. Applying Theorem 5.4 to Example 12, we can write

$$\mu_{2X-1} = 2\mu - 1.$$

Now,

$$\mu = E(X) = \sum_{x=4}^{9} xf(x)$$

$$= (4)(\tfrac{1}{12}) + (5)(\tfrac{1}{12}) + (6)(\tfrac{1}{4}) + (7)(\tfrac{1}{4}) + (8)(\tfrac{1}{6}) + (9)(\tfrac{1}{6})$$

$$= \tfrac{41}{6}.$$

Therefore,

$$\mu_{2X-1} = (2)(\tfrac{41}{6}) - 1 = \$12.67$$

as before.

Two additional properties that will be very useful in succeeding chapters involve the mean of sums and differences of random variables and the mean of the product of independent random variables. For the purpose of illustration, we shall restrict our proofs to the bivariate case in which X and Y are discrete random variables with joint probability distribution having the probabilities $f(x, y)$, where $x = x_1, x_2, \ldots, x_m$ and $y = y_1, y_2, \ldots, y_n$.

THEOREM 5.5

The mean of the sum or difference of two or more random variables is equal to the sum or difference of the means of the variables. That is,

$$\mu_{X+Y} = \mu_X + \mu_Y \quad \textit{and} \quad \mu_{X-Y} = \mu_X - \mu_Y.$$

Proof. By definition,

$$\mu_{X+Y} = E(X + Y)$$

$$= \sum_{i=1}^{m} \sum_{j=1}^{n} (x_i + y_j)f(x_i, y_j)$$

$$= \sum_{i=1}^{m} \sum_{j=1}^{n} x_i f(x_i, y_j) + \sum_{i=1}^{m} \sum_{j=1}^{n} y_j f(x_i, y_j) \quad \text{(by Exercise 12, page 19)}$$

$$= \mu_X + \mu_Y.$$

Identical steps can be applied to the difference of two random variables.

If X represents the daily production of an item from machine A and Y the daily production of the same kind of item from machine B, then $X + Y$ represents the total number of items produced daily from both machines. Theorem 5.5 states that the average daily production for both machines is equal to the sum of the average daily production of each machine.

THEOREM 5.6 ————————————————————————————————————

The mean of the product of two or more independent variables is equal to the product of the means of the variables. Therefore, if X and Y are independent,

$$\mu_{XY} = \mu_X \mu_Y.$$

————————————————————————————————————

Proof. By definition,

$$\mu_{XY} = E(XY) = \sum_{i=1}^{m} \sum_{j=1}^{n} x_i y_j f(x_i, y_j).$$

Since X and Y are independent, we may write

$$f(x, y) = g(x)h(y),$$

where $g(x)$ and $h(y)$ are the marginal distributions of X and Y, respectively. Hence

$$\mu_{XY} = \sum_{i=1}^{m} \sum_{j=1}^{n} x_i y_j g(x_i) h(y_j)$$

$$= \left[\sum_{i=1}^{m} x_i g(x_i) \right] \left[\sum_{j=1}^{n} y_j h(y_j) \right] \quad \text{(by Exercise 15 on page 19)}$$

$$= \mu_X \mu_Y.$$

Theorem 5.6 can be illustrated by tossing a green die and a red die. Let the random variable X represent the outcome on the green die and the random variable Y represent the outcome on the red die. Then XY represents the product of the numbers that occur on the pair of dice. In the long run, the average of the products of the numbers is equal to the product of the average number that occurs on the green die and the average number that occurs on the red die.

We conclude this section by proving three theorems that are useful in calculating variances or standard deviations.

THEOREM 5.7

If X is a random variable and b is a constant, then

$$\sigma^2_{X+b} = \sigma^2_X = \sigma^2.$$

Proof. By Theorem 5.3,

$$\sigma^2_{X+b} = E\{[(X + b) - \mu_{X+b}]^2\}.$$

Now, from Theorem 5.4, $\mu_{X+b} = \mu + b$. Therefore,

$$\sigma^2_{X+b} = E[(X + b - \mu - b)^2]$$
$$= E[(X - \mu)^2]$$
$$= \sigma^2.$$

This theorem states that the variance is unchanged if a constant is added to or subtracted from a random variable. The addition or subtraction of a constant simply shifts the values of X to the right or to the left but does not change their variability. Of course, we already verified this for finite populations in our discussion of coded data on page 37.

THEOREM 5.8

If X is a random variable and a is any constant, then

$$\sigma^2_{aX} = a^2\sigma^2_X = a^2\sigma^2.$$

Proof. By Theorem 5.3,

$$\sigma^2_{aX} = E\{[aX - \mu_{aX}]^2\}.$$

Now, by Corollary 2 of Theorem 5.4, $\mu_{aX} = a\mu$. Therefore,

$$\sigma^2_{aX} = E[(aX - a\mu)^2]$$
$$= a^2 E[(X - \mu)^2]$$
$$= a^2\sigma^2.$$

Therefore, if a random variable is multiplied or divided by a constant, the variance is multiplied or divided by the square of the constant. This too was verified on page 37 for finite populations.

THEOREM 5.9

The variance of the sum or difference of two or more independent random variables is equal to the sum of the variances of the variables. That is, if X and Y are independent,

$$\sigma^2_{X+Y} = \sigma^2_X + \sigma^2_Y \quad \text{and} \quad \sigma^2_{X-Y} = \sigma^2_X + \sigma^2_Y.$$

Proof. By extending Theorem 5.3 to the case of two variables, we have

$$\sigma^2_{X-Y} = E\{[(X - Y) - \mu_{X-Y}]^2\}.$$

Now, from Theorem 5.5, $\mu_{X-Y} = \mu_X - \mu_Y$. Therefore,

$$\begin{aligned}
\sigma^2_{X-Y} &= E\{[(X - Y) - (\mu_X - \mu_Y)]^2\} \\
&= E\{[(X - \mu_X) - (Y - \mu_Y)]^2\} \\
&= E[(X - \mu_X)^2] + E[(Y - \mu_Y)^2] - 2E[(X - \mu_X)(Y - \mu_Y)] \\
&= \sigma^2_X + \sigma^2_Y,
\end{aligned}$$

since for independent variables

$$\begin{aligned}
E[(X - \mu_X)(Y - \mu_Y)] &= E(XY - \mu_X Y - \mu_Y X + \mu_X \mu_Y) \\
&= E(X)E(Y) - \mu_X E(Y) - \mu_Y E(X) + E(\mu_X \mu_Y) \\
&= \mu_X \mu_Y - \mu_X \mu_Y - \mu_Y \mu_X + \mu_X \mu_Y \\
&= 0.
\end{aligned}$$

The proof for the variance of the sum of two independent random variables is left for the reader.

Example 22. If X and Y are independent random variables with variances $\sigma^2_X = 1$ and $\sigma^2_Y = 2$, find the variance of the random variable $Z = 3X - 2Y + 5$.

Solution. Treating $3X - 2Y$ as a single random variable, we can use Theorem 5.7 to write

$$\sigma^2_Z = \sigma^2_{3X - 2Y + 5} = \sigma^2_{3X - 2Y}.$$

Now, applying Theorem 5.9 and then Theorem 5.8 to the two independent variables $3X$ and $2Y$, we have

$$\sigma_Z^2 = \sigma_{3X}^2 + \sigma_{2Y}^2$$
$$= 9\sigma_X^2 + 4\sigma_Y^2$$
$$= (9)(1) + (4)(2)$$
$$= 17.$$

EXERCISES

1. A shipment of 7 television sets contains 2 defectives. A hotel makes a random purchase of 3 of the sets. If X is the number of defective sets purchased by the hotel, find the mean of X.

2. The probability distribution of the discrete random variable X is

$$f(x) = \binom{3}{x}\left(\frac{1}{4}\right)^x\left(\frac{3}{4}\right)^{3-x}, \qquad \text{for } x = 0, 1, 2, 3.$$

Find the mean of X.

3. Find the expected number of jazz records when 4 records are selected at random from a collection consisting of 5 jazz records, 2 classical records, and 3 polka records.

4. Find the mean of the random variable T representing the total of the 3 coins in Exercise 5 on page 125.

5. Let X be a random variable with the following probability distribution:

x	0	1	2	3
$P(X = x)$	$\frac{8}{27}$	$\frac{4}{9}$	$\frac{2}{9}$	$\frac{1}{27}$

Find the mean of X.

6. By investing in a particular stock, a person can make a profit in 1 year of $4000 with probability 0.4 or take a loss of $1000 with probability 0.7. What is this person's expected gain?

7. In a gambling game a man is paid $3 if he draws a jack or queen and $5 if he draws a king or ace from an ordinary deck of 52 playing cards. If he draws any other card, he loses. How much should he pay to play if the game is fair?

8. A race-car driver wishes to insure his car for the racing season for $50,000. The insurance company estimates a total loss may occur with probability 0.002, a 50% loss with probability 0.01, and a 25% loss with probability 0.1. Ignoring all

other partial losses, what premium should the insurance company charge each season to realize an average profit of $500?

9. Let X represent the outcome when a balanced die is tossed. Find $\mu_{g(X)}$, where $g(X) = 3X^2 + 4$.

10. Let X be a random variable with the following probability distribution:

x	-3	6	9
$P(X = x)$	$\frac{1}{6}$	$\frac{1}{2}$	$\frac{1}{3}$

Find $\mu_{g(X)}$, where $g(X) = (2X + 1)^2$.

11. Find the expected value of the random variable $g(X) = X^2$, where X has the probability distribution of Exercise 2.

12. Suppose that X and Y have the following joint probability function:

		x	
y		2	4
1		0.10	0.15
3		0.20	0.30
5		0.10	0.15

Find the expected value of $g(X, Y) = XY^2$.

13. Referring to the random variables whose joint probability distribution is given in Exercise 12, find the mean of $g(X, Y) = XY$.

14. From a sack of fruit containing 3 oranges, 2 apples, and 3 bananas, a random sample of 4 pieces of fruit is selected. If X is the number of oranges and Y is the number of apples in the sample, find $E(X^2Y - 2XY)$.

15. Suppose that X and Y have the following joint probability distribution:

		x		
y		1	2	3
1		0	$\frac{1}{6}$	$\frac{2}{12}$
2		$\frac{1}{5}$	$\frac{1}{9}$	0
3		$\frac{2}{15}$	$\frac{1}{4}$	$\frac{1}{18}$

Find μ_X and μ_Y.

16. Evaluate μ_X and μ_Y for the random variables of Exercise 3 on page 131.

17. Using the definition on page 140, find the variance of the random variable X of Exercise 1.

18. Let X be a random variable with the following probability distribution:

x	-2	3	5
$P(X = x)$	0.3	0.2	0.5

   Find the standard deviation of X.

19. From a group of 5 men and 3 women a committee of size 3 is selected at random. Calculate the interval $\mu \pm 2\sigma$ for the random variable X representing the number of women on the committee. According to Chebyshev's theorem, how does one interpret this interval?

20. The random variable X, representing the number of chocolate chips in a cookie, has the following probability distribution:

x	2	3	4	5	6
$P(X = x)$	0.01	0.25	0.4	0.3	0.04

   Using Theorem 5.2, find the variance of X.

21. Referring to Exercise 9, find $\sigma^2_{g(X)}$ for the function $g(X) = 3X^2 + 4$.

22. Find the standard deviation of the random variable $g(X) = (2X + 1)^2$ in Exercise 10.

23. Referring to Exercise 20, find the mean of the random variable $g(X) = 3X - 2$

    (a) by using Theorem 5.1;
    (b) by using Theorem 5.4.

24. Using Theorem 5.4, find the mean of the random variable $g(X) = 5X + 3$, where X has the probability distribution of Exercise 5.

25. Let X represent the number that occurs when a green die is tossed and Y the number that occurs when a red die is tossed. Find the variance of the random variable

    (a) $2X - Y$;   (b) $X + 3Y - 5$.

26. If X and Y are independent random variables with the variances $\sigma^2_X = 5$ and $\sigma^2_Y = 3$, find the variance of the random variable $Z = -2X + 4Y - 3$.

27. **Covariance:** The *covariance* of two random variables X and Y, denoted by σ_{XY}, is defined to be

$$\sigma_{XY} = E[(X - \mu_X)(Y - \mu_Y)].$$

The covariance will be positive when large values of X are associated with large values of Y and small values of X are associated with small values of Y. If small values of X are associated with large values of Y, and vice versa, then the covariance will be negative. Therefore, the covariance provides a measure of association between the values of the two random variables.

(a) Show that $\sigma_{XY} = E(XY) - \mu_X\mu_Y$.

(b) Find the covariance of the random variables X and Y in Exercise 15.

6 Some Discrete Probability Distributions

No matter whether a discrete probability distribution is represented graphically by a histogram, in tabular form, or by means of a formula, the behavior of a random variable is described. Often, the observations generated by different statistical experiments have the same general type of behavior. Consequently, discrete random variables associated with these experiments can be described by essentially the same probability distribution and therefore can be represented by a single formula. In fact, one needs only a handful of important discrete probability distributions to describe most random variables encountered in practice.

6.1

**Uniform
Distribution**

The simplest of all discrete probability distributions is one in which the random variable assumes all its values with equal probabilities. Such a probability distribution is called the **discrete uniform distribution.**

DEFINITION

Discrete Uniform Distribution. If the random variable X assumes the values $x_1, x_2, \ldots, x_k$, with equal probabilities, then the *discrete uniform distribution* is given by

$$f(x; k) = \frac{1}{k}, \qquad \text{for } x = x_1, x_2, \ldots, x_k.$$

We have used the notation $f(x; k)$ instead of $f(x)$ to indicate that the uniform distribution depends on the parameter k.

Example 1. When a die is tossed, each element of the sample space $S = \{1, 2, 3, 4, 5, 6\}$ occurs with probability $\frac{1}{6}$. Therefore, we have a uniform distribution, with

$$f(x; 6) = \tfrac{1}{6}, \qquad \text{for } x = 1, 2, 3, 4, 5, 6.$$

Example 2. Suppose that an employee is selected at random from a staff of 10 to supervise a certain project. Each employee has the same probability $\frac{1}{10}$ of being selected. If we assume that the employees have been numbered in some way from 1 to 10, the distribution is uniform with $f(x; 10) = \frac{1}{10}$, for $x = 1, 2, \ldots, 10$.

The graphic representation of the uniform distribution by means of a histogram always turns out to be a set of rectangles with equal heights. The histogram for Example 1 is shown in Figure 6.1.

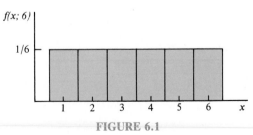

FIGURE 6.1
Histogram for the tossing of a die.

It is not difficult to see that the distribution of all possible subsets of size n from a finite sample space is itself uniform. Suppose that the sample space consists of the 4 clergymen, A, B, C, and D, in a small town, from which 2 are to be chosen at random for a community Thanksgiving Day service. The total number of possible combinations is $\binom{4}{2} = 6$, which we could list as follows: AB, AC, AD, BC, BD, and CD. Since each has the same probability of being drawn, the distribution of subsets is uniform, with

$$f(x; 6) = \tfrac{1}{6}, \qquad \text{for } x = 1, 2, \ldots, 6.$$

Therefore,

$$P(B \text{ and } D \text{ are chosen}) = P(X = 5) = f(5; 6) = \tfrac{1}{6}.$$

In general, the value of k in the formula of the uniform distribution is given by $\binom{N}{n}$ when selecting a subset of size n from a finite sample space of size N.

> **Example 3.** Find the uniform distribution for the subsets of months of size 3.
>
> **Solution.** Since there are 12 possible months, we may choose 3 at random in $\binom{12}{3} = 220$ ways. Numbering these subsets from 1 to 220 the probability distribution is given by
>
> $$f(x; 220) = \tfrac{1}{220}, \qquad \text{for } x = 1, 2, \ldots, 220.$$
>
> Thus the probability of choosing subset number 5 is
>
> $$f(5; 220) = \tfrac{1}{220}.$$

6.2
Binomial and Multinomial Distributions

An experiment often consists of repeated trials, each with two possible outcomes, which may be labeled **success** or **failure**. This is true in the flipping of a coin 5 times, where each trial may result in a head or a tail. We may choose to define either outcome as a success. It is also true if 5 cards are drawn in succession from an ordinary deck and each trial is labeled "success" or "failure," depending on whether the card is red or black. If each card is replaced and the deck shuffled before the next drawing, then the two experiments described above have similar properties, in that repeated trials are independent and the probability of a success remains constant, $\frac{1}{2}$, from trial to trial. Experiments of this type are known as **binomial experiments.** Observe in the card-drawing example that the probabilities of a success for the repeated trials change if the cards are not replaced. That is, the probability of selecting a red card on the first draw is $\frac{1}{2}$, but on the second draw it is a conditional probability having a value of $\frac{26}{51}$ or $\frac{25}{51}$, depending on the color that occurred on the first draw. This then would no longer be considered a binomial experiment.

A binomial experiment is one that possesses the following properties:

1. The experiment consists of n repeated trials.
2. Each trial results in an outcome that may be classified as a success or a failure.
3. The probability of a success, denoted by p, remains constant from trial to trial.
4. The repeated trials are independent.

Consider the binomial experiment where a coin is tossed 3 times and a head is designated a success. The number of successes is a random variable X assuming integral values from 0 through 3. The 8 possible outcomes and the corresponding values of X are as follows:

Outcome	x
TTT	0
THT	1
TTH	1
HTT	1
THH	2
HTH	2
HHT	2
HHH	3

Since the trials are independent with constant probability of success equal to $\frac{1}{2}$, the $P(HTH) = P(H)P(T)P(H) = (\frac{1}{2})(\frac{1}{2})(\frac{1}{2}) = \frac{1}{8}$. Similarly, each of the other possible outcomes occur with probability $\frac{1}{8}$. The probability distribution of X is therefore given by

x	0	1	2	3
$P(X = x)$	$\frac{1}{8}$	$\frac{3}{8}$	$\frac{3}{8}$	$\frac{1}{8}$

or by the formula

$$f(x) = \frac{\binom{3}{x}}{8}, \qquad \text{for } x = 0, 1, 2, 3.$$

The number X of successes in n trials of a binomial experiment is called a **binomial random variable.** The probability distribution of this discrete variable is called the **binomial distribution** and its values will be denoted by $b(x; n, p)$, since they depend on the number of trials and the probability of a success on a given trial. Thus, for the 3 tosses of a coin, the probability distribution should be rewritten

$$b(x; 3, \tfrac{1}{2}) = \frac{\binom{3}{x}}{8}, \qquad \text{for } x = 0, 1, 2, 3.$$

Let us now generalize the foregoing illustration to yield a formula for $b(x; n, p)$. That is, we wish to find a formula that gives the probability of x successes in n trials for a binomial experiment. First, consider the probability of x successes and $n - x$ failures in a specified order. Since the trials are independent, we can multiply all the probabilities corresponding to the different outcomes. Each success occurs with probability p and each failure with probability $q = 1 - p$. Therefore, the probability for the specified order is $p^x q^{n-x}$. We must now determine the total number of sample points in the experiment that have x successes and $n - x$ failures. This number is equal to the number of partitions of n outcomes into two groups with x in one group and $n - x$ in the other and is given by $\binom{n}{x}$. Because these partitions are mutually exclusive, we add the probabilities of all the different partitions to obtain the general formula, or simply multiply $p^x q^{n-x}$ by $\binom{n}{x}$.

DEFINITION

Binomial Distribution. If a binomial trial can result in a success with probability p and a failure with probability $q = 1 - p$, then the probability distribution of the binomial random variable X, the number of successes in n independent trials is

$$b(x; n, p) = \binom{n}{x} p^x q^{n-x}, \qquad \text{for } x = 0, 1, 2, \dots, n.$$

Note that when $n = 3$ and $p = \frac{1}{2}$,

$$b\left(x; 3, \frac{1}{2}\right) = \binom{3}{x}\left(\frac{1}{2}\right)^x \left(\frac{1}{2}\right)^{3-x} = \frac{\binom{3}{x}}{8},$$

which agrees with the answer above for the number of heads when three coins are tossed.

Example 4. Find the probability of obtaining exactly three 2's if an ordinary die is tossed 5 times.

Solution. The probability of a success on each of the 5 independent trials is $\frac{1}{6}$ and the probability of a failure is $\frac{5}{6}$. The outcome of 2 is considered a success here. Therefore,

$$b\left(3; 4, \frac{1}{6}\right) = \binom{5}{3}\left(\frac{1}{6}\right)^3 \left(\frac{5}{6}\right)^2$$

$$= \frac{5!}{3!2!} \cdot \frac{5^2}{6^5}$$

$$= 0.032.$$

Example 5. In a certain city district the need for money to buy drugs is given as the reason for 75% of all thefts. What is the probability that exactly 2 of the next 4 theft cases reported in this district resulted from the need for money to buy drugs?

Solution. Assuming that the thefts are independent and $p = \frac{3}{4}$ for each of the 4 thefts, then

$$b\left(2; 4, \frac{3}{4}\right) = \binom{4}{2}\left(\frac{3}{4}\right)^2\left(\frac{1}{4}\right)^2$$

$$= \frac{4!}{2!2!} \cdot \frac{3^2}{4^4}$$

$$= 0.211.$$

The binomial distribution derives its name from the fact that the $(n + 1)$ terms in the binomial expansion of $(q + p)^n$ correspond to the various values of $b(x; n, p)$ for $x = 0, 1, 2, \ldots, n$. That is,

$$(q + p)^n = \binom{n}{0}q^n + \binom{n}{1}pq^{n-1} + \binom{n}{2}p^2q^{n-2} + \cdots + \binom{n}{n}p^n$$

$$= b(0; n, p) + b(1; n, p) + b(2; n, p) + \cdots + b(n; n, p).$$

Since $p + q = 1$, we see that $\sum_{x=0}^{n} b(x; n, p) = 1$, a condition that must hold for any probability distribution.

Frequently, we are interested in problems in which it is necessary to find $P(X < r)$ or $P(a \leq X \leq b)$. Fortunately, binomial sums $\sum_{x=0}^{r} b(x; n, p)$ are available and are given in Table A.2 of the Appendix for samples of size $n = 1, 2, \ldots, 20$, and selected values of p from 0.1 to 0.9. We illustrate the use of Table A.2 with the following example.

Example 6. The probability that a patient recovers from a rare blood disease is 0.4. If 15 people are known to have contracted this disease, what is the probability that (a) at least 10 survive; (b) from 3 to 8 survive; and (c) exactly 5 survive?

Solution
(a) Let X be the number of people that survive. Then

$$P(X \geq 10) = 1 - P(X < 10)$$

$$= 1 - \sum_{x=0}^{9} b(x; 15, 0.4)$$

$$= 1 - 0.9662$$

$$= 0.0338.$$

(b) $P(3 \leq X \leq 8) = \displaystyle\sum_{x=3}^{8} b(x; 15, 0.4)$

$$= \sum_{x=0}^{8} b(x; 15, 0.4) - \sum_{x=0}^{2} b(x; 15, 0.4)$$

$$= 0.9050 - 0.0271$$

$$= 0.8779.$$

(c) $P(X = 5) = b(5; 15, 0.4)$

$$= \sum_{x=0}^{5} b(x; 15, 0.4) - \sum_{x=0}^{4} b(x; 15, 0.4)$$

$$= 0.4032 - 0.2173$$

$$= 0.1859.$$

THEOREM 6.1

The mean and variance of the binomial distribution $b(x; n, p)$ are

$$\mu = np \quad and \quad \sigma^2 = npq.$$

Proof. Let the outcome on the jth trial be represented by the random variable I_j, which assumes the values 0 and 1 with probabilities q and p, respectively. This is called a **Bernoulli variable** or perhaps more appropriately, an **indicator variable,** since $I_j = 0$ indicates a failure and $I_j = 1$ indicates a success.

Therefore, in a binomial experiment the number of successes can be written as the sum of the n independent indicator variables. Hence

$$X = I_1 + I_2 + \cdots + I_n.$$

The mean of any I_j is $E(I_j) = 0 \cdot q + 1 \cdot p = p$. Therefore, using Theorem 5.5 of page 146, the mean of the binomial distribution is

$$\mu = E(X) = E(I_1) + E(I_2) + \cdots + E(I_n)$$

$$= \underbrace{p + p + \cdots + p}_{n \text{ terms}}$$

$$= np.$$

The variance of any I_j is given by

$$\sigma_{I_j}^2 = E[(I_j - p)^2] = E(I_j^2) - p^2$$
$$= (0)^2 q + (1)^2 p - p^2$$
$$= p(1 - p) = pq.$$

Therefore, by Theorem 5.9 on page 149, the variance of the binomial distribution is

$$\sigma_X^2 = \sigma_{I_1}^2 + \sigma_{I_2}^2 + \cdots + \sigma_{I_n}^2$$
$$= \underbrace{pq + pq + \cdots + pq}_{n \text{ terms}}$$
$$= npq.$$

Example 7. Using Chebyshev's theorem, find and interpret the interval $\mu \pm 2\sigma$ for Example 6.

Solution. Since Example 6 was a binomial experiment with $n = 15$ and $p = 0.4$, by Theorem 6.1 we have

$$\mu = (15)(0.4) = 6 \quad \text{and} \quad \sigma^2 = (15)(0.4)(0.6) = 3.6.$$

Taking the square root of 3.6, we find that $\sigma = 1.897$. Hence the required interval is $6 \pm (2)(1.897)$, or from 2.206 to 9.794. Chebyshev's theorem states that the number of recoveries of 15 patients subjected to the given disease has a probability of at least $\frac{3}{4}$ of falling between 2.206 and 9.794.

The binomial experiment becomes a **multinomial experiment** if we let each trial have more than two possible outcomes. Hence the outcomes, when a pair of dice is tossed, might be recorded as a matching pair, a total of 7 or 11, or neither of these cases, and therefore constitutes a multinomial experiment. The drawing of a card from a deck *with replacement* is also a multinomial experiment if the 4 suits are the outcomes of interest.

In general, if a given trial can result in any one of k possible outcomes $E_1, E_2, \ldots, E_k$, with probabilities $p_1, p_2, \ldots, p_k$, then the **multinomial distribution** will give the probability that E_1 occurs x_1 times, E_2 occurs x_2 times, $\ldots$, E_k occurs x_k times in n independent trials, where

$$x_1 + x_2 + \cdots + x_k = n.$$

We shall denote this joint probability distribution by $f(x_1, x_2, \ldots, x_k; p_1, p_2, \ldots, p_k, n)$. Clearly, $p_1 + p_2 + \cdots + p_k = 1$, since the result of each trial must be one of the k possible outcomes.

To derive the general formula, we proceed as in the binomial case. Since the trials are independent, any specified order yielding x_1 outcomes for E_1, x_2 for E_2, $\ldots$, x_k for E_k will occur with probability $p_1^{x_1} p_2^{x_2} \cdots p_k^{x_k}$. The total number of orders yielding similar outcomes for the n trials is equal to the number of partitions of n items into k groups with x_1 in the first group, x_2 in the second group, $\ldots$, x_k in the kth group. This can be done in

$$\binom{n}{x_1, x_2, \ldots, x_k} = \frac{n!}{x_1! x_2! \cdots x_k!}$$

ways. Since the partitions are mutually exclusive and occur with equal probability, we obtain the multinomial distribution by multiplying the probability for a specified order by the total number of partitions.

DEFINITION

Multinomial Distribution. If a given trial can result in the k outcomes E_1, E_2, $\ldots$, E_k, with probabilities p_1, p_2, $\ldots$, p_k, then the probability distribution of the random variables X_1, X_2, $\ldots$, X_k, representing the number of occurrences for E_1, E_2, $\ldots$, E_k in n independent trials, is

$$f(x_1, x_2, \ldots, x_k; p_1, p_2, \ldots, p_k, n) = \binom{n}{x_1, x_2, \ldots, x_k} p_1^{x_1} p_2^{x_2} \cdots p_k^{x_k},$$

with $\sum_{i=1}^{k} x_i = n$ and $\sum_{i=1}^{k} p_i = 1$.

The multinomial distribution derives its name from the fact that the terms of the multinomial expansion of $(p_1 + p_2 + \cdots + p_k)^n$ correspond to all the possible values of $f(x_1, x_2, \ldots, x_k; p_1, p_2, \ldots, p_k, n)$.

Example 8. If a pair of dice is tossed 6 times, what is the probability of obtaining a total of 7 or 11 twice, a matching pair once, and any other combination 3 times?

Solution. We list the following possible events:

      E_1: a total of 7 or 11 occurs,

      E_2: a matching pair occurs,

      E_3: neither a pair nor a total of 7 or 11 occurs.

The corresponding probabilities for a given trial are $p_1 = \frac{2}{9}$, $p_2 = \frac{1}{6}$, and $p_3 = \frac{11}{18}$. These values remain constant for all 6 trials. Using the multinomial distribution with $x_1 = 2$, $x_2 = 1$, and $x_3 = 3$, we obtain the required probability:

$$f\left(2, 1, 3; \frac{2}{9}, \frac{1}{6}, \frac{11}{18}, 6\right) = \binom{6}{2, 1, 3}\left(\frac{2}{9}\right)^2\left(\frac{1}{6}\right)^1\left(\frac{11}{18}\right)^3$$

$$= \frac{6!}{2!1!3!} \cdot \frac{2^2}{9^2} \cdot \frac{1}{6} \cdot \frac{11^3}{18^3}$$

$$= 0.1127.$$

EXERCISES

1. Find a formula for the distribution of the random variable X representing the number on a tag drawn at random from a box containing 12 tags numbered 1 to 12. What is the probability that the number drawn is less than 4?

2. A roulette wheel is divided into 25 sectors of equal area numbered from 1 to 25. Find a formula for the probability distribution of X, the number that occurs when the wheel is spun.

3. Find the uniform distribution for the random samples of committees of size 4 chosen from 6 students.

4. A baseball player's batting average is 0.250. What is the probability that he gets exactly 1 hit in his next 5 times at bat?

5. If we define the random variable X to be equal to the number of heads that occur when a balanced coin is flipped once, find the probability distribution of X. What two well-known distributions describe the values of X?

6. A multiple-choice quiz has 15 questions, each with 4 possible answers of which only 1 is the correct answer. What is the probability that sheer guesswork yields from 5 to 10 correct answers?

7. The probability that a patient recovers from a delicate heart operation is 0.9. What is the probability that exactly 5 of the next 7 patients having this operation survive?

8. A study conducted at George Washington University and the National Institute of Health examined national attitudes about tranquilizers. The study revealed that approximately 70% believe "tranquilizers don't really cure anything, they just cover up the real trouble." According to this study, what is the probability that at least 3 of the next 5 people selected at random will be of the opinion that tranquilizers actually do cure the problem rather than just cover it up?

9. A survey of the residents in a United States city showed that 20% preferred a white telephone over any other color available. What is the probability that more than one-half of the next 20 telephones installed in this city will be white?

10. One-fourth of the female freshmen entering a Virginia college are out-of-state students. If the students are assigned at random to the dormitories, 3 to a room, what is the probability that in one room at most 2 of the 3 roommates are out-of-state students?

11. If X represents the number of people that believe tranquilizers cure the problem in Exercise 8 when 5 people are selected at random, find the probability distribution of X. Using Chebyshev's theorem, find and interpret the interval $\sigma \pm 2\sigma$.

12. Suppose that airplane engines operate independently in flight and fail with probability $q = \frac{1}{5}$. Assuming that a plane makes a safe flight if at least one-half of its engines run, determine whether a 4-engine plane or a 2-engine plane has the highest probability for a successful flight.

13. Repeat Exercise 12 for $q = \frac{1}{2}$ and $q = \frac{1}{3}$.

14. In Exercise 6, how many correct answers would you expect based on sheer guesswork? Using Chebyshev's theorem, find and interpret the interval $\mu \pm 2\sigma$.

15. If 64 coins are tossed a large number of times, how many heads can we expect on the average per toss? Using Chebyshev's theorem, between what two values would you expect the number of heads to fall at least $\frac{3}{4}$ of the time?

16. A card is drawn from a well-shuffled deck of 52 playing cards, the result recorded, and the card replaced. If the experiment is repeated 5 times, what is the probability of obtaining 2 spades and 1 heart?

17. The surface of a circular dart board has a small center circle called the bull's-eye and 20 pie-shaped regions numbered from 1 to 20. Each of the pie-shaped regions is further divided into 3 parts such that a person throwing a dart that lands on a specified number scores the value of the number, double the number, or triple the number, depending on which of the 3 parts the dart falls. If a person hits the bull's-eye with probability 0.01, hits a double with probability 0.10, a triple with probability 0.05, and misses the dart board with probability 0.02, what is the probability that 7 throws will result in no bull's-eyes, no triples, a double twice, and a complete miss once?

18. Find the probability of obtaining 2 ones, 1 two, 1 three, 2 fours, 3 fives, and 1 six in 10 rolls of a balanced die.

19. According to the theory of genetics a certain cross of guinea pigs will result in red, black, and white offspring in the ratio 8:4:4. Find the probability that among 8 such offspring 5 will be red, 2 black, and 1 white.

20. The probabilities are 0.4, 0.2, 0.3, and 0.1, respectively, that a delegate to a certain convention arrived by air, bus, automobile, or train. What is the probability that among 9 delegates randomly selected at this convention, 3 arrived by air, 3 arrived by bus, 1 arrived by automobile, and 2 arrived by train?

6.3

Hypergeometric Distribution

In Section 6.2 we saw that the binomial distribution did not apply if we wished to find the probability of observing 3 red cards in 5 draws from an ordinary deck of 52 playing cards, unless each card is replaced and the deck reshuffled before the next drawing is made. To solve the problem of sampling without replacement, let us restate the problem. If 5 cards are drawn at random, we are interested in the probability of selecting 3 red cards from the 26 available and 2 black cards from the 26 black cards available in the deck. There are $\binom{26}{3}$ ways of selecting 3 red cards; for each of these ways we can choose 2 black cards in $\binom{26}{2}$ ways. Therefore, the total number of ways to select 3 red and 2 black cards in 5 draws is the product $\binom{26}{3}\binom{26}{2}$. The total number of ways to select any 5 cards from the 52 that are available is $\binom{52}{5}$. Hence the probability of selecting 5 cards without replacement, of which 3 are red and 2 are black, is given by

$$\frac{\binom{26}{3}\binom{26}{2}}{\binom{52}{5}} = \frac{\dfrac{26!}{3!23!} \cdot \dfrac{26!}{2!24!}}{\dfrac{52!}{5!47!}} = 0.3251.$$

In general, we are interested in the probability of selecting x successes from the k items labeled "success" and $n - x$ failures from the $N - k$ items labeled "failures," when a random sample of size n is selected from a finite population of size N. This is known as a **hypergeometric experiment.**

A hypergeometric experiment is one that possesses the following two properties:

1. A random sample of size n is selected from a population of N items.
2. k of the N items may be classified as successes and $N - k$ classified as failures.

The number X of successes in a hypergeometric experiment is called a **hypergeometric random variable.** Accordingly, the probability distribution of the hypergeometric variable is called the **hypergeometric distribution** and its values will be denoted by $h(x; N, n, k)$, since they depend on the number of successes k in the set N from which we select n items.

Example 9. A committee of size 5 is to be selected at random from 3 women and 5 men. Find the probability distribution for the number of women on the committee.

Solution. Let the random variable X be the number of women on the committee. The two properties of a hypergeometric experiment are satisfied. Hence

$$P(X = 0) = h(0; 8, 5, 3) = \frac{\binom{3}{0}\binom{5}{5}}{\binom{8}{5}} = \frac{1}{56},$$

$$P(X = 1) = h(1; 8, 5, 3) = \frac{\binom{3}{1}\binom{5}{4}}{\binom{8}{5}} = \frac{15}{56},$$

$$P(X = 2) = h(2; 8, 5, 3) = \frac{\binom{3}{2}\binom{5}{3}}{\binom{8}{5}} = \frac{30}{56},$$

$$P(X = 3) = h(3; 8, 5, 3) = \frac{\binom{3}{3}\binom{5}{2}}{\binom{8}{5}} = \frac{10}{56}.$$

In tabular form the hypergeometric distribution of X is as follows:

x	0	1	2	3
$P(X = x)$	$\frac{1}{56}$	$\frac{15}{56}$	$\frac{30}{56}$	$\frac{10}{56}$

It is not difficult to see that the probability distribution can be given by the formula

$$h(x; 8, 5, 3) = \frac{\binom{3}{x}\binom{5}{5-x}}{\binom{8}{5}}, \qquad \text{for } x = 0, 1, 2, 3.$$

Let us now generalize Example 9 to find a formula for $h(x; N, n, k)$. The total number of samples of size n chosen from N items is $\binom{N}{n}$. These samples are assumed to be equally likely. There are $\binom{k}{x}$ ways of selecting x successes from the k that are available, and for each of these ways we can choose the $n - x$ failures in $\binom{N - k}{n - x}$ ways. Thus the total number of favorable samples among the $\binom{N}{n}$ possible samples is given by $\binom{k}{x}\binom{N - k}{n - x}$. Hence we have the following definition.

DEFINITION

Hypergeometric Distribution. If a population of size N contains k items labeled "success" and $N - k$ items labeled "failure," then the probability distribution of the hypergeometric random variable X, the number of successes in a random sample of size n, is

$$h(x; N, n, k) = \frac{\binom{k}{x}\binom{N - k}{n - x}}{\binom{N}{n}}, \qquad \text{for } x = 0, 1, 2, \ldots, n.$$

Example 10. If 5 cards are dealt from a standard deck of 52 playing cards, what is the probability that 3 will be hearts?

Solution. By using the hypergeometric distribution with $n = 5$, $N = 52$, $k = 13$, and $x = 3$, we find the probability of receiving 3 hearts to be

$$h(3; 52, 5, 13) = \frac{\binom{13}{3}\binom{39}{2}}{\binom{52}{5}} = 0.0815.$$

To find the mean and variance of the hypergeometric distribution, we can again write

$$X = I_1 + I_2 + \cdots + I_n,$$

where I_j assumes a value of 1 or 0, depending on whether we have a success or a failure on the jth draw. However, since the indicator variables are no longer independent, the problem of finding the mean and variance becomes much more complex. Therefore, we omit the proof and merely include the results in the following theorem.

THEOREM 6.2
_____

The mean and variance of the hypergeometric distribution $h(x; N, n, k)$ are

$$\mu = \frac{nk}{N},$$

$$\sigma^2 = \frac{N-n}{N-1} \cdot n \cdot \frac{k}{N}\left(1 - \frac{k}{N}\right).$$

_____

Example 11. By calculating the mean and variance for the probability distribution of Example 9, verify the formulas of Theorem 6.2.

Solution. For the hypergeometric distribution of Example 9 with $N = 8$, $n = 5$, and $k = 3$, we obtain

$$\mu = (0)\left(\frac{1}{56}\right) + (1)\left(\frac{15}{56}\right) + (2)\left(\frac{30}{56}\right) + (3)\left(\frac{10}{56}\right)$$

$$= \frac{15}{8} = \frac{nk}{N}.$$

Now

$$E(X^2) = (0)\left(\frac{1}{56}\right) + (1)\left(\frac{15}{56}\right) + (4)\left(\frac{30}{56}\right) + (9)\left(\frac{10}{56}\right)$$

$$= \frac{225}{56},$$

and therefore

$$\sigma^2 = \frac{225}{56} - \left(\frac{15}{8}\right)^2 = \frac{225}{448}$$

$$= \frac{N-n}{N-1} \cdot n \cdot \frac{k}{N}\left(1 - \frac{k}{N}\right).$$

Example 12. Using Chebyshev's theorem, find and interpret the interval $\mu \pm 2\sigma$ for Example 10.

Solution. Since Example 10 was a hypergeometric experiment with $N = 52$, $n = 5$, and $k = 13$, then by Theorem 6.2 we have

$$\mu = \frac{(5)(13)}{52} = \frac{5}{4} = 1.25,$$

$$\sigma^2 = \left(\frac{52 - 5}{51}\right)(5)\left(\frac{13}{52}\right)\left(1 - \frac{13}{52}\right)$$

$$= 0.8640.$$

Taking the square root of 0.8640, we find $\sigma = 0.93$. Hence the required interval is $1.25 \pm (2)(0.93)$, or from -0.61 to 3.11. Chebyshev's theorem states that the number of hearts obtained when 5 cards are dealt from an ordinary deck of 52 playing cards has a probability of at least $\frac{3}{4}$ of falling between -0.61 and 3.11; that is, at least $\frac{3}{4}$ of the time the 5 cards include less than 4 hearts.

If n is small relative to N, the probability for each drawing will change only slightly. Hence we essentially have a binomial experiment and can approximate the hypergeometric distribution by using the binomial distribution with $p = k/N$. The mean and variance can also be approximated by the formulas

$$\mu = np = \frac{nk}{N},$$

$$\sigma^2 = npq = n \cdot \frac{k}{N}\left(1 - \frac{k}{N}\right).$$

Comparing these formulas with those of Theorem 6.2, we see that the mean is the same, whereas the variance differs by a correction factor of $(N - n)/(N - 1)$. This is negligible when n is small relative to N.

Example 13. The telephone company reports that among 5000 telephones installed in a new subdivision 4000 have pushbuttons. If 10 people are called at random, what is the probability that exactly 3 will be talking on dial telephones?

Solution. Since the population size $N = 5000$ is large relative to the sample size $n = 10$, we shall approximate the desired probability by using the binomial distribution. The probability of calling someone with a dial telephone is 0.2. Therefore, the probability that exactly 3 people are called who have dial telephones is

$$h(3; 5000, 10, 1000) \simeq b(3; 10, 0.2)$$

$$= \sum_{x=0}^{3} b(x; 10, 0.2) - \sum_{x=0}^{2} b(x; 10, 0.2)$$

$$= 0.8791 - 0.6778$$

$$= 0.2013.$$

The hypergeometric distribution can be extended to treat the case in which the population can be partitioned into k cells $A_1, A_2, \ldots, A_k$, with a_1 elements in the first cell, a_2 elements in the second cell, $\ldots$, a_k elements in the kth cell. We are now interested in the probability that a random sample of size n yields x_1 elements from A_1, x_2 elements from A_2, $\ldots$, and x_k elements from A_k. Let us represent this probability by $f(x_1, x_2, \ldots, x_k; a_1, a_2, \ldots, a_k, N, n)$.

To obtain a general formula, we note that the total number of samples that can be chosen of size n from a population of size N is still $\binom{N}{n}$. There are $\binom{a_1}{x_1}$ ways of selecting x_1 items from the items in A_1, and for each of these we can choose x_2 items from the items in A_2 in $\binom{a_2}{x_2}$ ways. Therefore, we can select x_1 items from A_1 and x_2 items from A_2 in $\binom{a_1}{x_1}\binom{a_2}{x_2}$ ways. Continuing in this way, we can select all n items consisting of x_1 from A_1, x_2 from A_2, $\ldots$, and x_k from A_k in $\binom{a_1}{x_1}\binom{a_2}{x_2} \cdots \binom{a_k}{x_k}$ ways. The required probability distribution is now defined as follows.

DEFINITION

_____

Multivariate Hypergeometric Distribution. If a population of size N can be partitioned into the k cells $A_1, A_2, \ldots, A_k$, with $a_1, a_2, \ldots, a_k$ elements, respectively, then the probability distribution of the random variables $X_1, X_2, \ldots, X_k$, representing the number of elements selected from $A_1, A_2, \ldots, A_k$ in a random sample of size n, is

$$f(x_1, x_2, \ldots, x_k; a_1, a_2, \ldots, a_k, N, n) = \frac{\binom{a_1}{x_1}\binom{a_2}{x_2}\cdots\binom{a_k}{x_k}}{\binom{N}{n}},$$

with $\sum_{i=1}^{k} x_i = n$ and $\sum_{i=1}^{k} a_i = N$.

Example 14. A gardener wishes to landscape a piece of property by planting flowers across the front and back of the house. From a box containing 3 tulip bulbs, 4 daffodil bulbs, and 3 hyacinth bulbs he selects 5 at random to be planted at the front of the house, and the remaining 5 are planted at the rear of the house. What is the probability that 1 tulip plant, 2 daffodil plants, and 2 hyacinth plants bloom at the front of the house?

Solution. Using the multivariate hypergeometric distribution with $x_1 = 1$, $x_2 = 2$, $x_3 = 2$, $a_1 = 3$, $a_2 = 4$, $a_3 = 3$, $N = 10$, and $n = 5$, we find that the desired probability is

$$f(1, 2, 2; 3, 4, 3, 10, 5) = \frac{\binom{3}{1}\binom{4}{2}\binom{3}{2}}{\binom{10}{5}} = \frac{3}{14}.$$

EXERCISES

1. If 7 cards are dealt from an ordinary deck of 52 playing cards, what is the probability that
   (a) exactly 2 of them will be face cards?
   (b) at least 1 of them will be a queen?

2. A homeowner plants 5 bulbs selected at random from a box containing 5 tulip bulbs and 4 daffodil bulbs. What is the probability that he planted 2 daffodil bulbs and 3 tulip bulbs?

3. A random committee of size 3 is selected from 4 men and 2 women. Write a formula for the probability distribution of the random variable X representing the number of men on the committee. Find the $P(2 \leq X \leq 3)$.

4. From a lot of 12 missiles 5 are selected at random and fired. If the lot contains 3 defective missiles that will not fire, what is the probability that
   (a) all 5 will fire?
   (b) at most 2 will not fire?

5. What is the probability that a waitress will refuse to serve alcoholic beverages to only 2 minors if she randomly checks the I.D.s of 6 students from among 9 students of which 4 are not of legal age?

6. In Exercise 4 how many defective missiles might we expect to be included among the 5 that are selected? Use Chebyshev's theorem to describe the variability of the number of defective missiles included when 5 are selected from several lots each of size 12 containing 3 defective missiles.

7. If a person is dealt 13 cards from an ordinary deck of 52 playing cards several times, how many hearts per hand can he expect? Between what two values would you expect the number of hearts to fall at least 75% of the time?

8. It is estimated that 4000 of the 10,000 voting residents of a town are against a new sales tax. If 15 eligible voters are selected at random and asked their opinion, what is the probability that at most 7 favor the new tax?

9. An annexation suit is being considered against a county subdivision of 1200 residents by a neighboring city. If the occupants of one-half the residences object to being annexed, what is the probability that in a random sample of 10 at least 3 favor the annexation suit?

10. Among 150 IRS employees in a large city, only 30 are women. If 10 of the applicants are chosen at random to provide free tax assistance for the residents of this city, use the binomial approximation to the hypergeometric to find the probability that at least 3 women are selected.

11. A nationwide survey of 17,000 seniors by the University of Michigan reveals that almost 70% disapprove of daily pot smoking according to a report in _Parade,_ September 14, 1980. If 18 of these seniors are selected at random and asked their opinions, what is the probability that more than 9 but less than 14 disapprove of smoking pot?

12. Find the probability of being dealt a bridge hand of 13 cards containing 5 spades, 2 hearts, 3 diamonds, and 3 clubs.

13. A foreign student club lists as its members 2 Canadians, 3 Japanese, 5 Italians, and 2 Germans. If a committee of 4 is selected at random, find the probability that
    (a) all nationalities are represented;
    (b) all nationalities except the Italians are represented.

14. An urn contains 3 green balls, 2 blue balls, and 4 red balls. In a random sample of 5 balls, find the probability that both blue balls and at least 1 red ball are selected.

15. A car rental agency at a local airport has available 5 Fords, 7 Chevrolets, 4 Dodges, 3 Datsuns, and 4 Toyotas. If the agency randomly selects 9 of these cars to chauffeur delegates from the airport to the downtown convention center, find the probability that 2 Fords, 3 Chevrolets, 1 Dodge, 1 Datsun, and 2 Toyotas are used.

6.4

Negative Binomial and Geometric Distributions

Let us consider an experiment in which the properties are the same as those listed for a binomial experiment, with the exception that the trials will be repeated until a *fixed* number of successes occur. Therefore, instead of finding the probability of x successes in n trials, where n is fixed, we are now interested in the probability that the kth success occurs on the xth trial. Experiments of this kind are called **negative binomial experiments.**

As an illustration let us consider a controlled experiment in which a mouse is exposed to a contagious disease. The probability that a mouse will actually catch the disease is 0.6. We are interested in finding the probability that the seventh mouse exposed to this disease will be the fifth to catch it. In this experiment catching the disease will be considered a success. Designating a success by S and a failure by F, a possible order of achieving the desired result is *SFSSSFS*, which occurs with probability $(0.6)(0.4)(0.6)(0.6)(0.6)(0.4)(0.6) = (0.6)^5(0.4)^2$. We could list all possible orders by rearranging the Fs and Ss, except for the last outcome, which must be the fifth success. The total number of possible orders is equal to the number of partitions of the first 6 trials into two groups, with the 2 failures assigned to the one group and the 4 successes assigned to the other group. This can be done in

$$\binom{6}{4} = 15$$

mutually exclusive ways. Hence if X represents the trial on which the kth success occurs, then the probability that the seventh mouse exposed to the disease is the fifth to catch it is

$$P(X = 7) = \binom{6}{4}(0.6)^5(0.4)^2 = 0.1866.$$

The number X of trials to produce k successes in a negative binomial experiment is called a **negative binomial random variable** and its probability distribution is called the **negative binomial distribution.** Since its probabilities depend on the number of successes desired and the probability of a success on a given trial, we shall denote them by the symbol $b^*(x; k, p)$. To obtain the general formula for $b^*(x; k, p)$, consider the probability of a success on the xth trial preceded by $k - 1$ successes and $x - k$ failures in some specified order. Since the trials are independent, we can multiply all the probabilities corresponding to each desired outcome. Each success occurs with probability p and each failure with probability $q = 1 - p$. Therefore, the probability for the specified order, ending in a success, is $p^{k-1}q^{x-k}p =$

$p^k q^{x-k}$. The total number of sample points in the experiment ending in a success, after the occurrence of $k - 1$ successes and $x - k$ failures in any order, is equal to the number of partitions of $x - 1$ trials into two groups with $k - 1$ successes corresponding to one group and $x - k$ failures corresponding to the other group. This number is given by the term $\binom{x - 1}{k - 1}$, all mutually exclusive and occurring with equal probability $p^k q^{x-k}$. We obtain the general formula by multiplying $p^k q^{x-k}$ by $\binom{x - 1}{k - 1}$.

DEFINITION

_____

Negative Binomial Distribution. If repeated independent trials can result in a success with probability p and a failure with probability $q = 1 - p$, then the probability distribution of the random variable X, the number of the trial on which the kth success occurs is given by

$$b^*(x; k, p) = \binom{x - 1}{k - 1} p^k q^{x-k}, \qquad \text{for } x = k, k + 1, k + 2, \dots.$$

_____

Example 15. Find the probability that a person tossing 3 coins will get either all heads or all tails for the second time on the fifth toss.

Solution. Using the negative binomial distribution with $x = 5$, $k = 2$, and $p = \frac{1}{4}$, we have

$$b^*\left(5; 2, \frac{1}{4}\right) = \binom{4}{1}\left(\frac{1}{4}\right)^2\left(\frac{3}{4}\right)^3$$

$$= \frac{4!}{1!3!} \cdot \frac{3^3}{4^5}$$

$$= \frac{37}{256}.$$

The negative binomial distribution derives its name from the fact that each term in the expansion of $p^k(1 - q)^{-k}$ corresponds to the values of $b^*(x; k, p)$ for $x = k, k + 1, k + 2, \dots$.

If we consider the special case of the negative binomial distribution where $k = 1$, we have a probability distribution for the number of trials required for a single success. An example would be the tossing of a coin until a head occurs. We might be interested in the probability that the first head occurs on the fourth toss. The negative binomial distribution then reduces to the form $b*(x; 1, p) = pq^{x-1}$, $x = 1, 2, 3, \ldots$. Since the successive terms constitute a geometric progression, it is customary to refer to this special case as the **geometric distribution** and denote its values by $g(x; p)$.

DEFINITION

Geometric Distribution. If repeated independent trials can result in a success with probability p and a failure with probability $q = 1 - p$, then the probability distribution of the random variable X, the number of the trial on which the first success occurs, is given by

$$g(x; p) = pq^{x-1}, \qquad \text{for } x = 1, 2, 3, \ldots.$$

Example 16. Find the probability that a person flipping a balanced coin requires 4 tosses to get a head.

Solution. Using the geometric distribution with $x = 4$ and $p = \frac{1}{2}$, we have

$$g(4; \tfrac{1}{2}) = \tfrac{1}{2}(\tfrac{1}{2})^3 = \tfrac{1}{16}.$$

6.5
Poisson Distribution

Experiments yielding numerical values of a random variable X, the number of outcomes occurring during a given time interval or in a specified region, are often called **Poisson experiments.** The given time interval may be of any length, such as a minute, a day, a week, a month, or even a year. Hence a Poisson experiment might generate observations for the random variable X representing the number of telephone calls per hour received by an office, the number of days school is closed due to snow during the winter, or the number of postponed games due to rain during a baseball season. The specified region could be a line segment, an area, a volume, or perhaps a piece of material. In this case X might represent the number of field mice per acre, the number of bacteria in a given culture, or the number of typing errors per page.

A Poisson experiment is one that possesses the following properties:

1. The number of outcomes occurring in one time interval or specified region is independent of the number that occur in any other disjoint time interval or region of space.
2. The probability that a single outcome will occur during a very short time interval or in a small region is proportional to the length of the time interval

or the size of the region and does not depend on the number of outcomes occurring outside this time interval or region.

3. The probability that more than one outcome will occur in such a short time interval or fall in such a small region is negligible.

The number X of outcomes occurring in a Poisson experiment is called a **Poisson random variable** and its probability distribution is called the **Poisson distribution.** Since its probabilities depend only on μ, the average number of outcomes occurring in the given time interval or specified region, we shall denote them by the symbol $p(x; \mu)$. The derivation of the formula for $p(x; \mu)$, based on the properties for a Poisson experiment listed above, is beyond the scope of this text. We list the result in the following definition.

DEFINITION

Poisson Distribution. The probability distribution of the Poisson random variable X, representing the number of outcomes occurring in a given time interval or specified region, is

$$p(x; \mu) = \frac{e^{-\mu}\mu^x}{x!}, \qquad \text{for } x = 0, 1, 2, \ldots,$$

where μ is the average number of outcomes occurring in the given time interval or specified region and $e = 2.71828 \cdots$.

Table A.3 contains Poisson probability sums $\sum\limits_{x=0}^{r} p(x; \mu)$ for a few selected values of μ ranging from 0.1 to 18. We illustrate the use of this table with the following two examples.

Example 17. The average number of days school is closed due to snow during the winter in a certain city in the eastern part of United States is 4. What is the probability that the schools in this city will close for 6 days during a winter?

Solution. Using the Poisson distribution with $x = 6$ and $\mu = 4$, we find from Table A.3 that

$$p(6; 4) = \frac{e^{-4}4^6}{6!} = \sum_{x=0}^{6} p(x; 4) - \sum_{x=0}^{5} p(x; 4)$$

$$= 0.8893 - 0.7851 = 0.1042.$$

Example 18. The average number of field mice per acre in a 5-acre wheat field is estimated to be 10. Find the probability that a given acre contains more than 15 mice.

Solution. Let X be the number of field mice per acre. Then using Table A.3, we have

$$P(X > 15) = 1 - P(X \leq 15)$$

$$= 1 - \sum_{x=0}^{15} p(x; 10)$$

$$= 1 - 0.9513$$

$$= 0.0487.$$

The variance of the Poisson distribution can be shown to be equal to the mean. Thus in Example 17, where $\mu = 4$, we also have $\sigma^2 = 4$ and hence $\sigma = 2$. Using Chebyshev's theorem, we can state that our random variable has a probability of at least $\frac{3}{4}$ of falling in the interval $\mu \pm 2\sigma = 4 \pm (2)(2)$, or from 0 to 8. Therefore, we conclude that at least $\frac{3}{4}$ of the time the schools of the given city will be closed anywhere from 0 to 8 days during the winter season.

The Poisson and binomial distributions have histograms with approximately the same shape when n is large and p is close to zero. Hence, if these two conditions hold, the Poisson distribution, with $\mu = np$, can be used to approximate binomial probabilities. If p is close to 1, we can interchange what we have defined to be a success and a failure, thereby changing p to a value close to zero.

Example 19. Suppose that on the average 1 person in every 1000 is an alcoholic. Find the probability that a random sample of 8000 people will yield fewer than 7 alcoholics.

Solution. This is essentially a binomial experiment where $n = 8000$ and $p = 0.001$. Since p is very close to zero and n is quite large, we shall approximate with the Poisson distribution using $\mu = (8000)(0.001) = 8$. Hence, if X represents the number of alcoholics, we have

$$P(X < 7) = \sum_{x=0}^{6} b(x; 8000, 0.001)$$

$$\simeq \sum_{x=0}^{6} p(x; 8)$$

$$= 0.3134.$$

1. The probability that a person living in a certain city owns a dog is estimated to be 0.3. Find the probability that the tenth person randomly interviewed in this city is the fifth one to own a dog.

2. A scientist inoculates several mice, one at a time, with a disease germ until he finds 2 that have contracted the disease. If the possibility of contracting the disease is $\frac{1}{6}$, what is the probability that 8 mice are required?

3. Suppose the probability is 0.8 that any given person will believe a tale about life after death. What is the probability that
   (a) the sixth person to hear this tale is the fourth one to believe it?
   (b) the third person to hear this tale is the first one to believe it?

4. Find the probability that a person flipping a coin gets
   (a) the third head on the seventh flip;
   (b) the first head on the fourth flip.

5. Three people toss a coin and the odd man pays for the coffee. If the coins all turn up the same, they are tossed again. Find the probability that fewer than 4 tosses are needed.

6. According to a study published by a group of University of Massachusetts sociologists, about two-thirds of the 20 million persons in this country who take Valium are women. Assuming this figure to be a valid estimate, find the probability that on a given day the fifth prescription written by a doctor for Valium is
   (a) the first prescribing Valium for a woman;
   (b) the third prescribing Valium for a woman.

7. On the average a certain intersection results in 3 traffic accidents per month. What is the probability that in any given month at this intersection
   (a) exactly 5 accidents will occur?
   (b) less than 3 accidents will occur?
   (c) at least 2 accidents will occur?

8. A secretary makes 2 errors per page on the average. What is the probability that on the next page she makes
   (a) 4 or more errors?
   (b) no errors?

9. A certain area of the eastern United States is, on the average, hit by 6 hurricanes a year. Find the probability that in a given year this area will be hit by
   (a) fewer than 4 hurricanes;
   (b) anywhere from 6 to 8 hurricanes.

10. The average number of oil tankers arriving each day at a certain port city is known to be 10. The facilities at the port can handle at most 15 tankers per day. What is the probability that the port is unable to handle all the tankers that arrive
    (a) on a given day?
    (b) on one of the next 3 days?

11. A restaurant prepares a tossed salad containing on the average 5 vegetables. Find the probability that the salad contains more than 5 vegetables
    (a) on a given day;
    (b) on 3 of the next 4 days;
    (c) for the first time in April on April 5.

12. The probability that a person dies from a certain respiratory infection is 0.002. Find the probability that fewer than 5 of the next 2000 so infected will die.

13. Suppose that on the average 1 person in 1000 makes a numerical error in preparing his income tax return. If 10,000 forms are selected at random and examined, find the probability that 6, 7, or 8 of the forms will be in error.

14. The probability that a student fails the screening test for scoliosis (curvature of the spine) at a local high school is known to be 0.004. Of the next 1875 students who are screened for scoliosis, find the probability that
    (a) fewer than 5 fail the test;
    (b) 8, 9, or 10 fail the test.

15. Using Chebyshev's theorem, find and interpret the interval $\mu \pm 2\sigma$ for Exercise 12.

16. Using Chebyshev's theorem, find and interpret the interval $\mu \pm 3\sigma$ for Exercise 13.

7 Normal Distribution

Continuous random variables and their associated density functions arise whenever our experimental data are defined over a continuous sample space. Therefore, whenever we measure time intervals, weights, heights, volumes, and so forth, our underlying population is described by a continuous distribution. Just as there are several special discrete probability distributions there are also numerous types of continuous distributions whose graphs may display varying amounts of skewness or in some cases may be perfectly symmetric. Among these, by far the most important is the continuous distribution whose graph is a symmetric bell-shaped curve extending indefinitely in both directions. It is this distribution that provided a basis upon which much of the theory of statistical inference has been developed.

7.1

Normal Curve

The most important continuous probability distribution in the entire field of statistics is the **normal distribution.** Its graph, called the **normal curve,** is the bell-shaped curve of Figure 7.1 that describes so many sets of data that occur in nature, industry, and research. In 1733, DeMoivre derived the mathematical equation of the normal curve. The normal distribution is often referred to as the **Gaussian distribution** in honor of Gauss (1777–1855), who also derived its equation from a study of errors in repeated measurements of the same quantity.

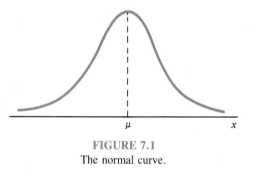

FIGURE 7.1
The normal curve.

A continuous random variable X having the bell-shaped distribution of Figure 7.1 is called a **normal random variable.** The mathematical equation for the probability distribution of the normal variable depends upon the two parameters μ and σ, its mean and standard deviation. Hence we denote the values of the density function of X by $n(x; \mu, \sigma)$.

DEFINITION

Normal Curve. If X is a normal random variable with mean μ and variance σ^2, then the equation of the *normal curve* is

$$n(x; \mu, \sigma) = \frac{1}{\sqrt{2\pi}\sigma} e^{-\frac{1}{2}\left(\frac{x-\mu}{\sigma}\right)^2}, \quad \text{for } -\infty < x < \infty,$$

where $\pi = 3.14159\ldots$ and $e = 2.71828\ldots.$

Once μ and σ are specified, the normal curve is completely determined. For example, if $\mu = 50$ and $\sigma = 5$, then the ordinates of $n(x; 50, 5)$ can easily be computed for various values of x and the curve drawn. In Figure 7.2 we have sketched two normal curves having the same standard deviation but

184

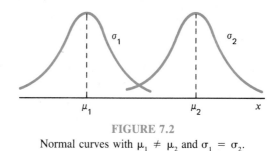

FIGURE 7.2
Normal curves with $\mu_1 \neq \mu_2$ and $\sigma_1 = \sigma_2$.

different means. The two curves are identical in form but are centered at different positions along the horizontal axis.

In Figure 7.3 we have sketched two normal curves with the same mean but different standard deviations. This time we see that the two curves are centered at exactly the same position on the horizontal axis, but the curve with the larger standard deviation is lower and spreads out farther. Remember that the area under a probability curve must be equal to 1, and therefore the more variable the set of observations, the lower and wider the corresponding curve will be.

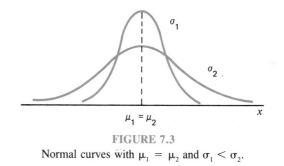

FIGURE 7.3
Normal curves with $\mu_1 = \mu_2$ and $\sigma_1 < \sigma_2$.

Figure 7.4 shows the results of sketching two normal curves having different means and different standard deviations. Clearly, they are centered at different positions on the horizontal axis and their shapes reflect the two different values of σ.

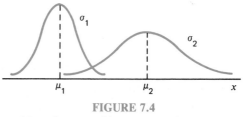

FIGURE 7.4
Normal curves with $\mu_1 \neq \mu_2$ and $\sigma_1 < \sigma_2$.

From an inspection of Figures 7.1 through 7.4, we list the following properties of the normal curve:

1. The *mode*, which is the point on the horizontal axis where the curve is a maximum, occurs at $x = \mu$.
2. The curve is symmetric about a vertical axis through the mean μ.
3. The normal curve approaches the horizontal axis asymptotically as we proceed in either direction away from the mean.
4. The total area under the curve and above the horizontal axis is equal to 1.

Many random variables have probability distributions that can be described adequately by the normal curve once μ and σ^2 are specified. In this chapter we assume that these two parameters are known, perhaps from previous investigations. Later, in Chapter 8, we consider methods of estimating μ and σ^2 from the available experimental data.

7.2
Areas Under the Normal Curve

The curve of any continuous probability distribution or density function is constructed so that the area under the curve bounded by the two ordinates $x = x_1$ and $x = x_2$ equals the probability that the random variable X assumes a value between $x = x_1$ and $x = x_2$. Thus, for the normal curve in Figure 7.5, $P(x_1 < X < x_2)$ is represented by the area of the shaded region.

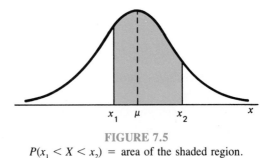

FIGURE 7.5
$P(x_1 < X < x_2)$ = area of the shaded region.

In Figures 7.2 through 7.4 we saw how the normal curve is dependent upon the mean and the standard deviation of the distribution under investigation. The area under the curve between any two ordinates must then also depend upon the values of μ and σ. This is evident in Figure 7.6, where we have shaded regions corresponding to $P(x_1 < X < x_2)$ for two curves with different means and variances. The $P(x_1 < X < x_2)$, where X is the random

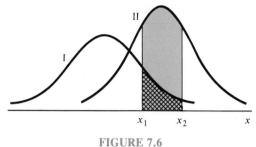

FIGURE 7.6
$P(x_1 < X < x_2)$ for different normal curves.

variable describing distribution I, is indicated by the crosshatched area. If X is the random variable describing distribution II, then the $P(x_1 < X < x_2)$ is given by the entire shaded region. Obviously, the two shaded regions are different in size; therefore, the probability associated with each distribution will be different.

It would be a hopeless task to attempt to set up separate tables of normal curve areas for every conceivable value of μ and σ. Yet we must use tables if we hope to avoid the use of integral calculus. Fortunately, we are able to transform all the observations of any normal random variable X to a new set of observations of a normal random variable Z with mean zero and variance 1. This can be done by means of the transformation

$$Z = \frac{X - \mu}{\sigma}.$$

The mean of Z is zero, since

$$E(Z) = \frac{1}{\sigma} E(X - \mu) = \frac{1}{\sigma}(\mu - \mu) = 0,$$

and the variance is

$$\sigma_Z^2 = \sigma_{(X-\mu)/\sigma}^2 = \sigma_{X/\sigma}^2 = \frac{1}{\sigma^2}\sigma_X^2 = \frac{\sigma^2}{\sigma^2} = 1.$$

DEFINITION

Standard Normal Distribution. The distribution of a normal random variable with mean zero and standard deviation equal to 1 is called a *standard normal distribution*.

Whenever X is between the values $x = x_1$ and $x = x_2$, the random variable Z will fall between the corresponding values

$$z_1 = \frac{x_1 - \mu}{\sigma} \quad \text{and} \quad z_2 = \frac{x_2 - \mu}{\sigma}.$$

The original and transformed distributions are illustrated in Figure 7.7. Since all the values of X falling between x_1 and x_2 have corresponding z values between z_1 and z_2, the area under the X curve between the ordinates $x = x_1$ and $x = x_2$ in Figure 7.7 equals the area under the Z curve between the transformed ordinates $z = z_1$ and $z = z_2$. Hence we have

$$P(x_1 < X < x_2) = P(z_1 < Z < z_2).$$

We have now reduced the required number of tables of normal-curve areas to one—that of the standard normal distribution. Table A.4 gives the area under the standard normal curve corresponding to $P(Z < z)$ for values of z from -3.49 to 3.49. To illustrate the use of this table, let us find the probability that Z is less than 1.74. First we locate a value of z equal to 1.7 in the left column and then move across the row to the column under 0.04, where we read 0.9591. Therefore, $P(Z < 1.74) = 0.9591$.

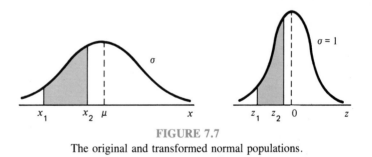

FIGURE 7.7
The original and transformed normal populations.

Occasionally, we are required to find a value of z corresponding to a specified probability that falls between values listed in Table A.4 (see Example 3). For convenience, we shall always choose the z value corresponding to the tabular value that comes closest to the specified probability. However, if the given probability falls midway between two tabular values, we shall choose for z the value falling midway between the corresponding values of z. For instance, to find the z value corresponding to a probability of 0.7975, which falls between 0.7967 and 0.7995 in Table A.4, we choose $z = 0.83$, since 0.7975 is closer to 0.7967. On the other hand, for a probability of 0.7981, which falls midway between 0.7967 and 0.7995, we take $z = 0.835$.

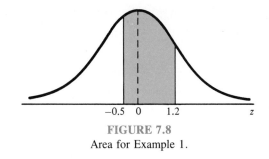

FIGURE 7.8
Area for Example 1.

Example 1. Given a normal distribution with $\mu = 50$ and $\sigma = 10$, find the probability that X assumes a value between 45 and 62.

Solution. The z values corresponding to $x_1 = 45$ and $x_2 = 62$ are

$$z_1 = \frac{45 - 50}{10} = -0.5,$$

$$z_2 = \frac{62 - 50}{10} = 1.2.$$

Therefore,

$$P(45 < X < 62) = P(-0.5 < Z < 1.2).$$

The $P(-0.5 < Z < 1.2)$ is given by the area of the shaded region in Figure 7.8. This area may be found by subtracting the area to the left of the ordinate $z = -0.5$ from the entire area to the left of $z = 1.2$. Using Table A.4, we have

$$P(45 < X < 62) = P(-0.5 < Z < 1.2)$$
$$= P(Z < 1.2) - P(Z < -0.5)$$
$$= 0.8849 - 0.3085$$
$$= 0.5764.$$

Example 2. Given a normal distribution with $\mu = 300$ and $\sigma = 50$, find the probability that X assumes a value greater than 362.

Solution. The normal probability distribution showing the desired area is given in Figure 7.9. To find the $P(X > 365)$, we need to evaluate the area under the normal curve to the right of $x = 362$. This can be done by

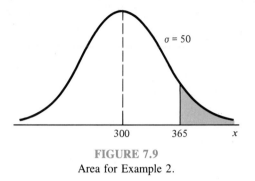

FIGURE 7.9
Area for Example 2.

transforming $x = 362$ to the corresponding z value, obtaining the area to the left of z from Table A.4, and then subtracting this area from 1. We find that

$$z = \frac{362 - 300}{50} = 1.24.$$

Hence

$$
\begin{aligned}
P(X > 362) &= P(Z < 1.24) \\
&= 1 - P(Z < 1.24) \\
&= 1 - 0.8925 \\
&= 0.1075.
\end{aligned}
$$

Example 3. Given a normal distribution with $\mu = 40$ and $\sigma = 6$, find the value x that has (a) 38% of the area below it and (b) 5% of the area above it.

Solution. The preceding two examples were solved by going first from a value of x to a z value and then computing the desired area. In this example we reverse the process and begin with a known area or probability, find the z value, and then determine x by rearranging the formula

$$z = \frac{x - \mu}{\sigma} \qquad \text{to give} \qquad x = \sigma z + \mu.$$

(a) An area of 0.38 to the left of the desired x value is shown shaded in Figure 7.10. We require a z value that leaves an area of 0.38 to the left. From Table A.4 we find $P(Z < -0.31) = 0.38$ so that the desired

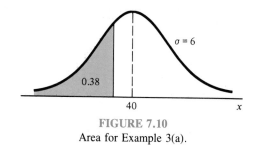

FIGURE 7.10
Area for Example 3(a).

z value is -0.31. Hence

$$x = (6)(-0.31) + 40$$
$$= 38.14.$$

(b) In Figure 7.11 we shade an area equal to 0.05 to the right of the desired x value. This time we require a z value that leaves an area of 0.05 to the right and hence an area of 0.95 to the left. Again from Table A.4 we find $P(Z < 1.645) = 0.95$ so that the desired z value is 1.645 and

$$x = (6)(1.645) + 40$$
$$= 49.87.$$

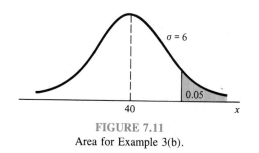

FIGURE 7.11
Area for Example 3(b).

The percentages associated with the Empirical Rule as stated in Section 3.3 were determined theoretically by using the normal-curve areas of Table A.4. For example, the probability of a normal random variable X with mean μ and variance σ^2 assuming a value between $x_1 = \mu - 2\sigma$ and $x_2 = \mu + 2\sigma$ is equal to the area under a standard normal curve between the ordinates $z = z_1$ and $z = z_2$, where

$$z_1 = \frac{(\mu - 2\sigma) - \mu}{\sigma} = -2$$

and

$$z_2 = \frac{(\mu + 2\sigma) - \mu}{\sigma} = 2.$$

Hence

$$
\begin{aligned}
P(\mu - 2\sigma < X < \mu + 2\sigma) &= P(-2 < Z < 2) \\
&= P(Z < 2) - P(Z < -2) \\
&= 0.9772 - 0.0228 \\
&= 0.9544,
\end{aligned}
$$

which is equivalent to stating that 95.44% of the measurements of a normal random variable fall within the interval $\mu \pm 2\sigma$.

7.3
Applications of the Normal Distribution

Some of the many problems in which the normal distribution is applicable are treated in the following examples. The use of the normal curve to approximate binomial probabilities will be considered in Section 7.4.

Example 4. A certain type of storage battery lasts on the average 3.0 years, with a standard deviation of 0.5 year. Assuming that the battery lives are normally distributed, find the probability that a given battery will last less than 2.3 years.

Solution. First construct a diagram such as Figure 7.12, showing the given distribution of battery lives and the desired area. To find the $P(X < 2.3)$, we need to evaluate the area under the normal curve to the left of 2.3. This is accomplished by finding the area to the left of the corresponding z value. Hence we find that

$$z = \frac{2.3 - 3}{0.5} = -1.4,$$

and then, using Table A.4, we have

$$
\begin{aligned}
P(X < 2.3) &= P(Z < -1.4) \\
&= 0.0808.
\end{aligned}
$$

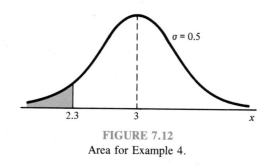

FIGURE 7.12
Area for Example 4.

Example 5. An electrical firm manufactures light bulbs that have a length of life that is normally distributed with mean equal to 800 hours and a standard deviation of 40 hours. Find the probability that a bulb burns between 778 and 834 hours.

Solution. The distribution of light bulbs is illustrated in Figure 7.13. The z values corresponding to $x_1 = 778$ and $x_2 = 834$ are

$$z_1 = \frac{778 - 800}{40} = -0.55,$$

$$z_2 = \frac{834 - 800}{40} = 0.85.$$

Hence

$$
\begin{aligned}
P(778 < X < 834) &= P(-0.55 < Z < 0.85) \\
&= P(Z < 0.85) - P(Z < -0.55) \\
&= 0.8023 - 0.2912 \\
&= 0.5111.
\end{aligned}
$$

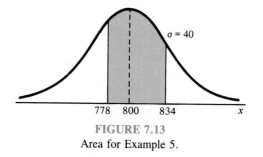

FIGURE 7.13
Area for Example 5.

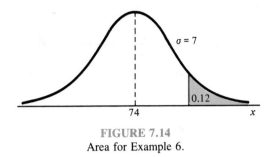

FIGURE 7.14
Area for Example 6.

Example 6. On an examination the average grade was 74 and the standard deviation was 7. If 12% of the class are given A's, and the grades are curved to follow a normal distribution, what is the lowest possible A and the highest possible B?

Solution. In this example we begin with a known area or probability, find the z value, and then determine x from the formula $x = \sigma z + \mu$. An area of 0.12, corresponding to the fraction of students receiving A's, is shaded in Figure 7.14. We require a z value that leaves 0.12 of the area to the right and hence an area of 0.88 to the left. From Table A.4, $P(Z < 1.175) = 0.88$ so that the desired z value is 1.175. Hence

$$x = (7)(1.175) + 74$$
$$= 82.225.$$

Therefore, the lowest A is 83 and the highest B is 82.

Example 7. Refer to Example 6 and find D_6.

Solution. The sixth decile, D_6, is the x value below which 60% of the area lies as shown by the shaded region in Figure 7.15. From Table A.4 we find $P(Z < 0.25) = 0.6$ so that the desired z value is 0.25. Now

$$x = (7)(0.25) + 74$$
$$= 75.75.$$

Hence $D_6 = 75.75$. That is, 60% of the grades are 75 or less.

Example 8. If the average height of miniature poodles is 30 centimeters, with a standard deviation of 4.1 centimeters, what percentage of miniature poodles exceeds 35 centimeters in height, assuming that the heights follow

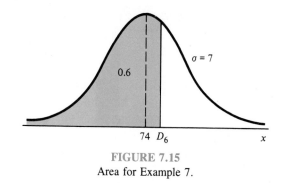

FIGURE 7.15
Area for Example 7.

a normal distribution and can be measured to any desired degree of accuracy?

Solution. A percentage is found by multiplying the relative frequency by 100%. Since the relative frequency for an interval is equal to the probability of falling in the interval, we must find the area to the right of $x = 35$ in Figure 7.16. This can be done by transforming $x = 35$ to the corresponding z value, obtaining the area to the left of z from Table A.4, and then subtracting this area from 1. We find that

$$z = \frac{35 - 30}{4.1} = 1.22.$$

Hence

$$P(X > 35) = P(Z > 1.22)$$
$$= 1 - P(Z < 1.22)$$
$$= 1 - 0.8888$$
$$= 0.1112.$$

Therefore, 11.12% of miniature poodles exceed 35 centimeters in height.

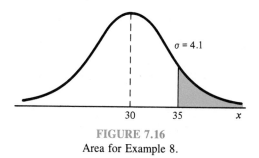

FIGURE 7.16
Area for Example 8.

Example 9. Find the percentage of miniature poodles exceeding 35 centimeters in Example 8 if the heights are all measured to the nearest centimeter.

Solution. This problem differs from Example 8 in that we now assign a measurement of 35 centimeters to all poodles whose heights are greater than 34.5 centimeters and less than 35.5 centimeters. We are actually approximating a discrete distribution by means of a continuous normal distribution. The required area is the region shaded to the right of 35.5 in Figure 7.17. We now find that

$$z = \frac{35.5 - 30}{4.1} = 1.34.$$

Hence

$$P(X > 35.5) = P(Z > 1.34)$$
$$= 1 - P(Z < 1.34)$$
$$= 1 - 0.9099$$
$$= 0.0901.$$

Therefore, 9.01% of miniature poodles exceed 35 centimeters in height when measured to the nearest centimeter. The difference of 2.11% between this answer and that of Example 8 represents all those poodles having a height greater than 35 and less than 35.5 centimeters that are now recorded as being 35 centimeters tall.

Example 10. The quality grade-point averages of 300 college freshmen follow approximately a normal distribution with a mean of 2.1 and a standard deviation of 0.8. How many of these freshmen would you expect to have a score between 2.5 and 3.5 inclusive if the point averages are computed to the nearest tenth?

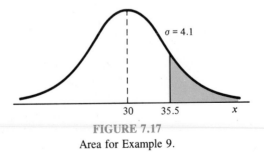

FIGURE 7.17
Area for Example 9.

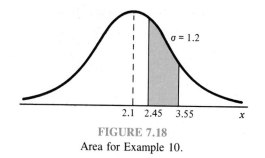

FIGURE 7.18
Area for Example 10.

Solution. Since the scores are recorded to the nearest tenth, we require the area between $x_1 = 2.45$ and $x_2 = 3.55$, as indicated in Figure 7.18. The corresponding z values are

$$z_1 = \frac{2.45 - 2.1}{0.8} = 0.44,$$

$$z_2 = \frac{3.55 - 2.1}{0.8} = 1.81.$$

Therefore,

$$P(2.45 < X < 3.55) = P(0.44 < Z < 1.81)$$
$$= P(Z < 1.81) - P(Z < 0.44)$$
$$= 0.9649 - 0.6700$$
$$= 0.2949.$$

Hence 29.49%, or approximately 88 of the 300 freshmen, should have a score between 2.5 and 3.5 inclusive.

EXERCISES

1. Given a normal distribution with $\mu = 40$ and $\sigma = 6$, find
   (a) the area below 32;
   (b) the area above 27;
   (c) the area between 42 and 51;
   (d) the x value that has 45% of the area below it;
   (e) the x value that has 13% of the area above it.

2. Given a normal distribution with $\mu = 200$ and $\sigma^2 = 100$, find
   (a) the area below 214;
   (b) the area above 179;
   (c) the area between 188 and 206;
   (d) the x value that has 80% of the area below it;
   (e) the two x values containing the middle 75% of the area.

3. Given the normally distributed variable X with mean 18 and standard deviation 2.5, find
   (a) $P(X < 15)$;
   (b) $P(17 < X < 21)$;
   (c) the value of k such that $P(X < k) = 0.2578$;
   (d) the value of k such that $P(X > k) = 0.1539$.

4. A soft-drink machine is regulated so that it discharges an average of 200 milliliters per cup. If the amount of drink is normally distributed with a standard deviation equal to 15 milliliters,
   (a) what fraction of the cups will contain more than 224 milliliters?
   (b) what is the probability that a cup contains between 191 and 209 milliliters?
   (c) how many cups will likely overflow if 230-milliliter cups are used for the next 1000 drinks?
   (d) below what value do we get the smallest 25% of the drinks?

5. The finished inside diameter of a piston ring is normally distributed with a mean of 10 centimeters and a standard deviation of 0.03 centimeter.
   (a) What proportion of rings will have inside diameters exceeding 10.075 centimeters?
   (b) What is the probability that a piston ring will have an inside diameter between 9.97 and 10.03 centimeters?
   (c) Below what value of inside diameter will 15% of the piston rings fall?

6. A lawyer commutes daily from his suburban home to his midtown office. On the average the trip one way takes 24 minutes, with a standard deviation of 3.8 minutes. Assume the distribution of trip times to be normally distributed.
   (a) What is the probability that a trip will take at least $\frac{1}{2}$ hour?
   (b) If the office opens at 9:00 A.M. and he leaves his house at 8:45 A.M. daily, what percentage of the time is he late for work?
   (c) If he leaves the house at 8:35 A.M. and coffee is served at the office from 8:50 A.M. until 9:00 A.M., what is the probability that he misses coffee?
   (d) Find the length of time above which we find the slowest 15% of the trips.
   (e) Find the probability that 2 of the next 3 trips will take at least $\frac{1}{2}$ hour.

7. If a set of grades on a statistics examination are approximately normally distributed with a mean of 74 and a standard deviation of 7.9, find
   (a) the lowest passing grade if the lowest 10% of the students are given F's;
   (b) the highest B if the top 5% of the students are given A's;
   (c) the lowest B if the top 10% of the students are given A's and the next 25% are given B's.

8. In a mathematics examination the average grade was 82 and the standard deviation was 5. All students with grades from 88 to 94 received a grade of B. If the grades are approximately normally distributed and 8 students received a B grade, how many students took the examination?

9. The heights of 1000 students are normally distributed with a mean of 174.5 centimeters and a standard deviation of 6.9 centimeters. Assuming that the heights are recorded to the nearest half of a centimeter, how many of these students would you expect to have heights
   (a) less than 160.0 centimeters?
   (b) between 171.5 and 182.0 centimeters inclusive?
   (c) equal to 175.0 centimeters?
   (d) greater than or equal to 188.0 centimeters?

10. A company pays its employees an average wage of $7.25 an hour with a standard deviation of 60 cents. If the wages are approximately normally distributed and paid to the nearest cent,
    (a) what percentage of the workers receive wages between $6.75 and $7.69 an hour inclusive?
    (b) the highest 5% of the employee hourly wages are greater than what amount?

11. The weights of a large number of miniature poodles are approximately normally distributed with a mean of 8 kilograms and a standard deviation of 0.9 kilogram. If measurements are recorded to the nearest tenth of a kilogram, find the fraction of these poodles with weights
    (a) over 9.5 kilograms;
    (b) at most 8.6 kilograms;
    (c) between 7.3 and 9.1 kilograms inclusive.

12. The tensile strength of a certain metal component is normally distributed with a mean of 10,000 kilograms per square centimeter and a standard deviation of 100 kilograms per square centimeter. Measurements are recorded to the nearest 50 kilograms per square centimeter.
    (a) What proportion of these components exceeds 10,150 kilograms per square centimeter in tensile strength?
    (b) If specifications require that all components have tensile strength between 9800 and 10,200 kilograms per square centimeter inclusive, what proportion of pieces would you expect to scrap?

13. If a set of observations is normally distributed, what percentage of the observations differs from the mean by
    (a) more than 1.3σ?
    (b) less than 0.52σ?

14. The IQs of 600 applicants to a certain college are approximately normally distributed with a mean of 115 and a standard deviation of 12. If the college requires an IQ of at least 95, how many of these students will be rejected on this basis regardless of their other qualifications?

15. The average rainfall, recorded to the nearest hundredth of a centimeter, in Roanoke, Virginia, for the month of March is 9.22 centimeters. Assuming a normal

distribution with a standard deviation of 2.83 centimeters, find the probability that next March Roanoke receives

(a) less than 1.84 centimeters of rain;

(b) more than 5 centimeters but not over 7 centimeters of rain;

(c) more than 13.8 centimeters of rain.

16. The average life of a certain type of small motor is 10 years, with a standard deviation of 2 years. The manufacturer replaces free all motors that fail while under guarantee. If he is willing to replace only 3% of the motors that fail, how long a guarantee should he offer? Assume that the lives of the motors follow a normal distribution.

7.4

Normal Approximation to the Binomial Distribution

Probabilities associated with binomial experiments are readily obtainable from the formula $b(x; n, p)$ of the binomial distribution or from Table A.2 when n is small. If n is not listed in any available table, we can compute the binomial probabilities by approximation procedures. In Section 6.5 we illustrated how the Poisson distribution can be used to approximate binomial probabilities when n is large and p is very close to zero or 1. Both the binomial and Poisson distributions are discrete. The first application of a continuous probability distribution to approximate probabilities over a discrete sample space was demonstrated in Section 7.3, Examples 9 and 10, where the normal curve was used. We now state a theorem that allows us to use areas under the normal curve to approximate binomial probabilities when n is sufficiently large.

THEOREM 7.1

If X is a binomial random variable with mean $\mu = np$ and variance $\sigma^2 = npq$, then the limiting form of the distribution of

$$Z = \frac{X - np}{npq},$$

as $n \to \infty$, is the standardized normal distribution.

It turns out that the proper normal distribution provides a very accurate approximation to the binomial distribution when n is large and p is close to $\frac{1}{2}$. In fact, even when n is small and p is not extremely close to zero or 1, the approximation is fairly good.

To investigate the normal approximation to the binomial distribution, we first draw the histogram for $b(x; 15, 0.4)$ and then superimpose the particular normal curve having the same mean and variance as the binomial variable X. Hence we draw a normal curve with

$$\mu = np = (15)(0.4) = 6,$$
$$\sigma^2 = npq = (15)(0.4)(0.6) = 3.6.$$

The histogram of $b(x; 15, 0.4)$ and the corresponding superimposed normal curve, which is completely determined by its mean and variance, are illustrated in Figure 7.19.

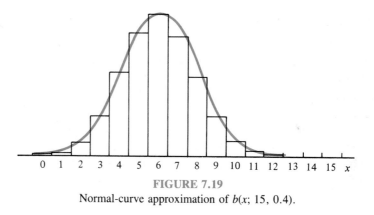

FIGURE 7.19
Normal-curve approximation of $b(x; 15, 0.4)$.

The exact probability of the binomial random variable X assuming a given value x is equal to the area of the rectangle whose base is centered at x. For example, the exact probability that X assumes the value 4 is equal to the area of the rectangle with base centered at $x = 4$. Using the formula for the binomial distribution, we find this area to be

$$b(4; 15, 0.4) = 0.1268.$$

This same probability is approximately equal to the area of the shaded region under the normal curve between the two ordinates $x_1 = 3.5$ and $x_2 = 4.5$ in Figure 7.20. Converting to z values, we have

$$z_1 = \frac{3.5 - 6}{1.9} = -1.316,$$

$$z_2 = \frac{4.5 - 6}{1.9} = -0.789.$$

If X is a binomial random variable and Z a standard normal variable, then

$$P(X = 4) = b(4; 15, 0.4)$$
$$\simeq P(-1.316 < Z < -0.789)$$
$$= P(Z < -0.789) - P(Z < -1.316)$$
$$= 0.2151 - 0.0941$$
$$= 0.1210.$$

This agrees very closely with the exact value of 0.1268.

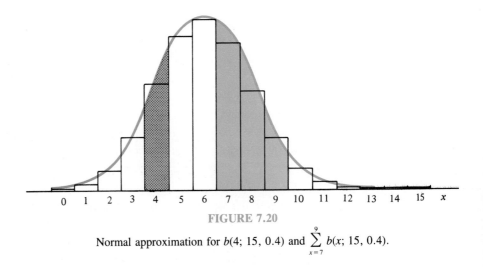

FIGURE 7.20

Normal approximation for $b(4; 15, 0.4)$ and $\sum_{x=7}^{9} b(x; 15, 0.4)$.

The normal approximation is most useful in calculating binomial sums for large values of n, which, without tables of binomial sums, is an impossible task. Referring to Figure 7.20, we might be interested in the probability that X assumes a value from 7 to 9 inclusive. The exact probability is given by

$$P(7 \leq X \leq 9) = \sum_{x=7}^{9} b(x; 15, 0.4)$$

$$= \sum_{x=0}^{9} b(x; 15, 0.4) - \sum_{x=0}^{6} b(x; 15, 0.4)$$

$$= 0.9662 - 0.6098$$

$$= 0.3564,$$

which is equal to the sum of the areas of the rectangles with bases centered at $x = 7$, 8, and 9. For the normal approximation we find the area of the shaded region under the curve between the ordinates $x_1 = 6.5$ and $x_2 = 9.5$ in Figure 7.20. The corresponding z values are

$$z_1 = \frac{6.5 - 6}{1.9} = 0.263,$$

$$z_2 = \frac{9.5 - 6}{1.9} = 1.842.$$

Now

$$\begin{aligned}
P(7 \le X \le 9) &\simeq P(0.263 < Z < 1.842) \\
&= P(Z < 1.842) - P(Z < 0.263) \\
&= 0.9673 - 0.6037 \\
&= 0.3636.
\end{aligned}$$

Once again the normal-curve approximation provides a value that agrees very closely with the exact value of 0.3564. The degree of accuracy, which depends on how well the curve fits the histogram, will increase as n increases. This is particularly true when p is not very close to $\frac{1}{2}$ and the histogram is no longer symmetric. Figures 7.21 and 7.22 show the histograms for $b(x; 6, 0.2)$ and $b(x; 15, 0.2)$, respectively. It is evident that a normal curve would fit the histogram when $n = 15$ considerably better than when $n = 6$.

In summary, we use the normal approximation to evaluate binomial prob-

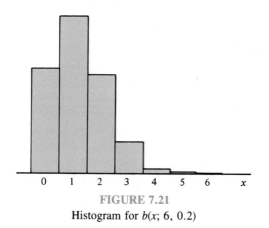

FIGURE 7.21

Histogram for $b(x; 6, 0.2)$

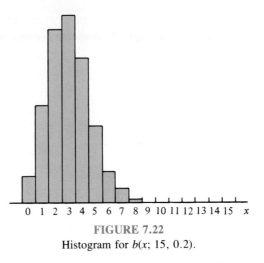

FIGURE 7.22

Histogram for $b(x; 15, 0.2)$.

abilities whenever p is not close to zero or 1. The approximation is excellent when n is large and fairly good for small values of n if p is reasonably close to $\frac{1}{2}$. One possible guide to determine when the normal approximation may be used is provided by calculating np and nq. If both np and nq are greater than 5, the approximation will be good.

Example 11. The probability that a patient recovers from a rare blood disease is 0.6. If 100 people are known to have contracted this disease, what is the probability that less than one-half survive?

Solution. Let the binomial variable X represent the number of patients that survive. Since $n = 100$, we should obtain fairly accurate results using the normal-curve approximation with

$$\mu = np = (100)(0.6) = 60,$$
$$\sigma = \sqrt{npq} = \sqrt{(100)(0.6)(0.4)} = 4.9.$$

To obtain the desired probability, we have to find the area to the left of $x = 49.5$. The z value corresponding to 49.5 is

$$z = \frac{49.5 - 60}{4.9} = -2.14,$$

and the probability of fewer than 50 of the 100 patients surviving is given

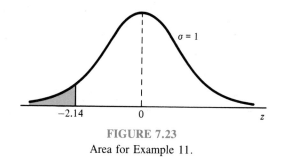

FIGURE 7.23
Area for Example 11.

by the area of the shaded region in Figure 7.23. Hence

$$P(X < 50) = \sum_{x=0}^{49} b(x; 100, 0.6)$$

$$\simeq P(Z < -2.14)$$

$$= 0.0162.$$

Example 12. A multiple-choice quiz has 200 questions, each with 4 possible answers, of which only 1 is the correct answer. What is the probability that sheer guesswork yields from 25 to 30 correct answers for 80 of the 200 problems about which the student has no knowledge?

Solution. The probability of a correct answer for each of the 80 questions is $p = \frac{1}{4}$. If X represents the number of correct answers due to guesswork, then

$$P(25 \leq X \leq 30) = \sum_{x=25}^{30} b(x; 80, \tfrac{1}{4}).$$

Using the normal-curve approximation with

$$\mu = np = (80)(\tfrac{1}{4}) = 20,$$
$$\sigma = \sqrt{npq} = \sqrt{(80)(\tfrac{1}{4})(\tfrac{3}{4})} = 3.87,$$

we need the area between $x_1 = 24.5$ and $x_2 = 30.5$. The corresponding z values are

$$z_1 = \frac{24.5 - 20}{3.87} = 1.16,$$

$$z_2 = \frac{30.5 - 20}{3.87} = 2.71.$$

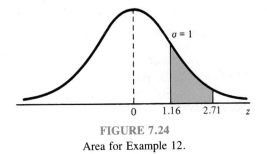

FIGURE 7.24
Area for Example 12.

The probability of correctly guessing from 25 to 30 questions is given by the area of the shaded region in Figure 7.24. From Table A.4 we find that

$$P(25 \leq X \leq 30) = \sum_{x=25}^{30} b(x; 80, \tfrac{1}{4})$$

$$\approx P(1.16 < Z < 2.71)$$
$$= P(Z < 2.71) - P(Z < 1.16)$$
$$= 0.9966 - 0.8770$$
$$= 0.1196.$$

EXERCISES

1. Find the error in approximating $\sum_{x=1}^{4} b(x; 20, 0.1)$ by the normal-curve approximation.

2. A coin is tossed 400 times. Use the normal-curve approximation to find the probability of obtaining
   (a) between 185 and 210 heads inclusive;
   (b) exactly 205 heads;
   (c) fewer than 176 or more than 227 heads.

3. A pair of dice is rolled 180 times. What is the probability that a total of 7 occurs
   (a) at least 25 times;
   (b) between 33 and 41 times inclusive;
   (c) exactly 30 times.

4. The probability that a patient recovers from a delicate heart operation is 0.9. Of the next 100 patients having this operation, what is the probability that
   (a) between 84 and 95 inclusive survive?
   (b) fewer than 86 survive?

5. A pheasant hunter claims that she brings down 75% of the birds she shoots at. Of the next 80 pheasants shot at by this hunter, what is the probability that
   (a) at least 50 escape?
   (b) at most 56 are brought down?

6. A drug manufacturer claims that a certain drug cures a blood disease on the average 80% of the time. To check the claim, government testers used the drug on a sample of 100 individuals and decide to accept the claim if 75 or more are cured.
   (a) What is the probability that the claim will be rejected when the cure probability is in fact 0.8?
   (b) What is the probability that the claim will be accepted by the government when the cure probability is as low as 0.7?

7. If 20% of the residents in a U.S. city prefer a white telephone over any other color available, what is the probability that among the next 1000 telephones installed in this city
   (a) between 170 and 185 inclusive will be white?
   (b) at least 210 but not more than 225 will be white?

8. One-sixth of the male freshmen entering a large state school are out-of-state students. If the students are assigned at random to the dormitories, 180 to a building, what is the probability that in a given dormitory at least one-fifth of the students are from out of state?

9. A certain pharmaceutical company knows that, on the average, 5% of a certain type of pill has an ingredient that is below the minimum strength and thus unacceptable. What is the probability that fewer than 10 in a sample of 200 pills will be unacceptable?

10. According to the May/June issue of *Consumers Digest*, census figures show that in 1978 almost 53% of all the households in the United States were composed of only one or two people. What is the probability that between 490 and 515 inclusive of the next 1000 randomly selected households in America will consist of either one or two people?

8 Sampling Theory

Statisticians, for the most part, are concerned with making statistical inferences concerning population parameters on the basis of partial or incomplete evidence. This incomplete evidence is provided by drawing samples from the populations and computing the values of appropriate statistics. Now, the value of a statistic will depend on the observed sample values, and therefore will vary from sample to sample. Before we can make reliable inferences concerning the value of a population parameter, it is essential that we understand the chance variations associated with the appropriate statistic for the particular sampling process being used.

The field of statistical inference is basically concerned with generalizations and predictions. For example, we might claim, based on the opinions of several people interviewed on the street, that in a forthcoming election 60% of the eligible voters in the city of Detroit favor a certain candidate. In this case we are dealing with a random sample of opinions from a very large finite population. As a second illustration we might state that the average cost to build a residence in Charleston, South Carolina, is between $65,000 and $70,000, based on the estimates of 3 contractors selected at random from the 30 now building in this city. The population being sampled here is again finite but very small. Finally, let us consider a soft-drink dispensing machine in which the average amount of drink dispensed is being held to 240 milliliters. A company official computes the mean of 40 drinks to obtain $\bar{x} = 236$ milliliters, and on the basis of this value decides that the machine is still dispensing drinks with an average content of $\mu = 240$ milliliters. The 40 drinks represent a sample from the infinite population of possible drinks that will be dispensed by this machine.

In each of the examples above we have computed a statistic from a sample selected from the population, and from these statistics we made various statements concerning the values of population parameters which may or may not be true. The company official made the decision that the soft-drink machine dispenses drinks with an average content of 240 milliliters, even though the sample mean was 236 milliliters, because he knows from sampling theory that such a sample value is likely to occur. In fact, if he ran similar tests, say every hour, he would expect the values of $\bar{x}$ to fluctuate above and below $\mu = 240$ milliliters. Only when the value of $\bar{x}$ is subantially different from 240 milliliters will the company official initiate action to adjust the machine.

Since many random samples are generally possible from the same population, we would expect every statistic, whether it be $\bar{x}$, $\tilde{x}$, or perhaps s^2, to vary somewhat from sample to sample. Hence a statistic is a *random variable* that depends only on the observed sample. In keeping with the convention adopted in Chapter 5 for representing random variables, we use a capital letter to denote the appropriate statistic and the corresponding small or lowercase letter for one of its values. For example, the sample mean discussed in the preceding paragraph is a statistic that is represented by $\bar{X}$. The value of the random variable $\bar{X}$ for a given sample is then denoted, as in Section 2.2, by $\bar{x}$. Similarly, the sample median $\tilde{X}$ and the sample variance S^2 are statistics whose values for a given sample are denoted as $\tilde{x}$ and s^2. Since a statistic is a random variable, it must have a probability distribution.

Sampling Distribution. The probability distribution of a statistic is called a *sampling distribution.*

The probability distribution of $\bar{X}$ is called the **sampling distribution of the mean**. It is customary to refer to the standard deviation of the sampling distribution as the **standard error** of the statistic. Therefore, the standard error of the mean is just the standard deviation of the sampling distribution of $\bar{X}$. Also, the standard error of the sample standard deviation for all possible samples of size n selected from a specified population is the standard deviation of the statistic S.

The sampling distribution of a statistic will depend on the size of the population, the size of the samples, and the method of choosing the samples. In this chapter we study several of the more important sampling distributions of frequently used statistics. Applications of these sampling distributions to problems of statistical inference are considered throughout most of the remaining chapters.

8.2
Sampling Distributions of the Mean

The first important sampling distribution to be considered is that of the mean $\bar{X}$. To illustrate, we sample from a discrete uniform population consisting of the values 0, 1, 2, and 3. Clearly, the four observations making up the population are values of a random variable X having the probability distribution

$$f(x) = \tfrac{1}{4}, \qquad \text{for } x = 0, 1, 2, 3,$$

with mean

$$\mu = E(X) = \sum_{x=0}^{3} xf(x)$$

$$= \frac{0 + 1 + 2 + 3}{4} = \frac{3}{2}$$

and variance

$$\sigma^2 = E[(X - \mu)^2] = \sum_{x=0}^{3} (x - \mu)^2 f(x)$$

$$= \frac{(0-\frac{3}{2})^2 + (1-\frac{3}{2})^2 + (2-\frac{3}{2})^2 + (3-\frac{3}{2})^2}{4}$$

$$= \frac{5}{4}.$$

The probability histogram of this discrete uniform population is shown in Figure 8.1.

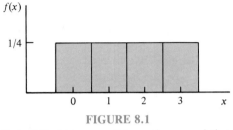

FIGURE 8.1

Probability histogram of the uniform population.

Suppose that we list all possible samples of size 2, *with replacement,* and then for each sample compute $\bar{x}$. The 16 possible samples and their means are given in Table 8.1. The statistic $\overline{X}$ clearly assumes values $\bar{x}$ fluctuating from 0 to 3. The sampling distribution of $\overline{X}$, which is just its relative frequency distribution, is given in Table 8.2, and the probability histogram or relative frequency histogram is shown in Figure 8.2.

TABLE 8.1

Means of Random Samples with Replacement

No.	Sample	$\bar{x}$	No.	Sample	$\bar{x}$
1	0, 0	0	9	2, 0	1.0
2	0, 1	0.5	10	2, 1	1.5
3	0, 2	1.0	11	2, 2	2.0
4	0, 3	1.5	12	2, 3	2.5
5	1, 0	0.5	13	3, 0	1.5
6	1, 1	1.0	14	3, 1	2.0
7	1, 2	1.5	15	3, 2	2.5
8	1, 3	2.0	16	3, 3	3.0

It can be seen that the probability histogram of the sampling distribution of $\overline{X}$ may be approximated very closely by a normal curve once we have determined the appropriate mean and variance. Let us designate this mean and variance of $\overline{X}$ by the symbols $\mu_{\overline{X}}$ and $\sigma_{\overline{X}}^2$, respectively. Then proceeding as in

TABLE 8.2
Sampling
Distribution of $\overline{X}$
with Replacement

$\overline{x}$	f	$f(\overline{x})$
0	1	$\frac{1}{16}$
0.5	2	$\frac{2}{16}$
1.0	3	$\frac{3}{16}$
1.5	4	$\frac{4}{16}$
2.0	3	$\frac{3}{16}$
2.5	2	$\frac{2}{16}$
3.0	1	$\frac{1}{16}$

Sections 5.5 and 5.6 with the data of Table 8.2, we obtain

$$\mu_{\overline{X}} = \sum \overline{x} f(\overline{x}) = \frac{3}{2} = \mu$$

and

$$\sigma_{\overline{X}}^2 = \sum \left(\overline{x} - \frac{3}{2}\right)^2 f(\overline{x}) = \frac{5}{8} = \frac{\frac{5}{4}}{2} = \frac{\sigma^2}{n}.$$

The mean and variance of $\overline{X}$ have been computed from the values in Table 8.2.

One could easily show that the sampling distribution of the means of 64 possible samples of size 3, selected with replacement, will be even more closely approximated by a normal curve with $\mu_{\overline{X}} = \frac{3}{2}$ and $\sigma_{\overline{X}}^2 = \frac{5}{12}$. The mean of the variable $\overline{X}$ is always equal to the mean of the population from which

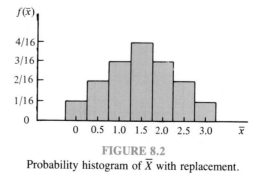

FIGURE 8.2
Probability histogram of $\overline{X}$ with replacement.

the random samples are chosen and in no way depends on the size of the sample. The variance of $\overline{X}$, however, does depend on the sample size and is equal to the original population variance σ^2 divided by n. Consequently, the larger the sample size, the smaller the standard error of $\overline{X}$, and the closer a particular $\overline{x}$ is likely to be to μ. Hence $\overline{x}$ could be used as an estimate of μ. The results of the foregoing example illustrate the following well-known theorem.

THEOREM 8.1

If all possible random samples of size n are drawn with replacement from a finite population of size N with mean μ and standard deviation σ, then for n sufficiently large the sampling distribution of the mean $\overline{X}$ will be approximately normally distributed with mean $\mu_{\overline{X}} = \mu$ and standard deviation $\sigma_{\overline{X}} = \sigma/\sqrt{n}$. Hence

$$z = \frac{\overline{x} - \mu}{\sigma/\sqrt{n}}$$

is a value of a standard normal variable Z.

Theorem 8.1 is valid for any finite population when $n \geq 30$. If $n < 30$, the results will be valid only if the population being sampled is not too different from a normal population. If the population is known to be bell-shaped, the sampling distribution of $\overline{X}$ will be approximately a normal distribution, regardless of the size of the sample.

Example 1. Given the population 1, 1, 1, 3, 4, 5, 6, 6, 6, and 7, find the probability that a random sample of size 36, selected with replacement, will yield a sample mean greater than 3.8 but less than 4.5 if the mean is measured to the nearest tenth.

Solution. The probability distribution of our population may be written as

x	1	3	4	5	6	7
$P(X = x)$	0.3	0.1	0.1	0.1	0.3	0.1

Calculating the mean and variance by standard procedures, we find $\mu = 4$ and $\sigma^2 = 5$. The sampling distribution of $\overline{X}$ may be approximated by the normal distribution with mean $\mu_{\overline{X}} = \mu = 4$ and variance $\sigma_{\overline{X}}^2 = \sigma^2/n = \frac{5}{36}$. Taking the square root, we find the standard deviation to be $\sigma_{\overline{X}} =$

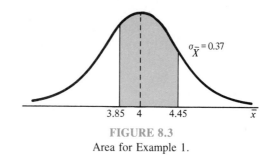

FIGURE 8.3
Area for Example 1.

0.373. The probability that $\overline{X}$ is greater than 3.8 and less than 4.5 is given by the area of the shaded region in Figure 8.3. The z values that correspond to $\overline{x}_1 = 3.85$ and $\overline{x}_2 = 4.45$ are

$$z_1 = \frac{3.85 - 4}{0.373} = -0.40,$$

$$z_2 = \frac{4.45 - 4}{0.373} = 1.21.$$

Therefore,

$$P(3.8 < \overline{X} < 4.5) \simeq P(-0.40 < Z < 1.21)$$
$$= P(Z < 1.21) - P(Z < -0.40)$$
$$= 0.8869 - 0.3446$$
$$= 0.5423.$$

To verify the result of Example 1, we could write the values of our population on tags and place them in a box from which we draw samples of size 36 with replacement. If we drew 100 samples each of size 36 and computed the sample means, we obtain what is known as an **experimental sampling distribution** of $\overline{X}$. The distribution given in Table 8.2 is often called the **theoretical sampling distribution** of $\overline{X}$. From the experimental sampling distribution we should find that approximately 54%, or 54 of our 100 sample means, fall within the interval from 3.85 to 4.45.

Suppose that we now draw all possible samples of size 2 from our uniform population, *without replacement,* and then for each compute $\overline{x}$. The 12 possible samples and their means are given in Table 8.3. The statistic $\overline{X}$ now assumes values that fluctuate from 0.5 to 2.5. The sampling distribution of $\overline{X}$ is given in Table 8.4 and the histogram of the population of sample means is shown in Figure 8.4.

TABLE 8.3
Means of Random Samples Without Replacement

No.	Sample	$\overline{x}$	No.	Sample	$\overline{x}$
1	0, 1	0.5	7	1, 0	0.5
2	0, 2	1.0	8	2, 0	1.0
3	0, 3	1.5	9	3, 0	1.5
4	1, 2	1.5	10	2, 1	1.5
5	1, 3	2.0	11	3, 1	2.0
6	2, 3	2.5	12	3, 2	2.5

TABLE 8.4
Sampling Distribution of $\overline{X}$
Without Replacement

$\overline{x}$	f	$f(\overline{x})$
0.5	2	$\frac{1}{6}$
1.0	2	$\frac{1}{6}$
1.5	4	$\frac{1}{3}$
2.0	2	$\frac{1}{6}$
2.5	2	$\frac{1}{6}$

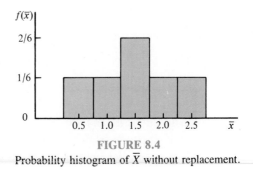

FIGURE 8.4
Probability histogram of $\overline{X}$ without replacement.

We cannot expect the sampling distribution of $\overline{X}$ to approximate the normal distribution very closely when the samples are selected without replacement from a small finite population unless the population is bell-shaped. However, the mean and variance of $\overline{X}$ are

$$\mu_{\overline{X}} = \sum \overline{x} f(\overline{x}) = \frac{3}{2} = \mu,$$

$$\sigma_{\overline{X}}^2 = \sum \left(\overline{x} - \frac{3}{2} \right)^2 f(\overline{x}) = \frac{5}{12} = \frac{\frac{5}{4}}{2} \left(\frac{4-2}{4-1} \right) = \frac{\sigma^2}{n} \left(\frac{N-n}{N-1} \right),$$

regardless of the size or form of the original population. When $n \geq 30$ and the population size is at least twice the sample size, we may apply the following theorem.

THEOREM 8.2 _____

If all possible random samples of size n are drawn, without replacement, from a finite population of size N with mean μ and standard deviation σ, then the sampling distribution of the sample mean $\overline{X}$ will be approximately normally distributed with a mean and standard deviation given by

$$\mu_{\overline{X}} = \mu,$$

$$\sigma_{\overline{X}} = \frac{\sigma}{\sqrt{n}} \sqrt{\frac{N-n}{N-1}}.$$

_____

Example 2. Given the population 1, 1, 1, 3, 4, 5, 6, 6, 6, and 7, find the mean and standard deviation for the sampling distribution of means for samples of size 4 selected at random without replacement. Between what two values would you expect at least $\frac{3}{4}$ of the sample means to fall?

Solution. From Example 1 we know that $\mu = 4$ and $\sigma^2 = 5$. Using Theorem 8.2, we find that the mean and standard deviation for the sampling distribution of means are

$$\mu_{\overline{X}} = 4,$$

$$\sigma_{\overline{X}} = \frac{\sqrt{5}}{\sqrt{4}} \sqrt{\frac{10-4}{10-1}} = 0.85.$$

Applying Chebyshev's theorem, we would expect at least $\frac{3}{4}$ of the sample means to fall in the interval $\mu_{\bar{X}} \pm 2\sigma_{\bar{X}} = 4 \pm (2)(0.85)$, or between 2.3 and 5.7.

The factor $[(N - n)/(N - 1)]^{1/2}$ in the formula for the standard deviation of $\bar{X}$ in Theorem 8.2 is called the **finite population correction factor.** For large N, relative to the sample size n, this correction factor will be close to 1, and $\sigma_{\bar{X}}^2$ will be approximately σ^2/n. Hence, for large or infinite populations, whether discrete or continuous, we state the following well-known theorem, called the **Central Limit Theorem.**

THEOREM 8.3

_____

Central Limit Theorem. If random samples of size n are drawn from a large or infinite population with mean μ *and variance* σ^2, *then the sampling distribution of the sample mean* $\bar{X}$ *is approximately normally distributed with mean* $\mu_{\bar{X}} = \mu$ *and standard deviation* $\sigma_{\bar{X}} = \sigma/\sqrt{n}$. *Hence*

$$z = \frac{\bar{x} - \mu}{\sigma/\sqrt{n}}$$

is a value of a standard normal variable Z.

_____

The normal approximation in Theorem 8.3 will be good if $n \geq 30$ regardless of the shape of the population. If $n < 30$, the approximation is good only if the population is not too different from a normal population. If the population is known to be normal, the sampling distribution of $\bar{X}$ will follow a normal distribution exactly, no matter how small the size of the samples.

Example 3. An electrical firm manufactures light bulbs that have a length of life that is approximately normally distributed, with mean equal to 800 hours and a standard deviation of 40 hours. Find the probability that a random sample of 16 bulbs will have an average life of less than 775 hours.

Solution. The sampling distribution of $\bar{X}$ will be approximately normal, with $\mu_{\bar{X}} = 800$ and $\sigma_{\bar{X}} = 40/\sqrt{16} = 10$. The desired probability is given by the area of the shaded region in Figure 8.5. Corresponding to $\bar{x} = 775$, we find that

$$z = \frac{775 - 800}{10} = -2.5$$

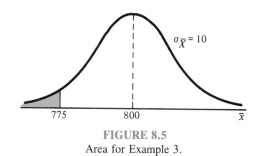

FIGURE 8.5
Area for Example 3.

and, therefore,

$$P(\bar{X} < 775) = P(Z < -2.5)$$
$$= 0.0062$$

8.3
t Distribution

Most of the time we are not fortunate enough to know the variance of the population from which we select our random samples. For samples of size $n \geq 30$, a good estimate of σ^2 is provided by calculating s^2. What then happens to the distribution of the z values in the Central Limit Theorem if we replace σ^2 by s^2? As long as s^2 is a good estimate of σ^2 and does not vary much from sample to sample, which is usually the case for $n \geq 30$, the values $(\bar{x} - \mu)/(s/\sqrt{n})$ are still approximately distributed as a standard normal variable, and the Central Limit Theorem is valid.

If the sample size is small ($n < 30$), the values of s^2 fluctuate considerably from sample to sample and the distribution of the values $(\bar{x} - \mu)(s/\sqrt{n})$ is no longer a standard normal distribution. We are now dealing with the distribution of a statistic that we shall call T, whose values are given by

$$t = \frac{\bar{x} - \mu}{s/\sqrt{n}}.$$

In 1908, W. S. Gosset published a paper in which he derived the equation of the probability distribution of T. At the time, Gosset was employed by an Irish brewery that disallowed publication of research by members of its staff. To circumvent this restriction, he published his work secretly under the name "Student." Consequently, the distribution of T is usually called the **Student-t distribution**, or simply the **t distribution**. In deriving the equation of this distribution, Gosset assumed that the samples were selected from a normal population. Although this would seem to be a very restrictive assumption, it can be shown that the sampling distribution of T for samples selected from nonnormal but bell-shaped distributions still approximates the t distribution very closely.

The actual mathematical formula is omitted here, since the areas under the curve have been tabulated in sufficient detail to meet the requirements of most problems. However, to evaluate probabilities associated with the t distribution we need to understand some of the characteristics of the t curve.

The distribution of T is similar to the distribution of Z, in that they both are symmetrical about a mean of zero. Both distributions are bell-shaped, but the t distribution is more variable, owing to the fact that the t values depend on the fluctuations of two quantities, $\bar{x}$ and s^2, whereas the z values depend only on the changes of $\bar{x}$ from sample to sample. The distribution of T differs from that of Z in that the variance depends on the sample size n and is always greater than 1. Only when the sample size $n \rightarrow \infty$ will the two distributions become the same.

The divisor, $n - 1$, that appears in the formula for s^2 is called the number of **degrees of freedom** associated with s^2. If $\bar{x}$ and s^2 are computed from samples of size n, the corresponding values of t are said to belong to a t distribution with v degrees of freedom, where $v = n - 1$. Thus we have a different t curve or t distribution for each possible sample size, a curve that becomes more and more like the standard normal curve as $n \rightarrow \infty$. In Figure 8.6 we show the relationship between a standard normal distribution ($v = \infty$) and t distributions with 2 and 5 degrees of freedom. The curve for $v = 5$ represents the sampling distribution of all t values computed from repeated random samples of size 6 taken from a normal population. Similarly, the curve for $v = 2$ represents the sampling distribution of all t values computed from samples of size 3.

THEOREM 8.4 _____

If $\bar{x}$ and s^2 are the mean and variance, respectively, of a random sample of size n taken from a population that is normally distributed with mean μ and variance σ^2, then

$$t = \frac{\bar{x} - \mu}{s/\sqrt{n}}$$

is a value of a random variable T having the t distribution with $v = n - 1$ degrees of freedom.

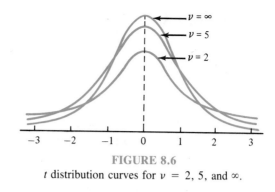

FIGURE 8.6

t distribution curves for $\nu = 2$, 5, and ∞.

The probability that a random sample produces a *t* value falling between any two specified values is equal to the area under the curve of the *t* distribution between the two ordinates corresponding to the specified values. It would be a tedious task to attempt to set up separate tables giving the areas between every conceivable pair of ordinates for all values of $n < 30$. Table A.5 gives only those *t* values above which we find a specified area α, where α is 0.1, 0.05, 0.025, 0.01, or 0.005. This table is set up differently from the table of normal-curve areas in that the areas are now the column headings and the entries are the *t* values. The left column gives the degrees of freedom. It is customary to let t_α represent the *t* value above which we find an area equal to α. Hence the *t* value with 10 degrees of freedom leaving an area of 0.025 to the right is $t_{0.025} = 2.228$. Since the *t* distribution is symmetric about a mean of zero, we have $t_{1-\alpha} = -t_\alpha$; that is, the *t* value leaving an area of $1 - \alpha$ to the right and therefore an area of α to the left is equal to the negative *t* value that leaves an area of α in the right tail of the distribution (see Figure 8.7). For a *t* distribution with 10 degrees of freedom we have $t_{0.975} = -t_{0.025} = -2.228$. This means that the *t* value of a random sample of size 11, selected from a normal population, will fall between -2.228 and 2.228, with probability equal to 0.95.

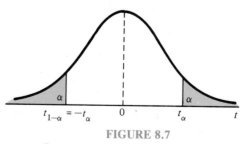

FIGURE 8.7

Symmetry property of the *t* distribution.

Example 4. Find $P(-t_{0.025} < T < t_{0.05})$.

Solution. Since $t_{0.05}$ leaves an area of 0.05 to the right, and $-t_{0.025}$ leaves an area of 0.025 to the left, we find a total area of

$$1 - 0.05 - 0.025 = 0.925$$

between $-t_{0.025}$ and $t_{0.05}$. Hence

$$P(-t_{0.025} < T < t_{0.05}) = 0.925.$$

Example 5. Find k such that $P(k < T < -1.761) = 0.045$, for a random sample of size 15 selected from a normal distribution.

Solution. From Table A.5 we note that 1.761 corresponds to $t_{0.05}$ when $v = 14$. Therefore, $-t_{0.05} = -1.761$. Since k in the original probability statement is to the left of $-t_{0.05} = -1.761$, let $k = -t_\alpha$. Then, from Figure 8.8, we have

$$0.045 = 0.05 - \alpha$$

or

$$\alpha = 0.005.$$

Hence, from Table A.5 with $v = 14$,

$$k = -t_{0.005} = -2.977$$

and

$$P(-2.977 < T < -1.761) = 0.045.$$

Exactly 95% of the values of a t distribution with $v = n - 1$ degrees of freedom lie between $-t_{0.025}$ and $t_{0.025}$. Of course, there are other t values that contain 95% of the distribution, such as $-t_{0.02}$ and $t_{0.03}$, but these values do not appear in Table A.5, and in the following discussion it is preferable to choose t values that leave exactly the same area in the two tails of our distribution. A t value that falls below $-t_{0.025}$ or above $t_{0.025}$ would tend to make one believe that either a rare event has taken place or our assumption about μ is in error. We shall make the latter decision and claim that our

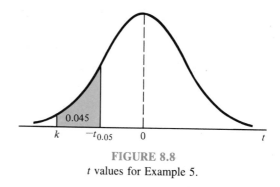

FIGURE 8.8

t values for Example 5.

assumed value of μ is in error. In fact, a *t* value falling below $-t_{0.01}$ or above $t_{0.01}$ would provide even stronger evidence that our assumed value of μ should be rejected. General procedures for testing claims concerning the value of the parameter μ, as well as many other parameters, will be discussed in Chapter 10.

Example 6. A manufacturer of light bulbs claims that his bulbs will burn on the average 500 hours. To maintain this average, he tests 25 bulbs each month. If the computed *t* value falls between $-t_{0.05}$ and $t_{0.05}$, he is satisfied with his claim. What conclusion should he draw from a sample that has a mean $\bar{x} = 518$ hours and a standard deviation $s = 40$ hours? Assume the distribution of burning times to be approximately normal.

Solution. From Table A.5 we find that $t_{0.05} = 1.711$ for 24 degrees of freedom. Therefore, the manufacturer is satisfied with his claim if a sample of 25 bulbs yields a *t* value between -1.711 and 1.711. If $\mu = 500$, then

$$t = \frac{518 - 500}{40/\sqrt{25}} = 2.25,$$

a value well above 1.711. The probability of obtaining a *t* value, with $\nu = 24$, equal to or greater than 2.25 is approximately 0.02. If $\mu > 500$, the value of *t* computed from the sample would be more reasonable. Hence the manufacturer is likely to conclude that his bulbs are a better product than he thought.

8.4

Sampling Distribution of the Differences of Means

Suppose that we now have two populations, the first with mean μ_1 and variance σ_1^2, and the second with mean μ_2 and variance σ_2^2. Let the values of the variable $\overline{X}_1$ represent the means of random samples of size n_1 drawn from the first population, and the values of $\overline{X}_2$ represent the means of random samples of size n_2 drawn from the second population, independent of the samples from the first population. The distribution of the differences $\overline{x}_1 - \overline{x}_2$ between the two sets of independent sample means is called the *sampling distribution* of the statistic $\overline{X}_1 - \overline{X}_2$.

To illustrate, let the first population be 3, 5, and 7, with the mean

$$\mu_1 = \frac{3 + 5 + 7}{3} = 5$$

and the variance

$$\sigma_1^2 = \frac{(3 - 5)^2 + (5 - 5)^2 + (7 - 5)^2}{3} = \frac{8}{3}.$$

The second population consists of the two values 0 and 3, with the mean

$$\mu_2 = \frac{0 + 3}{2} = \frac{3}{2}$$

and variance

$$\sigma_2^2 = \frac{(0 - \frac{3}{2})^2 + (3 - \frac{3}{2})^2}{2} = \frac{9}{4}.$$

From the first population we draw all possible samples of size $n_1 = 2$ with replacement, and, for each sample, the mean $\overline{x}_1$ is computed. Similarly, for the second population, all possible samples of size $n_2 = 3$ are drawn with replacement, and, for each of these we compute the mean $\overline{x}_2$. The two sets of possible samples and their means are given in Table 8.5. The 72 possible differences $\overline{x}_1 - \overline{x}_2$ are given in Table 8.6, and the frequency distribution of $\overline{X}_1 - \overline{X}_2$ is given in Table 8.7 with the corresponding probability histogram shown in Figure 8.9.

It is evident that the probability histogram of the sampling distribution of $\overline{X}_1 - \overline{X}_2$ may be approximated by a normal curve, the approximation improving as n_1 and n_2 increase. Applying Theorem 5.5 of Section 5.7 and

TABLE 8.5
Means of Random Samples with Replacement
from Two Finite Populations

From Population 1			From Population 2		
No.	Sample	$\bar{x}_1$	No.	Sample	$\bar{x}_2$
1	3, 3	3	1	0, 0, 0	0
2	3, 5	4	2	0, 0, 3	1
3	3, 7	5	3	0, 3, 0	1
4	5, 3	4	4	3, 0, 0	1
5	5, 5	5	5	0, 3, 3	2
6	5, 7	6	6	3, 0, 3	2
7	7, 3	5	7	3, 3, 0	2
8	7, 5	6	8	3, 3, 3	3
9	7, 7	7			

then Theorem 8.1 of Section 8.2, we find the mean of the differences of independent sample means to be

$$\mu_{\bar{X}_1 - \bar{X}_2} = \mu_{\bar{X}_1} - \mu_{\bar{X}_2} = \mu_1 - \mu_2.$$

Also, using successively Theorems 5.8 and 5.7 of Section 5.7 and then Theorem 8.1 of Section 8.2, we find the variance of the differences of independent sample means to be

$$\sigma_{\bar{X}_1 - \bar{X}_2}^2 = \sigma_{\bar{X}_1}^2 + \sigma_{\bar{X}_2}^2 = \frac{\sigma_1^2}{n_1} + \frac{\sigma_2^2}{n_2}.$$

TABLE 8.6
Differences of Independent Means

$\bar{x}_2$	$\bar{x}_1$								
	3	4	5	4	5	6	5	6	7
0	3	4	5	4	5	6	5	6	7
1	2	3	4	3	4	5	4	5	6
1	2	3	4	3	4	5	4	5	6
1	2	3	4	3	4	5	4	5	6
2	1	2	3	2	3	4	3	4	5
2	1	2	3	2	3	4	3	4	5
2	1	2	3	2	3	4	3	4	5
3	0	1	2	1	2	3	2	3	4

TABLE 8.7
Sampling Distribution of $\overline{X}_1 - \overline{X}_2$
with Replacement

$\overline{x}_1 - \overline{x}_2$	f	$f(\overline{x}_1 - \overline{x}_2)$
0	1	$\frac{1}{72}$
1	5	$\frac{5}{72}$
2	12	$\frac{12}{72}$
3	18	$\frac{18}{72}$
4	18	$\frac{18}{72}$
5	12	$\frac{12}{72}$
6	5	$\frac{5}{72}$
7	1	$\frac{1}{72}$

Therefore, in our illustration, the mean and variance of the sampling distribution of Table 8.7 are

$$\mu_{\overline{X}_1 - \overline{X}_2} = 5 - 1.5 = 3.5$$

and

$$\sigma^2_{\overline{X}_1 - \overline{X}_2} = \frac{\frac{8}{3}}{2} + \frac{\frac{9}{4}}{3} = \frac{25}{12}.$$

Of course, the same results could also have been found by proceeding directly as in Sections 5.5 and 5.6 with the data of Table 8.7.

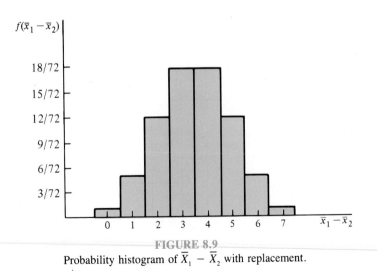

FIGURE 8.9
Probability histogram of $\overline{X}_1 - \overline{X}_2$ with replacement.

The results obtained for the sampling distribution of $\overline{X}_1 - \overline{X}_2$ by sampling with replacement from a finite population are also valid for infinite populations, continuous or discrete, and for finite populations when sampling is without replacement provided the population sizes, N_1 and N_2, are large relative to the sample sizes, n_1 and n_2, respectively. However, if the populations are small and sampling is without replacement, then we must compute $\sigma_{\overline{X}_1}$ and $\sigma_{\overline{X}_2}$ by the formula for $\sigma_{\overline{X}}$ in Theorem 8.2.

In this text we concern ourselves with the sampling distribution of the differences of independent means only when the size of the populations from which the samples are selected is large.

THEOREM 8.5

_____

If independent samples of size n_1 and n_2 are drawn from two large or infinite populations, discrete or continuous, with means μ_1 and μ_2 and variances σ_1^2 and σ_2^2, respectively, then the sampling distribution of the differences of means, $\overline{X}_1 - \overline{X}_2$, is approximately normally distributed with mean and standard deviation given by

$$\mu_{\overline{X}_1 - \overline{X}_2} = \mu_1 - \mu_2, \qquad and \qquad \sigma_{\overline{X}_1 - \overline{X}_2} = \sqrt{\frac{\sigma_1^2}{n_1} + \frac{\sigma_2^2}{n_2}}.$$

Hence

$$z = \frac{(\overline{x}_1 - \overline{x}_2) - (\mu_1 - \mu_2)}{\sqrt{(\sigma_1^2/n_1) + (\sigma_2^2/n_2)}}$$

is a value of a standard normal variable Z.

_____

If both n_1 and n_2 are greater than or equal to 30, the normal approximation for the distribution of $\overline{X}_1 - \overline{X}_2$ is very good regardless of the shapes of the two populations. However, even when n_1 and n_2 are less than 30, the normal approximation is reasonably good except when the populations are decidedly nonnormal.

Example 7. A sample of size $n_1 = 5$ is drawn at random from a population that is normally distributed with mean $\mu_1 = 50$ and variance $\sigma_1^2 = 9$, and the sample mean $\overline{x}_1$ is recorded. A second random sample of size $n_2 = 4$ is selected, independent of the first sample, from a different population that is also normally distributed, with mean $\mu_2 = 40$ and variance $\sigma^2 = 4$, and the sample mean $\overline{x}_2$ is recorded. What is the $P(\overline{X}_1 - \overline{X}_2 < 8.2)$?

Solution. From the sampling distribution of $\overline{X}_1 - \overline{X}_2$ we know that

$$\mu_{\overline{X}_1 - \overline{X}_2} = \mu_1 - \mu_2$$

$$= 50 - 40 = 10$$

and

$$\sigma^2_{\overline{X}_1 - \overline{X}_2} = \frac{\sigma_1^2}{n_1} + \frac{\sigma_2^2}{n_2}$$

$$= \frac{9}{5} + \frac{4}{4} = 2.8.$$

The desired probability is given by the area of the shaded region in Figure 8.10. Corresponding to the value $\bar{x}_1 - \bar{x}_2 = 8.2$, we find that

$$z = \frac{8.2 - 10}{\sqrt{2.8}} = -1.08,$$

and hence

$$P(\overline{X}_1 - \overline{X}_2 < 8.2) = P(Z < -1.08)$$

$$= 0.1401.$$

Example 8. The television picture tubes of manufacturer A have a mean lifetime of 6.5 years and a standard deviation of 0.9 year, while those of manufacturer B have a mean lifetime of 6.0 years and a standard deviation of 0.8 year. What is the probability that a random sample of 36 tubes from manufacturer A will have a mean lifetime that is at least 1 year more than the mean lifetime of a sample of 49 tubes from manufacturer B?

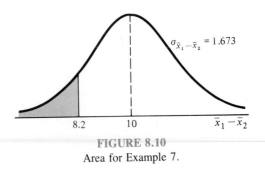

FIGURE 8.10
Area for Example 7.

Solution. We are given the following information.

Population 1	Population 2
$\mu_1 = 6.5$	$\mu_2 = 6.0$
$\sigma_1 = 0.9$	$\sigma_2 = 0.8$
$n_1 = 36$	$n_2 = 49$

If we use Theorem 8.5, the sampling distribution of $\overline{X}_1 - \overline{X}_2$ will have a mean and standard deviation given by

$$\mu_{\overline{X}_1 - \overline{X}_2} = 6.5 - 6.0 = 0.5,$$

$$\sigma_{\overline{X}_1 - \overline{X}_2} = \sqrt{\frac{0.81}{36} + \frac{0.64}{49}} = 0.189.$$

The probability that the mean of 36 tubes from manufacturer A will be at least 1 year longer than the mean of 49 tubes from manufacturer B is given by the area of the shaded region in Figure 8.11. Corresponding to the value $\overline{x}_1 - \overline{x}_2 = 1.0$, we find that

$$z = \frac{1.0 - 0.5}{0.189} = 2.65,$$

and hence

$$P(\overline{X}_1 - \overline{X}_2 \geq 1.0) = P(Z \geq 2.65)$$
$$= 1 - P(Z < 2.65)$$
$$= 1 - 0.9960$$
$$= 0.0040.$$

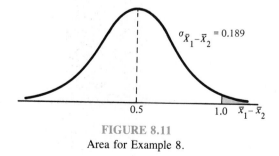

FIGURE 8.11
Area for Example 8.

The distribution of $\overline{X}_1 - \overline{X}_2$ in Theorem 8.5 suggests a more general result, which we state without proof in the following theorem.

THEOREM 8.6

───

If the random variables X and Y are independent and normally distributed with means μ_X and μ_Y and variances σ_X^2 and σ_Y^2, respectively, then the distribution of the difference X − Y is normally distributed with mean

$$\mu_{X-Y} = \mu_X - \mu_Y$$

and variance

$$\sigma_{X-Y}^2 = \sigma_X^2 + \sigma_Y^2.$$

───

EXERCISES

1. Random samples of size 4 are drawn, with replacement, from the finite population 2, 4, and 6.
   (a) Assuming that the 81 possible samples are all equally likely to occur, construct the sampling distribution of $\overline{X}$.
   (b) Construct a probability histogram for the sampling distribution of $\overline{X}$.
   (c) Verify that $\mu_{\overline{X}} = \mu$ and $\sigma_{\overline{X}}^2 = \sigma^2/n$.
   (d) Between what two values would you expect the middle 68% of the sample means to fall?

2. If, in Exercise 1, a sample of size 54 is drawn with replacement, what is the probability that the sample mean will be greater than 4.1 but less than 4.4? Assume the means to be measured to the nearest tenth.

3. A finite population consists of the numbers 2, 2, 4, 6, and 6 written on 5 tags, each of a different color.
   (a) Assuming that the 25 possible samples of size 2 that can be selected at random, with replacement, are all equally likely to occur, construct the sampling distribution of $\overline{X}$.
   (b) Construct a probability histogram for the sampling distribution of $\overline{X}$.
   (c) Verify that $\mu_{\overline{X}} = \mu$ and $\sigma_{\overline{X}}^2 = \sigma_X^2/n$.

4. Random samples of size 2 are drawn, without replacement, from the finite population 1, 1, 1, 2, 2, 3, and 4.
   (a) Assuming that the 42 possible samples are all equally likely to occur, construct the sampling distribution of $\overline{X}$.
   (b) Verify that $\mu_{\overline{X}} = \mu$ and $\sigma_{\overline{X}} = (\sigma/\sqrt{n})\sqrt{(N-n)/(N-1)}$.
   (c) Between what two values would you expect at least $\frac{8}{9}$ of the sample means to fall?

5. A certain type of thread is manufactured with a mean tensile strength of 78.3 kilograms and a standard deviation of 5.6 kilograms. Assuming that the population is infinite, how is the standard error of the mean changed when the sample size is

   (a) increased from 64 to 196?
   (b) decreased from 784 to 49?

6. If the standard error of the mean for the sampling distribution of random samples of size 36 from a large or infinite population is 2, how large must the size of the sample become if the standard error is to be reduced to 1.2?

7. Find the value of the finite population correction factor when

   (a) $n = 2$ and $N = 5$;
   (b) $n = 10$ and $N = 1000$;
   (c) $n = 40$ and $N = 10,000$.

8. If all possible samples of size 16 are drawn from a normal population with mean equal to 50 and standard deviation equal to 5, what is the probability that a sample mean $\overline{X}$ will fall in the interval from $\mu_{\overline{X}} - 1.9\sigma_{\overline{X}}$ to $\mu_{\overline{X}} - 0.4\sigma_{\overline{X}}$? Assume that the sample means can be measured to any degree of accuracy.

9. A soft-drink machine is being regulated so that the amount of drink dispensed averages 240 milliliters with a standard deviation of 15 milliliters. Periodically, the machine is checked by taking a sample of 40 drinks and computing the average content. If the mean of the 40 drinks is a value within the interval $\mu_{\overline{X}} \pm 2\sigma_{\overline{X}}$, the machine is thought to be operating satisfactorily; otherwise, adjustments are made. In Section 8.1, the company official found the mean of 40 drinks to be $\overline{x} = 236$ milliliters and concluded that the machine needed no adjustment. Was this a reasonable decision?

10. The heights of 1000 students are approximately normally distributed with a mean of 174.5 centimeters and a standard deviation of 6.9 centimeters. If 200 random samples of size 25 are drawn from this population and the means recorded to the nearest tenth of a centimeter, determine

    (a) the mean and standard error of the sampling distribution of $\overline{X}$;
    (b) the number of sample means that fall between 172.5 and 175.8 centimeters inclusive;
    (c) the number of sample means falling below 172.0 centimeters.

11. The random variable X, representing the number of cherries in a cherry puff, has the following probability distribution:

x	4	5	6	7
$P(X = x)$	0.2	0.4	0.3	0.1

   (a) Find the mean μ and the variance σ^2 of X.
   (b) Find the mean $\mu_{\overline{X}}$ and the variance $\sigma_{\overline{X}}^2$ of the mean $\overline{X}$ for random samples of 36 cherry puffs.

    (c) Find the probability that the average number of cherries in 36 cherry puffs will be less than 5.5.

12. If a certain machine makes electrical resistors having a mean resistance of 40 ohms and a standard deviation of 2 ohms, what is the probability that a random sample of 36 of these resistors will have a combined resistance of more than 1458 ohms?

13. (a) Find $t_{0.025}$ when $v = 14$.
    (b) Find $-t_{0.01}$ when $v = 10$.
    (c) Find $t_{0.995}$ when $v = 7$.

14. (a) Find $P(T < 2.365)$ when $v = 7$.
    (b) Find $P(T > 1.318)$ when $v = 24$.
    (c) Find $P(-1.356 < T < 2.179)$ when $v = 12$.
    (d) Find $P(T > -2.567)$ when $v = 17$.

15. (a) Find $P(-t_{0.005} < T < t_{0.01})$.
    (b) Find $P(T > -t_{0.025})$.

16. Given a random sample of size 24 from a normal distribution, find k such that
    (a) $P(-2.069 < T < k) = 0.965$.
    (b) $P(k < T < 2.807) = 0.095$.
    (c) $P(-k < T < k) = 0.90$.

17. A manufacturing firm claims that the batteries used in their electronic games will last an average of 30 hours. To maintain this average, 16 batteries are tested each month. If the computed t value falls between $-t_{0.025}$ and $t_{0.025}$, the firm is satisfied with its claim. What conclusion should the firm draw from a sample that has a mean $\bar{x} = 27.5$ hours and a standard deviation $s = 5$ hours? Assume the distribution of battery lives to be approximately normal.

18. A normal population with unknown variance is believed to have a mean of 20. Is one likely to obtain a random sample of size 9 from this population that has a mean $\bar{x} = 24$ and a standard deviation of $s = 4.1$? If not, what conclusion would you draw?

19. A cigarette manufacturer claims that his cigarettes have an average nicotine content of 1.83 milligrams. If a random sample of 8 cigarettes of this type have nicotine contents of 2.0, 1.7, 2.1, 1.9, 2.2, 2.1, 2.0, and 1.6 milligrams, would you agree with the manufacturer's claim?

20. Let $\bar{X}_1$ represent the mean of a sample of size $n_1 = 2$, selected with replacement, from the finite population -2, 0, 2, and 4. Similarly, let $\bar{X}_2$ represent the mean of a sample of size $n_2 = 2$, selected with replacement, from the population -1 and 1.
    (a) Assuming that the 64 possible differences $\bar{x}_1 - \bar{x}_2$ are equally likely to occur, construct the sampling distribution of $\bar{X}_1 - \bar{X}_2$.

(b) Construct a probability histogram for the sampling distribution of $\overline{X}_1 - \overline{X}_2$.

(c) Verify that $\mu_{\overline{X}_1 - \overline{X}_2} = \mu_1 - \mu_2$ and $\sigma^2_{\overline{X}_1 - \overline{X}_2} = (\sigma^2_1/n_1) + (\sigma^2_2/n_2)$.

21. Let $\overline{X}_1$ represent the mean of a sample of size $n_1 = 2$, selected with replacement, from the finite population 2, 3, and 7. Similarly, let $\overline{X}_2$ represent the mean of a sample of size $n_2 = 2$, selected with replacement, from the population 1, 1, and 3.

    (a) Assuming that the 81 possible differences $\overline{x}_1 - \overline{x}_2$ are equally likely to occur, construct the sampling distribution of $\overline{X}_1 - \overline{X}_2$.

    (b) Construct a probability histogram for the sampling distribution of $\overline{X}_1 - \overline{X}_2$.

    (c) Verify that $\mu_{\overline{X}_1 - \overline{X}_2} = \mu_1 - \mu_2$ and $\sigma^2_{\overline{X}_1 - \overline{X}_2} = (\sigma^2_1/n_1) + (\sigma^2_2/n_2)$.

22. If, in Exercise 21, samples of size $n_1 = 100$ and $n_2 = 25$ are drawn with replacement, what is the probability that $\overline{X}_1 - \overline{X}_2$ will be greater than 1.24 but less than 1.63? Assume that the sample means can be measured to any degree of accuracy.

23. A random sample of size 25 is taken from a normal population having a mean of 80 and a standard deviation of 5. A second random sample of size 36 is taken from a different normal population having a mean of 75 and a standard deviation of 3. Find the probability that the sample mean computed from the 25 measurements will exceed the sample mean computed from the 36 measurements by at least 3.4 but less than 5.9. Assume the means to be measured to the nearest tenth.

24. The distribution of heights of a certain breed of terrier dogs has a mean height of 72 centimeters and a standard deviation of 10 centimeters, whereas the distribution of heights of a certain breed of poodles has a mean height of 28 centimeters with a standard deviation of 5 centimeters. Assuming that the sample means can be measured to any degree of accuracy, find the probability that the sample mean for a random sample of heights of 64 terriers exceeds the sample mean for a random sample of 100 poodles by at most 44.2 centimeters.

25. The mean score for freshmen on an aptitude test, at a certain college, is 540, with a standard deviation of 50. What is the probability that two groups of students selected at random, consisting of 32 and 50 students, respectively, will differ in their mean scores by (a) more than 20 points; (b) an amount between 5 and 10 points? Assume the means to be measured to any degree of accuracy.

8.5

Simulated Experiments

In many areas of science and business, techniques have been used to *simulate* the values of a random variable associated with a statistical experiment. Such techniques are particularly useful in the study of inventory problems, traffic flows, collisions of particles in nuclear physics, certain medical studies, and in a variety of other studies where sequences of random outcomes of random variables are needed.

Simulating the outcomes of a random variable is accomplished by randomly assigning numbers to the values of the random variable in such a way so as not to change the probabilities associated with each possible outcome. These numbers might be assigned by various gambling devices, by tables of random numbers, or, more often today, by computer programs that generate random numbers. As a simple illustration, consider the experiment of tossing a balanced coin. Since the probabilities associated with a head and a tail are both $\frac{1}{2}$, we might simulate the outcomes, without ever tossing the coin, simply by choosing a certain column in Table A.12 and letting a 0, 1, 2, 3, or 4 represent heads and 5, 6, 7, 8, or 9 represent tails as we read down this column of random numbers. Note that this assignment preserves the probabilities since the probability of obtaining one of the five digits 0, 1, 2, 3, or 4 from the 10 digits is $\frac{1}{2}$, as is the probability of obtaining one of the five digits 5, 6, 7, 8, or 9. If we start with the fifth column in Table A.12, beginning with line 8, we get the digits 5, 1, 7, 9, 8, 0, 4, 0, 3, 8, ... as we read down the column. Hence, by simulating an experiment in which a coin is tossed, we have generated the following sequence of random outcomes: tail, head, tail, tail, tail, head, head, head, head, tail, ...

There are many ways that one can assign numbers to simulate outcomes of a random variable. In the illustration of the preceding paragraph, we simulated the outcomes of tossing a coin by using single-digit numbers from Table A.12. In general, we prefer to express probabilities as decimals rounded to one, two, three, or more places and then allocate, respectively, one, two, three, or more digit random numbers to the various values of the random variable. As we shall see in the next two examples, this method of assigning random numbers can be facilitated if we first construct a **cumulative probability distribution** giving the probabilities $P(X \leq x)$ for each value of x.

Example 9. Suppose the probabilities are, respectively, 0.03, 0.09, 0.14. 0.22, 0.27, 0.18, and 0.07 that a pharmacist will fill 0, 1, 2, 3, 4, 5, or 6 prescriptions for Valium tablets on any one day. Simulate the demand for this product for the next 15 days.

Solution. Let the random variable X represent the number of prescriptions filled for Valium tablets on any given day. To simulate values of X, we first determine the cumulative probabilities and then assign two-digit random numbers from Table A.12 to each of the x values according to the following scheme:

x	f(x)	P(X ≤ x)	Assigned Numbers
0	0.03	0.03	00–02
1	0.09	0.12	03–11
2	0.14	0.26	12–25
3	0.22	0.48	26–47
4	0.27	0.75	48–74
5	0.18	0.93	75–92
6	0.07	1.00	93–99

Because the probabilities are given to two decimal places, we accordingly assign two-digit random numbers to the values of X. Of the 100 random two-digit numbers from 00 to 99, we have assigned 3 of them to $x = 0$, 9 to $x = 1$, 14 to $x = 2$, and so forth. Note that this scheme preserves the probabilities since, for example, the probability that X is assigned the value 3 is still 0.22 because 22 of the 100 equally likely random numbers correspond to this value. Now, by starting on line 15 of Table A.12 and reading down columns 23 and 24, the first 15 two-digit numbers are 76, 99, 08, 11, 22, 59, 35, 11, 03, 04, 98, 36, 44, 27, and 40, to which we assign the values 5, 6, 1, 1, 2, 4, 3, 1, 1, 1, 6, 3, 3, 3, and 3, for the simulated numbers of prescriptions filled by this pharmacist for Valium tablets over the next 15 days.

Example 10. The probability that a person recovers from a rare blood disease is 0.4. Simulate the number of persons, in each of 10 groups of 5 people, who will recover from this disease.

Solution. The random variable X, representing the number of persons that survive when 5 people contract the disease, is a binomial variable with probabilities

$$b(x; 5, 0.4) = \binom{5}{x}(0.4)^x(0.6)^{5-x}$$

for $x = 0, 1, 2, 3, 4$, and 5. We assign three-digit random numbers from Table A.12 to each value of X according to the following scheme:

x	b(x; 5, 0.4)	P(X ≤ x)	Assigned Numbers
0	0.078	0.078	000–077
1	0.259	0.337	078–336
2	0.346	0.683	337–682
3	0.230	0.913	683–912
4	0.077	0.990	913–989
5	0.010	1.000	990–999

In this example we assign three-digit random numbers to the values of X because the probabilities have been rounded to three decimals. Of the 1000 equally likely random three-digit numbers from 000 to 999, we have assigned 78 of them to $x = 0$, 259 to $x = 1$, 346 to $x = 2$, and so forth. Once again, the probabilities have been preserved. Now, by starting on line 47 of Table A.12 and reading down columns 11, 12, and 13, and then continuing at line 1 with columns 14, 15, and 16, we get the 10 three-digit numbers 472, 492, 974, 762, 888, 818, 107, 368, 818, and 267, to which we assign the values 2, 2, 4, 3, 3, 3, 1, 2, 3, and 1 for the simulated numbers of people in 10 groups of 5 who will survive this blood disease.

8.6
Sampling Procedures

So far we have restricted our discussion of sampling distributions to simple random samples as defined in Chapter 1. There are, however, numerous other sampling procedures which in certain situations may be more efficient in the sense that they offer more information concerning the population at no more cost or at least the same information at less cost. In some cases the very nature of the experiment precludes the possibility of selecting a truly simple random sample. One widely employed sampling procedure involves the systematic selection of elements from the population such that every 10th item is inspected, or every 25th person listed in the telephone directory is selected, and so on.

DEFINITION

Systematic Sampling. _Systematic sampling selects every kth element in the population for the sample, with the starting point determined at random from the first k elements._

Systematic samples are very easy to obtain and are often used as if they were random samples. In fact, some systematic samples can lead to more precise inferences concerning population parameters simply because the sample values spread evenly over the entire population. However, a real danger in systematic sampling exists if one happens to choose a sampling interval that corresponds to any hidden periodicity. For example, in sampling average monthly gasoline sales, one should not sample every 12th month, since the sample would then include sales always for the same month and this might be a consistently high summer month for gasoline sales.

Another sampling procedure, which may be much more efficient than simple random sampling, is carried out by dividing the population into a number of mutually exclusive subpopulations, or strata, and then selecting a simple random sample from each stratum. The totality of these simple random samples from all the strata constitutes a **stratified random sample.**

DEFINITION

Stratified Random Sampling. Stratified random sampling selects simple ran-dom samples from mutually exclusive subpopulations, or strata, of the population.

In stratified random sampling the population is divided into strata such that the data of interest are fairly homogeneous within a given stratum. There-fore, in a study of annual incomes of medical doctors in a large city, it might be desirable to stratify the population according to the type of practice or specialty and then sample within these specialties. We would then expect the variability of the incomes within each specialty to be considerably less than within the entire population of incomes. This reduced variability within each stratum will yield a more precise estimate of the population mean so that a stratified random sample of a given size will be more efficient than a simple random sample of the same size. To achieve this homogeneity within the different strata, the stratification must be formed in such a way that there is some relationship between being in a certain stratum and the characteristic under study. In our illustration, the specialty of a doctor is related to annual income.

Stratification of a population results in strata of various sizes. Consider-ation must therefore be given to the sizes of the random samples selected from these strata. One procedure, using **proportional allocation,** chooses sample sizes proportional to the sizes of the different strata.

THEOREM 8.7

Sample Sizes for Proportional Allocation. If we divide a population of size N into k strata of sizes N_1, N_2, ..., N_k, and select samples of size n_1, n_2, ..., n_k, respectively, from the k strata, the allocation is proportional if

$$n_i = \left(\frac{N_i}{N}\right)n, \qquad \text{for } i = 1, 2, ..., k,$$

where n is the total size of the stratified random sample.

Proof. We can write

$$\frac{n_i}{N_i} = \frac{n}{N}$$

for $i = 1, 2, \ldots, k$, or equivalently

$$\frac{n_1}{N_1} = \frac{n_2}{n_2} = \cdots = \frac{n_k}{N_k} = \frac{n}{N},$$

and therefore the allocations are proportional.

Example 11. At a small private college the students may be classified according to the following scheme:

Classification	Number of Students
Senior	150
Junior	163
Sophomore	195
Freshman	220

If we use proportional allocation to select a stratified random sample of size $n = 40$, how large a sample must we take from each stratum?

Solution. From Theorem 8.7, with $n = 40$, $N_1 = 150$, $N_2 = 163$, $N_3 = 195$, $N_4 = 220$, and $N = 728$, it follows that

$$n_1 = \left(\frac{150}{728}\right)40 = 8, \qquad n_2 = \left(\frac{163}{728}\right)40 = 9,$$

$$n_3 = \left(\frac{195}{728}\right)40 = 11, \qquad n_4 = \left(\frac{220}{728}\right)40 = 12.$$

In all cases, we have rounded our calculations to the nearest integer.

In many statistical studies we can improve our efficiency over simple random sampling by randomly selecting groups or clusters of elements from a population and then sampling some or all of the elements within the selected cluster. For example, if a shipment of automotive parts consists of 5000 boxes, each containing 10 fuel pumps, and we wish to examine a random sample of

100 of these fuel pumps, it would be difficult to obtain a simple random sample without opening all 5000 boxes. A less costly procedure would be to select perhaps 10 of the boxes at random and examine all 10 fuel pumps in each of these boxes, or we might even select 50 of the 5000 boxes at random and then randomly pick 2 of the 10 fuel pumps from each of the 50 boxes for inspection. Sampling in this manner is referred to as **cluster sampling.**

DEFINITION

Cluster Sampling. Cluster sampling selects a sample containing either all, or a random selection, of the elements from clusters that have themselves been selected randomly from the population.

Cluster sampling has the advantage of being more cost efficient when the population is widely scattered. For example, in studying the investment habits of working adults in a given state, it is much cheaper to interview and collect data from individuals living close together in several randomly selected clusters or regions than to select a simple random sample from the entire state. When the clusters are geographic areas, such as regions of a state as we have here, or subdivisions of a large city, this kind of sampling is also called **area sampling.**

Several or all of the sampling procedures discussed so far may be used in the same study. For instance, if the members of a statistical group for the federal government wish to study voter opinion on the construction of additional nuclear power plants, they might let the voting districts within each of the 50 states represent clusters and then use proportional allocation to select a stratified random sample of voting districts. Then they might use simple random sampling or systematic sampling from the voter registration lists to sample voter opinions from within the selected districts.

EXERCISES

1. Use Table A.12 to simulate sums for 20 tosses of a pair of dice.

2. Suppose the probabilities are, respectively, 0.5, 0.3, 0.1, and 0.1 that a certain region of the country will be hit by 0, 1, 2, or 3 hurricanes during a given hurricane season. Entering Table A.12 on line 10 and reading down column 16, simulate the number of hurricanes to hit this region during the next 5 seasons.

3. Suppose the probabilities are, respectively, 0.08, 0.29, 0.35, 0.15, 0.09, 0.02, and 0.02 that there will be 0, 1, 2, 3, 4, 5, or 6 power outages in a large industrial city during the month of January. Entering Table A.12 on line 25 and reading down columns 39 and 40, simulate the number of power outages for this city during the month of January for the next 10 years.

4. Expressing all probabilities to three decimal places, simulate 7 random outcomes of the binomial random variable X for $n = 10$ and $p = 0.6$. Enter Table A.12 on line 18 and read down columns 21, 22, and 23.

5. Expressing all probabilities to four decimal places, simulate 15 random outcomes of a random variable having a Poisson distribution with the parameter $\mu = 4$. Enter Table A.12 on line 7 and read down the first four columns.

6. **Experimental sampling distribution:** A frequency distribution of a set of sample means obtained experimentally from repeated samples, is called an *experimental sampling distribution of the mean*. To simulate an experimental sampling distribution of the mean, we first select repeated samples of size n by grouping successive sets of n random digits from Table A.12, each set constituting a random sample, and then calculate the mean for each of these samples.

   (a) Use the scheme of Example 9 on page 234 to simulate 40 random samples of size $n = 5$ for the number of prescriptions filled from Monday through Friday for Valium tablets. Enter Table A.12 at line 30 and read down successive pairs of columns beginning with columns 4 and 5. Finally, calculate the mean for each of the 40 random samples.

   (b) Construct an experimental sampling distribution of the 40 means generated in part (a) having the 7 class intervals 1.6–2.0, 2.1–2.5, ..., 4.6–5.0.

   (c) Calculate the mean and the standard deviation of the random sample of 40 means generated in part (a) and compare with the corresponding values expected in accordance with the Central Limit Theorem.

7. The following are the numbers of hospitals located in 20 cities: 3, 5, 8, 5, 2, 4, 6, 9, 5, 7, 3, 12, 8, 3, 8, 5, 5, 10, 7, and 6. List the 4 possible systematic samples of size $n = 5$ that can be selected from this list by starting with one of the first four numbers and then taking each fourth number on the list.

8. At a large city university, students are classified according to the following scheme:

Housing	Number of Students
Campus dormitory	2100
Fraternity house	720
Private residence	3400

   If one uses proportional allocation to select a stratified random sample of size $n = 200$, how large a sample must be taken from each stratum?

9. Among the 250 employees of the local office of an international insurance company, 182 are whites, 51 are blacks, and 17 are Orientals. If we use proportional allocation to select a stratified random grievance committee of 15 employees, how many employees must we take from each race?

10. **Optimum allocation:** Proportional allocation is simple to use and is very effective when the variances of the various strata are not too different. If we let σ_1, σ_2, ..., σ_k denote the standard deviations of the k strata, we can improve our efficiency with stratified random sampling if we use *optimum allocation*. With optimum allocation the strata sample sizes are given by

$$n_i = \left(\frac{N_i \sigma_i}{\displaystyle\sum_{i=1}^{k} N_i \sigma_i} \right) n, \qquad \text{for } i = 1, 2, ..., k.$$

(a) Referring to the stratified data of Exercise 8, suppose that we wish to investigate grade-point averages. If the strata standard deviations for students housed in campus dormitories, fraternity houses, and private residences are, respectively, $\sigma_1 = 0.55$, $\sigma_2 = 0.48$, and $\sigma_3 = 0.62$, use optimum allocation to allocate the sample of size $n = 200$ to these three strata.

(b) Suppose that we wish to investigate wage discrimination in Exercise 9. If the strata standard deviations for monthly salaries for whites, blacks, and Orientals are, respectively, $\sigma_1 = \$200$, $\sigma_2 = \$120$, and $\sigma_3 = \$100$, use optimum allocation to allocate the sample of 15 employees to these three strata.

9 Estimation of Parameters

Research scientists, administrators in the fields of education, business, or government, and political pollsters are all interested in problems of estimation. Whether one is estimating the effectiveness of a new chemotherapy drug, the number of students entering college during the next decade, or the proportion of voters favoring each of three candidates in an upcoming state election for a seat in the U.S. Senate, a statistical inference must be made concerning a population parameter. Procedures for estimating the values of unknown population parameters from information provided by sample data are based on the theory of sampling distributions discussed in Chapter 8. These sampling distributions enable us to associate a specific level of confidence with each statistical inference as an expression of how much faith we can place in the sample statistic correctly estimating the population parameter.

9.1
Statistical Inference

The theory of **statistical inference** consists of those methods by which one makes inferences or generalizations about a population. The trend of today is to distinguish between the **classical method** of estimating a population parameter, whereby inferences are based strictly on information obtained from a random sample selected from the population, and the **Bayesian method,** which utilizes prior subjective knowledge about the probability distribution of the unknown parameters in conjunction with the information provided by the sample data. Throughout most of this chapter we shall use classical methods to estimate unknown population parameters such as the mean, proportion, and the variance by computing statistics from random samples and applying the theory of sampling distributions, much of which was covered in Chapter 8. For completeness, a brief discussion of the Bayesian approach to statistical decision theory is presented in Sections 9.9 and 9.10.

Statistical inference may be divided into two major areas: **estimation** and **tests of hypotheses.** We treat these two areas separately, dealing with the theory of estimation in this chapter and the theory of hypothesis testing in Chapter 10. To distinguish clearly between the two areas, consider the following examples. A candidate for public office may wish to estimate the true proportion of voters favoring him by obtaining the opinions from a random sample of 100 eligible voters. The fraction of voters in the sample favoring the candidate could be used as an estimate of the true proportion of the population of voters. A knowledge of the sampling distribution of a proportion enables one to establish the degree of accuracy of our estimate. This problem falls in the area of estimation.

Now consider the case in which a housewife is interested in finding out whether brand A floor wax is more scuff-resistant than brand B floor wax. She might hypothesize that brand A is better than brand B and, after proper testing, accept or reject this hypothesis. In this example we do not attempt to estimate a parameter, but instead we try to arrive at a correct decision about a prestated hypothesis. Once again we are dependent upon sampling theory to provide us with some measure of accuracy for our decision.

9.2
Classical Methods of Estimation

A **point estimate** of some population parameter θ is a single value $\hat{\theta}$ of a statistic $\hat{\Theta}$. For example, the value $\bar{x}$ of the statistic $\bar{X}$, computed from a sample of size n, is a point estimate of the population parameter μ. Similarly, $\hat{p} = x/n$, is a point estimate of the true proportion p for a binomial experiment.

244

The statistic that one uses to obtain a point estimate is called an **estimator** or a **decision function.** Hence the decision function S^2, which is a function of the random sample, is an estimator of σ^2 and the estimate s^2 is the "*action*" taken. Different samples will generally lead to different actions or estimates.

DEFINITION

Action Space. The set of all possible actions that can be taken in an estimation problem is called the *action space* or *decision space*.

An estimator is not expected to estimate the population parameter without error. We do not expect $\overline{X}$ to estimate μ exactly, but we certainly hope that it is not too far off. For a particular sample it is possible to obtain a closer estimate of μ by using the median $\tilde{X}$ as an estimator. Consider, for instance, a sample consisting of the values 2, 5, and 11 from a population whose mean is 4 but supposedly unknown. We would estimate μ to be $\bar{x} = 6$, using the sample mean as our estimate, or $\tilde{x} = 5$, using the median as our estimate. In this case the estimator $\tilde{X}$ produces an estimate closer to the true parameter than that of the estimator $\overline{X}$. On the other hand, if our random sample contains the values, 2, 6, and 7, then $\bar{x} = 5$ and $\tilde{x} = 6$, so that $\overline{X}$ is now the better estimator. Not knowing the true value of μ, we must decide in advance whether to use $\overline{X}$ or $\tilde{X}$ as our estimator.

What are the desirable properties of a "good" decision function that would influence us to choose one estimator rather than another? Let $\hat{\Theta}$ be an estimator whose value $\hat{\theta}$ is a point estimate of some unknown population parameter θ. Certainly, we would like the sampling distribution of $\hat{\Theta}$ to have a mean equal to the parameter estimated. In our definition of the sample variance on page 35, it was necessary to divide by $n - 1$, rather than n, if we were to have $E(S^2) = \sigma^2$. An estimator possessing this property is said to be **unbiased.**

DEFINITION

Unbiased Estimator. A statistic $\hat{\Theta}$ is said to be an *unbiased estimator* of the parameter θ if $\mu_{\hat{\Theta}} = E(\hat{\Theta}) = \theta$.

Although S^2 is an unbiased estimator of σ^2, S, on the other hand, is a biased estimator of σ with the bias becoming insignificant for large samples.

If $\hat{\Theta}_1$ and $\hat{\Theta}_2$ are two unbiased estimators of the same population parameter θ, we would choose the estimator whose sampling distribution has the smallest variance. Hence, if $\sigma^2_{\hat{\Theta}_1} < \sigma^2_{\hat{\Theta}_2}$, we say that $\hat{\Theta}_1$, is a **more efficient** estimator of θ than $\hat{\Theta}_2$.

DEFINITION

Most Efficient Estimator. If we consider all possible unbiased estimators of some parameter θ, the one with the smallest variance is the *most efficient estimator* of θ.

In Figure 9.1 we illustrate the sampling distributions of three different estimators $\hat{\Theta}_1$, $\hat{\Theta}_2$, and $\hat{\Theta}_3$, all estimating θ. It is clear that only $\hat{\Theta}_1$ and $\hat{\Theta}_2$ are unbiased, since their distributions are centered at θ. The estimator $\hat{\Theta}_1$ has a smaller variance than $\hat{\Theta}_2$ and is therefore more efficient. Hence our choice for an estimator of θ, among the three considered, would be $\hat{\Theta}_1$.

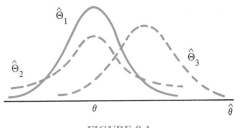

FIGURE 9.1

Sampling distributions of different estimators of θ.

For normal populations one can show that both $\overline{X}$ and $\tilde{X}$ are unbiased estimators of the population mean μ, but the variance of $\overline{X}$ is smaller than the variance of $\tilde{X}$. Thus both estimates $\overline{x}$ and $\tilde{x}$ will, on the average, equal the population mean μ, but $\overline{x}$ is likely to be closer to μ for a given sample.

Even the most efficient unbiased estimator is unlikely to estimate the population parameter exactly. It is true that our accuracy increases with large samples, but there is still no reason why we should expect a point estimate from a given sample to be exactly equal to the population parameter it is supposed to estimate. Perhaps it would be more desirable to determine an interval within which we would expect to find the value of the parameter. Such an interval is called an **interval estimate.**

An interval estimate of a population parameter θ is an interval of the form $\hat{\theta}_1 < \theta < \hat{\theta}_2$, where $\hat{\theta}_1$ and $\hat{\theta}_2$ depend on the value of the statistic $\hat{\Theta}$ for a particular sample and also on the sampling distribution of $\hat{\Theta}$. Thus a random sample of SAT verbal scores for students of the entering freshman class might produce an interval from 530 to 550 within which we expect to find the true average of all SAT verbal scores for the freshman class. The values of the end points, 530 and 550, will depend on the computed sample mean $\overline{x}$ and the sampling distribution of $\overline{X}$. As the sample size increases, we know that

$\sigma_{\bar{X}}^2 = \sigma^2/n$ decreases, and consequently our estimate is likely to be closer to the parameter μ, resulting in a shorter interval. Thus the interval estimate indicates, by its length, the accuracy of the point estimate.

Since different samples will generally yield different values of $\hat{\Theta}$ and, therefore, different values of $\hat{\theta}_1$ and $\hat{\theta}_2$, these end points of the interval are values of corresponding random variables $\hat{\Theta}_1$ and $\hat{\Theta}_2$. From the sampling distribution of $\hat{\Theta}$ we shall be able to determine $\hat{\theta}_1$ and $\hat{\theta}_2$ such that the $P(\hat{\Theta}_1 < \theta < \hat{\Theta}_2)$ is equal to any positive fractional value we care to specify. If, for instance, we find $\hat{\theta}_1$ and $\hat{\theta}_2$ such that

$$P(\hat{\Theta}_1 < \theta < \hat{\Theta}_2) = 1 - \alpha,$$

for $0 < \alpha < 1$, then we have a probability of $1 - \alpha$ of selecting a random sample that will produce an interval containing θ. The interval $\hat{\theta}_1 < \theta < \hat{\theta}_2$, computed from the selected sample, is then called a $(1 - \alpha)100\%$ **confidence interval,** the fraction $1 - \alpha$ is called the **confidence coefficient** or the **degree of confidence,** and the end points, $\hat{\theta}_1$ and $\hat{\theta}_2$, are called the lower and upper **confidence limits.** Thus, when $\alpha = 0.05$, we have a 95% confidence interval, and when $\alpha = 0.01$, we obtain a wider 99% confidence interval. The wider the confidence interval is, the more confident we can be that the given interval contains the unknown parameter. Of course, it is better to be 95% confident that the average life of a certain television transistor is between 6 and 7 years than to be 99% confident that it is between 3 and 10 years. Ideally, we prefer a short interval with a high degree of confidence. Sometimes, restrictions on the size of our sample prevent us from achieving short intervals without sacrificing some of our degree of confidence.

9.3

Estimating the Mean

A point estimator of the population mean μ is given by the statistic $\bar{X}$. The sampling distribution of $\bar{X}$ is centered at μ, and in most applications the variance is smaller than that of any other estimator. Thus the sample mean $\bar{x}$ will be used as a point estimate for the population mean μ. Now recall that $\sigma_{\bar{X}}^2 = \sigma^2/n$ so that a large sample will yield a value of $\bar{X}$ that comes from a sampling distribution with a small variance. Hence $\bar{x}$ is likely to be a very accurate estimate of μ when n is large.

Let us now consider the interval estimate of μ. If our sample is selected from a normal population or, failing this, if n is sufficiently large, we can establish a confidence interval for μ by considering the sampling distribution of $\bar{X}$. According to Theorems 8.1 and 8.3, we can expect the sampling dis-

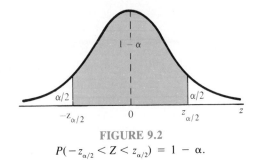

FIGURE 9.2
$$P(-z_{\alpha/2} < Z < z_{\alpha/2}) = 1 - \alpha.$$

tribution of $\overline{X}$ to be approximately normally distributed with mean $\mu_{\overline{X}} = \mu$ and standard deviation $\sigma_{\overline{X}} = \sigma/\sqrt{n}$. Writing $z_{\alpha/2}$ for the z value for which the area to its right under the standard normal curve is $\alpha/2$, we can see from Figure 9.2 that

$$P(-z_{\alpha/2} < Z < z_{\alpha/2}) = 1 - \alpha,$$

where

$$Z = \frac{\overline{X} - \mu}{\sigma/\sqrt{n}}.$$

Hence

$$P\left(-z_{\alpha/2} < \frac{\overline{X} - \mu}{\sigma/\sqrt{n}} < z_{\alpha/2}\right) = 1 - \alpha.$$

Multiplying each term in the inequality by $\sigma/\sqrt{n}$, and then subtracting $\overline{X}$ from each term and multiplying by -1, we have

$$P\left(\overline{X} - z_{\alpha/2}\frac{\sigma}{\sqrt{n}} < \mu < \overline{X} + z_{\alpha/2}\frac{\sigma}{\sqrt{n}}\right) = 1 - \alpha.$$

A random sample of size n is selected from a population whose variance σ^2 is known and the mean $\bar{x}$ is computed to give the following $(1 - \alpha)100\%$ confidence interval for μ.

Confidence Interval for μ; σ *Known.* If $\bar{x}$ is the mean of a random sample of size n from a population with known variance σ^2, a $(1 - \alpha)100\%$ confidence interval for μ is given by

$$\bar{x} - z_{\alpha/2}\frac{\sigma}{\sqrt{n}} < \mu < \bar{x} + z_{\alpha/2}\frac{\sigma}{\sqrt{n}},$$

where $z_{\alpha/2}$ is the z value leaving an area of $\alpha/2$ to the right.

For small samples selected from nonnormal populations, we cannot expect our degree of confidence to be accurate. However, for samples of size $n \geq 30$, regardless of the shape of most populations, sampling theory guarantees good results.

Clearly, the values of the random variables $\hat{\Theta}_1$ and $\hat{\Theta}_2$, defined in Section 9.2, are the confidence limits

$$\hat{\theta}_1 = \bar{x} - z_{\alpha/2}\frac{\sigma}{\sqrt{n}} \qquad \text{and} \qquad \hat{\theta}_2 = \bar{x} + z_{\alpha/2}\frac{\sigma}{\sqrt{n}}.$$

Different samples will yield different values of $\bar{x}$ and therefore produce different interval estimates of the parameter μ as shown in Figure 9.3. The circular dots at the center of each interval indicate the position of the point estimate $\bar{x}$ for each random sample. Most of the intervals are seen to contain μ, but not in every case. Note that all of these intervals are of the same width, since their widths depend only on the choice of $z_{\alpha/2}$ once $\bar{x}$ is determined. The larger the value we choose for $z_{\alpha/2}$, the wider we make all the intervals and the more confident we can be that the particular sample selected will produce an interval that contains the unknown parameter μ.

To compute a $(1 - \alpha)100\%$ confidence interval for μ, we have assumed

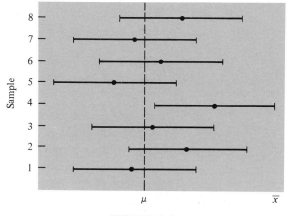

FIGURE 9.3
Interval estimates of μ for different samples.

that σ is known. Since this is generally not the case, we shall replace σ by the sample standard deviation s, provided that $n \geq 30$.

Example 1. The mean and standard deviation for the quality grade-point averages of a random sample of 36 college seniors are calculated to be 2.6 and 0.3, respectively. Find the 95% and 99% confidence intervals for the mean of the entire senior class.

Solution. The point estimate of μ is $\bar{x} = 2.6$. Since the sample size is large, the standard deviation σ can be approximated by $s = 0.3$. The z value, leaving an area of 0.025 to the right and therefore an area of 0.975 to the left, is $z_{0.025} = 1.96$ (Table A.4). Hence the 95% confidence interval is

$$2.6 - (1.96)(0.3/\sqrt{36}) < \mu < 2.6 + (1.96)(0.3/\sqrt{36}),$$

which reduces to

$$2.50 < \mu < 2.70.$$

To find a 99% confidence interval, we find the z value leaving an area of 0.005 to the right and 0.995 to the left. Therefore, using Table A.4 again, $z_{0.005} = 2.575$, and the 99% confidence interval is

$$2.6 - (2.575)(0.3/\sqrt{36}) < \mu < 2.6 + (2.575)(0.3/\sqrt{36}),$$

or simply

$$2.47 < \mu < 2.73.$$

We now see that a longer interval is required to estimate μ with a higher degree of accuracy.

The $(1 - \alpha)100\%$ confidence interval provides an estimate of the accuracy of our point estimate. If μ is actually the center value of the interval, then $\bar{x}$ estimates μ without error. Most of the time, however, $\bar{x}$ will not be exactly equal to μ and the point estimate is in error. The size of this error will be the absolute value of the difference between μ and $\bar{x}$, and we can be $(1 - \alpha)100\%$ confident that this difference will not exceed $z_{\alpha/2}\sigma/\sqrt{n}$. We can readily see this if we draw a diagram of a hypothetical confidence interval as in Figure 9.4.

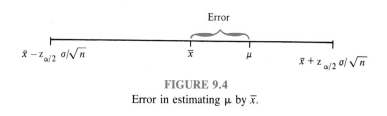

FIGURE 9.4
Error in estimating μ by $\bar{x}$.

THEOREM 9.1

Error in Estimating μ. *If* $\bar{x}$ *is used as an estimate of* μ, *we can then be* $(1 - \alpha)100\%$ *confident that the error will not exceed* $z_{\alpha/2}\sigma/\sqrt{n}$.

In Example 1 we are 95% confident that the sample mean $\bar{x} = 2.6$ differs from the true mean μ by no more than 0.1 and 99% confident that the difference is no more than 0.13.

Frequently, we wish to know how large a sample is necessary to ensure that the error in estimating μ will not exceed a specified amount e. By Theorem 9.1 this means that we must choose n such that $z_{\alpha/2}\sigma/\sqrt{n} = e$.

THEOREM 9.2

Sample Size for Estimating μ. *If* $\bar{x}$ *is used as an estimate of* μ, *we can be* $(1 - \alpha)100\%$ *confident that the error will not exceed a specified amount* e *when the sample size is*

$$n = \left(\frac{z_{\alpha/2}\sigma}{e}\right)^2.$$

When solving for the sample size, n, all fractional values are *rounded up* to the next whole number. By adhering to this principle, we can be sure that our degree of confidence never falls below $(1 - \alpha)100\%$.

Strictly speaking, the formula in Theorem 9.2 is applicable only if we know the variance of the population from which we are to select our sample. If we lack this information, a preliminary sample of size $n \geq 30$ could be taken to provide an estimate of σ, then, using Theorem 9.2, we could determine approximately how many observations are needed to provide the desired degree of accuracy.

Example 2. How large a sample is required in Example 1 if we want to be 95% confident that our estimate of μ is not off by more than 0.05?

Solution. The sample standard deviation $s = 0.3$ obtained from the preliminary sample of size 36 will be used for σ. Then, by Theorem 9.2,

$$n = \left[\frac{(1.96)(0.3)}{0.05} \right]^2 = 138.3.$$

Therefore, we can be 95% confident that a random sample of size 139 will provide an estimate $\bar{x}$ differing from μ by an amount not to exceed 0.05.

Frequently, we are attempting to estimate the mean of a population when the variance is unknown and it is impossible to obtain a sample of size $n \geq 30$. Cost can often be a factor that limits our sample size. As long as our population is approximately bell-shaped, confidence intervals can be computed when σ^2 is unknown and the sample size is small by using the sampling distribution of T, where

$$T = \frac{\bar{X} - \mu}{S/\sqrt{n}}.$$

The procedure is the same as for large samples except that we use the t distribution in place of the standard normal.

Referring to Figure 9.5, we can assert that

$$P(-t_{\sigma/2} < T < t_{\alpha/2}) = 1 - \alpha,$$

where $t_{\alpha/2}$ is the t value with $n - 1$ degrees of freedom, above which we find an area of $\alpha/2$. Owing to symmetry, an equal area of $\alpha/2$ will fall to the left of $-t_{\alpha/2}$. Substituting for T, we write

$$P\left(-t_{\alpha/2} < \frac{\bar{X} - \mu}{S/\sqrt{n}} < t_{\alpha/2} \right) = 1 - \alpha.$$

Multiplying each term in the inequality by $S/\sqrt{n}$, and then subtracting $\bar{X}$ from each term and multiplying by -1, we obtain

$$P\left(\bar{X} - t_{\alpha/2}\frac{S}{\sqrt{n}} < \mu < \bar{X} + t_{\alpha/2}\frac{S}{\sqrt{n}} \right) = 1 - \alpha.$$

For our particular random sample of size n, the mean $\bar{x}$ and standard deviation s are computed and the following $(1 - \alpha)100\%$ confidence interval for μ is obtained.

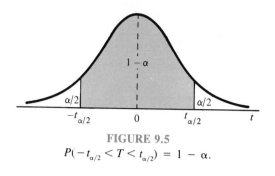

FIGURE 9.5
$P(-t_{\alpha/2} < T < t_{\alpha/2}) = 1 - \alpha.$

Small-Sample Confidence Interval for μ; σ *Unknown.* If $\bar{x}$ and s are the mean and standard deviation of a random sample of size $n < 30$ from an approximate normal population with unknown variance σ^2, a $(1 - \alpha)100\%$ confidence interval for μ is given by

$$\bar{x} - t_{\alpha/2}\frac{s}{\sqrt{n}} < \mu < \bar{x} + t_{\alpha/2}\frac{s}{\sqrt{n}},$$

where $t_{\alpha/2}$ is the t value with $\nu = n - 1$ degrees of freedom, leaving an area of $\alpha/2$ to the right.

Example 3. The contents of 7 similar containers of sulfuric acid are 9.8, 10.2, 10.4, 9.8, 10.0, 10.2, and 9.6 liters. Find a 95% confidence interval for the mean content of all such containers, assuming an approximate normal distribution for container contents.

Solution. The sample mean and standard deviation for the given data are

$$\bar{x} = 10.0 \quad \text{and} \quad s = 0.283.$$

Using Table A.5, we find $t_{0.025} = 2.447$ for $\nu = 6$ degrees of freedom. Hence the 95% confidence interval for μ is

$$10.0 - (2.447)(0.283/\sqrt{7}) < \mu < 10.0 + (2.447)(0.283/\sqrt{7}),$$

which reduces to

$$9.74 < \mu < 10.26.$$

9.4

Estimating the Difference Between Two Means

If we have two populations with means μ_1 and μ_2 and variances σ_1^2 and σ_2^2, respectively, a point estimator of the difference between μ_1 and μ_2 is given by the statistic $\bar{X}_1 - \bar{X}_2$. Therefore, to obtain a point estimate of $\mu_1 - \mu_2$, we shall select two independent random samples, one from each population, of size n_1 and n_2, and compute the difference, $\bar{x}_1 - \bar{x}_2$, of the sample means.

If our independent samples are selected from normal populations, or failing this, if n_1 and n_2 are both greater than 30, we can establish a confidence interval for $\mu_1 - \mu_2$ by considering the sampling distribution of $\bar{X}_1 - \bar{X}_2$.

According to Theorem 8.5 on page 227, we can expect the sampling distribution of $\bar{X}_1 - \bar{X}_2$ to be approximately normally distributed with mean $\mu_{\bar{X}_1 - \bar{X}_2} = \mu_1 - \mu_2$ and standard deviation $\sigma_{\bar{X}_1 - \bar{X}_2} = \sqrt{(\sigma_1^2/n_1) + (\sigma_2^2/n_2)}$. Then, we can assert with a probability of $1 - \alpha$ that the standard normal variable

$$Z = \frac{(\bar{X}_1 - \bar{X}_2) - (\mu_1 - \mu_2)}{\sqrt{(\sigma_1^2/n_1) + (\sigma_2^2/n_2)}}$$

will fall between $-z_{\alpha/2}$ and $z_{\alpha/2}$. Referring once again to Figure 9.2, we write

$$P(-z_{\alpha/2} < Z < z_{\alpha/2}) = 1 - \alpha.$$

Substituting for Z, we state equivalently that

$$P\left[-z_{\alpha/2} < \frac{(\bar{X}_1 - \bar{X}_2) - (\mu_1 - \mu_2)}{\sqrt{(\sigma_1^2/n_1) + (\sigma_2^2/n_2)}} < z_{\alpha/2} \right] = 1 - \alpha,$$

which leads to the following $(1 - \alpha)100\%$ confidence interval for $\mu_1 - \mu_2$.

Confidence Interval for $\mu_1 - \mu_2$; σ_1^2 and σ_2^2 Known. If $\bar{x}_1$ and $\bar{x}_2$ are the means of independent random samples of size n_1 and n_2 from populations with known variances σ_1^2 and σ_2^2, respectively, a $(1 - \alpha)100\%$ confidence interval for $\mu_1 - \mu_2$ is given by

$$(\bar{x}_1 - \bar{x}_2) - z_{\alpha/2} \sqrt{\frac{\sigma_1^2}{n_1} + \frac{\sigma_2^2}{n_2}} < \mu_1 - \mu_2 < (\bar{x}_1 - \bar{x}_2) + z_{\alpha/2} \sqrt{\frac{\sigma_1^2}{n_1} + \frac{\sigma_2^2}{n_2}},$$

where $z_{\alpha/2}$ is the z value leaving an area of $\alpha/2$ to the right.

The degree of confidence is exact when samples are selected from normal populations. For nonnormal populations we obtain an approximate confidence interval that is very good when both n_1 and n_2 exceed 30. As before, if σ_1^2 and σ_2^2 are unknown and our samples are sufficiently large, we may replace σ_1^2 by s_1^2 and σ_2^2 by s_2^2 without appreciably affecting the confidence interval.

Example 4. A standardized chemistry test was given to 50 girls and 75 boys. The girls made an average grade of 76 with a standard deviation of 6, while the boys made an average grade of 82 with a standard deviation of 8. Find a 96% confidence interval for the difference $\mu_1 - \mu_2$, where μ_1 is the mean score of all boys and μ_2 is the mean score of all girls who might take this test.

Solution. The point estimate of $\mu_1 - \mu_2$ is $\bar{x}_1 - \bar{x}_2 = 82 - 76 = 6$. Since n_1 and n_2 are both large, we can substitute $s_1 = 8$ for σ_1 and $s_2 = 6$ for σ_2. Using $\alpha = 0.04$, we find $z_{0.02} = 2.05$ from Table A.4. Hence substitution in the formula

$$(\bar{x}_1 - \bar{x}_2) - z_{\alpha/2} \sqrt{\frac{\sigma_1^2}{n_1} + \frac{\sigma_2^2}{n_2}} < \mu_1 - \mu_2 < (\bar{x}_1 - \bar{x}_2) + z_{\alpha/2} \sqrt{\frac{\sigma_1^2}{n_1} + \frac{\sigma_2^2}{n_2}}$$

yields the 96% confidence interval

$$6 - 2.05 \sqrt{\frac{64}{75} + \frac{36}{50}} < \mu_1 - \mu_2 < 6 + 2.05 \sqrt{\frac{64}{75} + \frac{36}{50}}$$

or

$$3.43 < \mu_1 - \mu_2 < 8.57.$$

The foregoing procedure for estimating the difference between two means is applicable if σ_1^2 and σ_2^2 are known or can be estimated from large samples. If the sample sizes are small, we must again resort to the t distribution to provide confidence intervals that are valid when the populations are approximately normally distributed.

Let us now assume that σ_1^2 and σ_2^2 are unknown and n_1 and n_2 are small (<30). If $\sigma_1^2 = \sigma_2^2 = \sigma^2$, we obtain a standard normal variable in the form

$$Z = \frac{(\bar{X}_1 - \bar{X}_2) - (\mu_1 - \mu_2)}{\sqrt{\sigma^2[(1/n_1) + (1/n_2)]}},$$

where σ^2 is to be estimated by combining or pooling the sample variances. Denoting the pooled estimator by S_p^2, we write

$$S_p^2 = \frac{(n_1 - 1)S_1^2 + (n_2 - 1)S_2^2}{n_1 + n_2 - 2}.$$

Substituting S_p^2 for σ_2 and noting that the divisor in S_p^2 is $n_1 + n_2 - 2$, we obtain the statistic

$$T = \frac{(\overline{X}_1 - \overline{X}_2) - (\mu_1 - \mu_2)}{S_p \sqrt{(1/n_1) + (1/n_2)}},$$

which has a t distribution with $\nu = n_1 + n_2 - 2$ degrees of freedom.
We can see by Figure 9.5 that

$$P(-t_{\alpha/2} < T < t_{\alpha/2}) = 1 - \alpha,$$

where $t_{\alpha/2}$ is the t value with $n_1 + n_2 - 2$ degrees of freedom, above which we find an area of $\alpha/2$.
Substituting for T in the inequality, we write

$$P\left[-t_{\alpha/2} < \frac{(\overline{X}_1 - \overline{X}_2) - (\mu_1 - \mu_2)}{S_p \sqrt{(1/n_1) + (1/n_2)}} < t_{\alpha/2}\right] = 1 - \alpha.$$

After performing the usual mathematical manipulations, the difference of the sample means $\overline{x}_1 - \overline{x}_2$ and the pooled variance

$$s_p^2 = \frac{(n_1 - 1)s_1^2 + (n_2 - 1)s_2^2}{n_1 + n_2 - 2}$$

are computed and then the following $(1 - \alpha)100\%$ confidence interval for $\mu_1 - \mu_2$ is obtained.

Small-Sample Confidence Interval for $\mu_1 - \mu_2$; $\sigma_1^2 = \sigma_2^2$ *Unknown.* If $\overline{x}_1$ and $\overline{x}_2$ are the means of small independent random samples of size n_1 and n_2, respectively, from approximate normal populations with unknown but equal variances, a $(1 - \alpha)100\%$ confidence interval for $\mu_1 - \mu_2$ is given by

$$(\overline{x}_1 - \overline{x}_2) - t_{\alpha/2}s_p \sqrt{\frac{1}{n_1} + \frac{1}{n_2}} < \mu_1 - \mu_2 < (\overline{x}_1 - \overline{x}_2) + t_{\alpha/2}s_p \sqrt{\frac{1}{n_1} + \frac{1}{n_2}},$$

where s_p is the pooled estimate of the population standard deviation and $t_{\alpha/2}$ is the t value with $\nu = n_1 + n_2 - 2$ degrees of freedom, leaving an area of $\alpha/2$ to the right.

Example 5. A course in mathematics is taught to 12 students by the conventional classroom procedure. A second group of 10 students was given the same course by means of programed materials. At the end of the semester the same examination was given to each group. The 12 students meeting in the classroom made an average grade of 85 with a standard deviation of 4, while the 10 students using programed materials made an average of 81 with a standard deviation of 5. Find a 90% confidence interval for the difference between the population means, assuming the populations are approximately normally distributed with equal variances.

Solution. Let μ_1 and μ_2 represent the average grades of all students who might take this course by the classroom and programed presentations, respectively. We wish to find a 90% confidence interval for $\mu_1 - \mu_2$. Our point estimate of $\mu_1 - \mu_2$ is $\bar{x}_1 - \bar{x}_2 = 85 - 81 = 4$. The pooled estimate, s_p^2, of the common variance, σ^2, is

$$s_p^2 = \frac{(11)(16) + (9)(25)}{12 + 10 - 2} = 20.05.$$

Upon taking the square root, $s_p = 4.478$. Using $\alpha = 0.1$, we find in Table A.5 that $t_{0.05} = 1.725$ for $\nu = n_1 + n_2 - 2 = 20$ degrees of freedom. Therefore, the 90% confidence interval for $\mu_1 - \mu_2$ is

$$4 - (1.725)(4.478)\sqrt{\frac{1}{12} + \frac{1}{10}} < \mu_1 - \mu_2$$

$$< 4 + (1.725)(4.478)\sqrt{\frac{1}{12} + \frac{1}{10}},$$

which simplifies to

$$0.69 < \mu_1 - \mu_2 < 7.31.$$

Hence we are 90% confident that the interval from 0.69 to 7.31 contains the true difference of the average grades for the two methods of instruction. The fact that both confidence limits are positive indicates that the classroom method of learning this particular mathematics course is superior to the method using programed materials.

The procedure for constructing confidence intervals for $\mu_1 - \mu_2$ from small samples assumes the populations to be normal and the population variances to be equal. Slight departures from either of these assumptions do not seriously alter the degree of confidence for our interval. A procedure is presented in Chapter 10 for testing the equality of two unknown population variances based on the information provided by the sample variances. If the population variances are considerably different, we still obtain good results when the populations are normal, provided $n_1 = n_2$. Therefore, in a planned experiment, one should make every effort to equalize the size of the samples.

Let us now consider the problem of finding an interval estimate of $\mu_1 - \mu_2$ for small samples when the unknown population variances are not likely to be equal, and it is impossible to select samples of equal size. The statistic most often used in this case is

$$T' = \frac{(\bar{X}_1 - \bar{X}_2) - (\mu_1 - \mu_2)}{\sqrt{(S_1^2/n_1) + (S_2^2/n_2)}},$$

which has approximately a t distribution with v degrees of freedom, where

$$v = \frac{(s_1^2/n_1 + s_2^2/n_2)^2}{[(s_1^2/n_1)^2/(n_1 - 1)] + [(s_2^2/n_2)^2/(n_2 - 1)]}.$$

Since v is seldom an integer, we round it off to the nearest whole number. Using the statistic T', we write

$$P(-t_{\alpha/2} < T' < t_{\alpha/2}) \simeq 1 - \alpha,$$

where $t_{\alpha/2}$ is the t value with v degrees of freedom, above which we find an area of $\alpha/2$. Substituting for T' in the inequality, and following the exact steps as before, we state the final result.

_____

Small-Sample Confidence Interval for $\mu_1 - \mu_2$; $\sigma_1^2 \neq \sigma_2^2$ *Unknown.* If $\bar{x}_1$ and s_1^2, and $\bar{x}_2$ and s_2^2, are the means and variances of small independent samples of size n_1 and n_2, respectively, from approximate normal distributions with unknown and unequal variances, an approximate $(1 - \alpha)100\%$ confidence interval for $\mu_1 - \mu_2$ is given by

$$(\bar{x}_1 - \bar{x}_2) - t_{\alpha/2}\sqrt{\frac{s_1^2}{n_1} + \frac{s_2^2}{n_2}} < \mu_1 - \mu_2 < (\bar{x}_1 - \bar{x}_2) + t_{\alpha/2}\sqrt{\frac{s_1^2}{n_1} + \frac{s_2^2}{n_2}},$$

where $t_{\alpha/2}$ is the t value with

$$v = \frac{(s_1^2/n_1 + s_2^2/n_2)^2}{[(s_1^2/n_1)^2/(n_1 - 1)] + [(s_2^2/n_2)^2/(n_2 - 1)]}$$

degrees of freedom, leaving an area $\alpha/2$ to the right.

Example 6. Records for the past 15 years have shown the average rainfall in a certain region of the country for the month of May to be 4.93 centimeters, with a standard deviation of 1.14 centimeters. A second region of the country has had an average rainfall in May of 2.64 centimeters with a standard deviation of 0.66 centimeter during the past 10 years. Find a 95% confidence interval for the difference of the true average rainfalls in these two regions, assuming that the observations come from normal populations with different variances.

Solution. For the first region we have $\bar{x}_1 = 4.93$, $s_1 = 1.14$, and $n_1 = 15$. For the second region $\bar{x}_2 = 2.64$, $s_2 = 0.66$, and $n_2 = 10$. We wish to find a 95% confidence interval for $\mu_1 - \mu_2$. Since the population variances are assumed to be unequal and our sample sizes are not the same, we can only find an approximate 95% confidence interval based on the t distribution with v degrees of freedom, where

$$v = \frac{(1.14^2/15 + 0.66^2/10)^2}{[(1.14^2/15)^2/14] + [(0.66^2/10)^2/9]}$$
$$= 22.7 \simeq 23.$$

Our point estimate of $\mu_1 - \mu_2$ is $\bar{x}_1 - \bar{x}_2 = 4.93 - 2.64 = 2.29$. Using $\alpha = 0.05$, we find in Table A.5 that $t_{0.025} = 2.069$ for $v = 23$ degrees of freedom. Therefore, the 95% confidence interval for $\mu_1 - \mu_2$ is

$$2.29 - 2.069 \sqrt{\frac{1.14^2}{15} + \frac{0.66^2}{10}} < \mu_1 - \mu_2$$
$$< 2.29 + 2.069 \sqrt{\frac{1.14^2}{15} + \frac{0.66^2}{10}},$$

which simplifies to

$$2.02 < \mu_1 - \mu_2 < 2.56.$$

Hence we are 95% confident that the interval from 2.02 to 2.56 centimeters contains the difference of the true average rainfalls for the two regions.

We conclude this section by considering estimation procedures for the difference of two means when the samples are not independent and the variances of the two populations are not necessarily equal. This will be true if the observations in the two samples occur in pairs so that the two observations are related. For instance, if we run a test on a new diet using 15 individuals, the weights before and after completion of the test form our two samples. Observations in the two samples made on the same individual are related and hence form a pair. To determine if the diet is effective, we must consider the differences, d_i, of paired observations. These differences are the values of a random sample $d_1, d_2,...,d_n$ from a population that we shall assume to be normal with mean μ_D and unknown variance σ_D^2. We estimate σ_D^2 by s_d^2, the variance of the differences constituting the sample. Therefore, s_d^2 is a value of the statistic S_d^2 that fluctuates from sample to sample. The point estimate of $\mu_1 - \mu_2 = \mu_D$ is given by $\bar{d}$.

Since the variance of $\bar{D}$ for paired observations is less than the variance of $\bar{X}_1 - \bar{X}_2$ for independent random samples, any statistical inference concerning $\mu_1 - \mu_2$ based on $\bar{D}$ will be more sensitive. For this reason it is often desirable to plan an experiment in which pairing has been purposely introduced. This may be achieved by recording the X_1 and X_2 measurements for each of n different subjects, whether they be individuals, animals, or plants. Thus the random variables might represent the weights of each of n individuals before and after a controlled diet experiment. We could also choose n pairs of subjects with each pair having a similar characteristic such as IQ, age, breed, and so on, and then for each pair one member is selected at random to yield a value of X_1 leaving the other member to provide the value of X_2. In this case X_1 and X_2 might represent the grades obtained by two individuals of equal IQ when one of the individuals is assigned at random to a class using the conventional lecture approach while the other individual is assigned to a class using programed materials.

A $(1 - \alpha)100\%$ confidence interval for μ_D can be established by writing

$$P(-t_{\alpha/2} < T < t_{\alpha/2}) = 1 - \alpha,$$

where

$$T = \frac{\bar{D} - \mu_D}{S_d/\sqrt{n}},$$

and $t_{\alpha/2}$, as before, is a value of the t distribution with $n - 1$ degrees of freedom.

It is now a routine procedure to replace T, by its definition, in the inequality above and carry out the mathematical steps that lead to the following $(1 - \alpha)100\%$ confidence interval for $\mu_1 - \mu_2 = \mu_D$.

Confidence Interval for $\mu_D = \mu_1 - \mu_2$ *for Paired Observations.* If $\bar{d}$ and s_d are the mean and standard deviation of the differences of n random pairs of measurements, a $(1 - \alpha)100\%$ confidence interval for $\mu_D = \mu_1 - \mu_2$ is

$$\bar{d} - t_{\alpha/2} \frac{s_d}{\sqrt{n}} < \mu_D < \bar{d} + t_{\alpha/2} \frac{s_d}{\sqrt{n}},$$

where $t_{\alpha/2}$ is the t value with $\nu = n - 1$ degrees of freedom, leaving an area of $\alpha/2$ to the right.

Example 7. Twenty college freshmen were divided into 10 pairs, each member of the pair having approximately the same IQ. One of each pair was selected at random and assigned to a mathematics section using programed materials only. The other member of each pair was assigned to a section in which the professor lectured. At the end of the semester each group was given the same examination and the following results were recorded.

Pair	Programed Materials	Lectures	d
1	76	81	−5
2	60	52	8
3	85	87	−2
4	58	70	−12
5	91	86	5
6	75	77	−2
7	82	90	−8
8	64	63	1
9	79	85	−6
10	88	83	5

Find a 98% confidence interval for the true difference in the two learning procedures.

Solution. We wish to find a 98% confidence interval for $\mu_1 - \mu_2$, where μ_1 and μ_2 represent the average grades of all students by the programed and lecture method of presentation, respectively. Since the observations are paired, $\mu_1 - \mu_2 = \mu_D$. The point estimate of μ_D is given by $\bar{d} = -1.6$. The variance, s_d^2, of the sample differences is

$$s_d^2 = \frac{n \sum d_i^2 - \left(\sum d_i \right)^2}{n(n-1)} = \frac{(10)(392) - (-16)^2}{(10)(9)} = 40.7.$$

By taking the square root, $s_d = 6.38$. Using $\alpha = 0.02$, we find in Table A.5 that $t_{0.01} = 2.821$ for $v = n - 1 = 9$ degrees of freedom. Therefore, our 98% confidence interval for μ_D is

$$-1.6 - (2.821)(6.38/\sqrt{10}) < \mu_D < -1.6 + (2.821)(6.38/\sqrt{10}),$$

or simply

$$-7.29 < \mu_D < 4.09.$$

Hence we are 98% confident that the interval from -7.29 to 4.09 contains the true difference of the average grades for the two methods of instruction. Since this interval allows for the possibility of μ_D being equal to zero, we are unable to state that one method of instruction is better than the other, even though this particular sample of differences shows the lecture procedure to be superior.

EXERCISES

1. (a) Find the parameters μ and σ^2 for the finite population 3, 5, and 2.
    (b) Set up a sampling distribution similar to Table 8.2 for $\bar{X}$ when samples of size 2 are selected at random, with replacement. Show that $\bar{X}$ is an unbiased estimator of μ.
    (c) By computing s^2 for each sample, obtain the sampling distribution S^2 and then show that S^2 is an unbiased estimator of σ^2.

2. Consider the statistic S'^2 whose values are computed from the formula $s'^2 = \sum_{i=1}^{n} (x_i - \bar{x})^2/n$. By computing s'^2 for each sample in Exercise 1, obtain the sampling distribution of S'^2. Show that $E(S'^2) \neq \sigma^2$ and hence that S'^2 is a biased estimator for σ^2.

3. An electrical firm manufactures light bulbs that have a length of life that is approximately normally distributed, with a standard deviation of 40 hours. If a random sample of 30 bulbs has an average life of 780 hours, find a 96% confidence interval for the population mean of all bulbs produced by this firm.

4. A soft-drink machine is regulated so that the amount of drink dispensed is approximately normally distributed with a standard deviation equal to 1.5 deciliters. Find a 95% confidence interval for the mean of all drinks dispensed by this machine if a random sample of 36 drinks had an average content of 22.5 deciliters.

5. The heights of a random sample of 50 college students showed a mean of 174.5 centimeters and a standard deviation of 6.9 centimeters.
   (a) Construct a 98% confidence interval for the mean height of all college students.
   (b) What can we assert with 98% confidence about the possible size of our error if we estimate the mean height of all college students to be 174.5?

6. A random sample of 100 automobile owners shows that an automobile is driven on the average 23,500 kilometers per year, in the state of Virginia, with a standard deviation of 3900 kilometers.
   (a) Construct a 99% confidence interval for the average number of miles an automobile is driven annually in Virginia.
   (b) What can we assert with 99% confidence about the possible size of our error if we estimate the average number of miles driven by car owners in Virginia to be 23,500 kilometers per year?

7. How large a sample is needed in Exercise 3 if we wish to be 96% confident that our sample mean will be within 10 hours of the true mean?

8. How large a sample is needed in Exercise 4 if we wish to be 95% confident that our sample mean will be within 0.3 ounce of the true mean?

9. An efficiency expert wishes to determine the average time that it takes to drill 3 holes in a certain metal clamp. How large a sample will he need to be 95% confident that his sample mean will be within 15 seconds of the true mean? Assume that it is known from previous studies that $\sigma = 40$ seconds.

10. Regular consumption of presweetened cereals contribute to tooth decay, heart disease, and other degenerative diseases, according to studies by Dr. W. H. Bowen of the National Institutes of Health and Dr. J. Yudben, Professor of Nutrition and Dietetics at the University of London. In a random sample of 20 similar single servings of Alpha-Bits, the average sugar content was 11.3 grams with a standard deviation of 2.45 grams. Assuming that the sugar contents are normally distributed, construct a 95% confidence interval for the mean sugar content for single servings of Alpha-Bits.

11. The contents of 10 similar containers of a commercial soap are 10.2, 9.7, 10.1, 10.3, 10.1, 9.8, 9.9, 10.4, 10.3, and 9.8 liters. Find a 99% confidence interval for the mean soap content of all such containers, assuming an approximate normal distribution.

12. A random sample of 8 cigarettes of a certain brand has an average nicotine content of 3.6 milligrams and a standard deviation of 0.9 milligrams. Construct a 99%

confidence interval for the true average nicotine content of this particular brand of cigarettes, assuming an approximate normal distribution.

13. A random sample of 12 female students in a certain dormitory showed an average weekly expenditure of $8.00 for snack foods, with a standard deviation of $1.75. Construct a 90% confidence interval for the average amount spent each week on snack foods by female students living in this dormitory, assuming the expenditures to be approximately normally distributed.

14. A random sample of size $n_1 = 25$ taken from a normal population with a standard deviation $\sigma_1 = 5$ has a mean $\bar{x}_1 = 80$. A second random sample of size $n_2 = 36$, taken from a different normal population with a standard deviation $\sigma_2 = 3$, has a mean $\bar{x}_2 = 75$. Find a 94% confidence interval for $\mu_1 - \mu_2$.

15. Two kinds of thread are being compared for strength. Fifty pieces of each type of thread are tested under similar conditions. Brand A had an average tensile strength of 78.3 kilograms with a standard deviation of 5.6 kilograms, while brand B had an average tensile strength of 87.2 kilograms with a standard deviation of 6.3 kilograms. Construct a 95% confidence interval for the difference of the population means.

16. A study was made to estimate the difference in salaries of college professors in the private and state colleges of Virginia. A random sample of 100 professors in private colleges showed an average 9-month salary of $25,000 with a standard deviation of $1200. A random sample of 200 professors in state colleges showed an average salary of $26,000 with a standard deviation of $1400. Find a 98% confidence interval for the difference between the average salaries of professors teaching in state and private colleges of Virginia.

17. Given two random samples of size $n_1 = 9$ and $n_2 = 16$, from two independent normal populations, with $\bar{x}_1 = 64$, $\bar{x}_2 = 59$, $s_1 = 6$, and $s_2 = 5$, find a 95% confidence interval for $\mu_1 - \mu_2$, assuming that $\sigma_1 = \sigma_2$.

18. Students may choose between a 3-semester-hour course in physics without labs and a 4-semester-hour course with labs. The final written examination is the same for each section. If 12 students in the section with labs made an average examination grade of 84 with a standard deviation of 4, and 18 students in the section without labs made an average grade of 77 with a standard deviation of 6, find a 99% confidence interval for the difference between the average grades for the two courses. Assume the populations to be approximately normally distributed with equal variances.

19. A taxi company is trying to decide whether to purchase brand A or brand B tires for its fleet of taxis. To estimate the difference in the two brands, an experiment is conducted using 12 of each brand. The tires are run until they wear out. The results are

Brand A: $\bar{x}_1 = 36{,}300$ kilometers, $s_1 = 5000$ kilometers.
Brand B: $\bar{x}_2 = 38{,}100$ kilometers, $s_2 = 6100$ kilometers.

Compute a 95% confidence interval for $\mu_1 - \mu_2$, assuming the populations to be approximately normally distributed.

20. The following data represent the running times of films produced by two motion-picture companies:

	Time (minutes)						
Company I	103	94	110	87	98		
Company II	97	82	123	92	175	88	118

Compute a 90% confidence interval for the difference between the average running times of films produced by the two companies. Assume that the running times for each of the companies are approximately normally distributed with unequal variances.

21. The government awarded grants to the agricultural departments of nine universities to test the yield capabilities of two new varieties of wheat. Each variety was planted on plots of equal area at each university and the yields, in kilograms per plot, recorded as follows:

	University								
	1	2	3	4	5	6	7	8	9
Variety 1	38	23	35	41	44	29	37	31	38
Variety 2	45	25	31	38	50	33	36	40	43

Find a 95% confidence interval for the mean difference between the yields of the two varieties assuming the distributions of yields to be approximately normal. Explain why pairing is necessary in this problem.

22. Referring to Exercise 19, find a 99% confidence interval for $\mu_1 - \mu_2$ if a tire from each company is assigned at random to the rear wheels of 8 taxis and the following distances in kilometers, recorded:

Taxi	Brand A	Brand B
1	34,400	36,700
2	45,500	46,800
3	36,700	37,700
4	32,000	31,100
5	48,400	47,800
6	32,800	36,400
7	38,100	38,900
8	30,100	31,500

23. It is claimed that a new diet will reduce a person's weight by 4.5 kilograms on the average in a period of 2 weeks. The weights of 7 women who followed this diet were recorded before and after a 2-week period:

	Woman						
	1	2	3	4	5	6	7
Weight before	58.5	60.3	61.7	69.0	64.0	62.6	56.7
Weight after	60.0	54.9	58.1	62.1	58.5	59.9	54.4

Test the manufacturer's claim by computing a 95% confidence interval for the mean difference in the weight. Assume the distribution of weights to be approximately normal.

9.5
Estimating a Proportion

A point estimator of the proportion p in a binomial experiment is given by the statistic $\hat{P} = X/n$, where X represents the number of successes in n trials. Therefore, the sample proportion $\hat{p} = x/n$ will be used as the point estimate of the parameter p.

If the unknown proportion p is not expected to be too close to zero or 1, we can establish a confidence interval for p by considering the sampling distribution of $\hat{P}$, which of course is a constant multiple of the random variable X. Hence, by Theorem 7.1 on page 200, for n sufficiently large, the distribution of $\hat{P}$ is approximately normally distributed with mean

$$\mu_{\hat{P}} = E(\hat{P}) = E\left(\frac{X}{n}\right) = \frac{np}{n} = p$$

and variance

$$\sigma_{\hat{P}}^2 = \sigma_{X/n}^2 = \frac{\sigma_X^2}{n^2} = \frac{npq}{n^2} = \frac{pq}{n}.$$

Therefore, we can assert that

$$P(-z_{\alpha/2} < Z < z_{\alpha/2}) = 1 - \alpha,$$

where

$$Z = \frac{\hat{P} - p}{\sqrt{pq/n}}$$

and $z_{\alpha/2}$ is the value of the standard normal curve above which we find an area of $\alpha/2$. Substituting for Z, we write

$$P\left(-z_{\alpha/2} < \frac{\hat{P} - p}{\sqrt{pq/n}} < z_{\alpha/2}\right) = 1 - \alpha.$$

Multiplying each term of the inequality by $\sqrt{pq/n}$, and then subtracting $\hat{P}$ and multiplying by -1, we obtain

$$P\left(\hat{P} - z_{\alpha/2}\sqrt{\frac{pq}{n}} < p < \hat{P} + z_{\alpha/2}\sqrt{\frac{pq}{n}}\right) = 1 - \alpha.$$

It is difficult to manipulate the inequalities to obtain a random interval whose end points are independent of p, the unknown parameter. When n is large, very little error is introduced by substituting the point estimate $\hat{p} = x/n$ for the p under the radical sign. Then we can write

$$P\left(\hat{P} - z_{\alpha/2}\sqrt{\frac{\hat{p}\hat{q}}{n}} < p < \hat{P} + z_{\alpha/2}\sqrt{\frac{\hat{p}\hat{q}}{n}}\right) \simeq 1 - \alpha.$$

For our particular random sample of size n, the sample proportion $\hat{p} = x/n$ is computed, and the following approximate $(1 - \alpha)100\%$ confidence interval for p is obtained.

Large-Sample Confidence Interval for p. If $\hat{p}$ is the proportion of successes in a random sample of size n, and $\hat{q} = 1 - \hat{p}$, an approximate $(1 - \alpha)100\%$ confidence interval for the binomial parameter p is given by

$$\hat{p} - z_{\alpha/2}\sqrt{\frac{\hat{p}\hat{q}}{n}} < p < \hat{p} + z_{\alpha/2}\sqrt{\frac{\hat{p}\hat{q}}{n}},$$

where $z_{\alpha/2}$ is the z value leaving an area of $\alpha/2$ to the right.

The method for finding a confidence interval for the binomial parameter p is also applicable when the binomial distribution is being used to approximate the hypergeometric distribution, that is, when n is small relative to N, as illustrated in Example 8.

Example 8. In a random sample of 500 people eating lunch at a hospital cafeteria on various Fridays, it was found that $x = 160$ preferred seafood. Find a 95% confidence interval for the actual proportion of people who eat seafood on Fridays at this cafeteria.

Solution. The point estimate of p is $\hat{p} = 160/500 = 0.32$. Using Table A.4, we find $z_{0.025} = 1.96$. Therefore, substituting in the formula

$$\hat{p} - z_{\alpha/2}\sqrt{\frac{\hat{p}\hat{q}}{n}} < p < \hat{p} + z_{\alpha/2}\sqrt{\frac{\hat{p}\hat{q}}{n}},$$

we obtain the 95% confidence interval

$$0.32 - 1.96\sqrt{\frac{(0.32)(0.68)}{500}} < p < 0.32 + 1.96\sqrt{\frac{(0.32)(0.68)}{500}},$$

which simplifies to

$$0.28 < p < 0.36.$$

If p is the center value of a $(1 - \alpha)100\%$ confidence interval, then $\hat{p}$ estimates p without error. Most of the time, however, $\hat{p}$ will not be exactly equal to p and the point estimate is in error. The size of this error will be the positive difference that separates p and $\hat{p}$, and we can be $(1 - \alpha)100\%$ confident that this difference will not exceed $z_{\alpha/2}\sqrt{\hat{p}\hat{q}/n}$. We can readily see this if we draw a diagram of a typical confidence interval as in Figure 9.6.

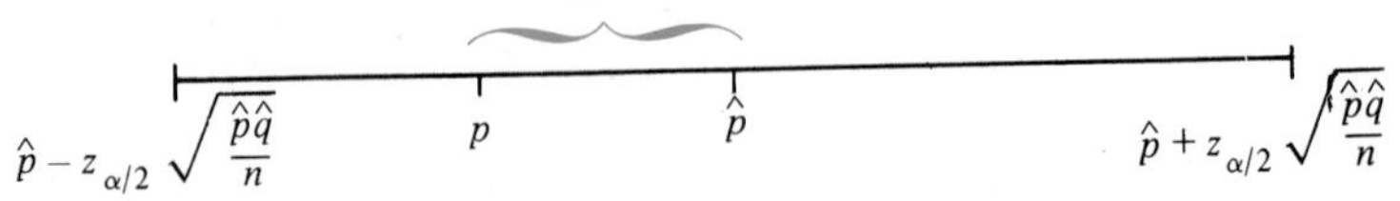

FIGURE 9.6

Error in estimating p by $\hat{p}$.

THEOREM 9.3

Error in Estimating p. If $\hat{p}$ is used as an estimate of p, then we can be $(1 - \alpha)100\%$ confident that the error will not exceed $z_{\alpha/2}\sqrt{\hat{p}\hat{q}/n}$.

In Example 8 we are 95% confident that the sample proportion $\hat{p} = 0.32$ differs from the true proportion p by an amount less than 0.04.

Let us now determine how large a sample is necessary to ensure that the error in estimating p will be less than a specified amount e. By Theorem 9.3 this means we must choose n such that $z_{\alpha/2}\sqrt{\hat{p}\hat{q}/n} = e$.

THEOREM 9.4

Sample Size for Estimating p. If $\hat{p}$ is used as an estimate of p, then we can be $(1 - \alpha)100\%$ confident that the error will not exceed a specified amount e when the sample size is

$$n = \frac{z_{\alpha/2}^2\hat{p}\hat{q}}{e^2}.$$

Theorem 9.4 is somewhat misleading in that we must use $\hat{p}$ to determine the sample size n, but $\hat{p}$ is computed from the sample. If a crude estimate of p can be made without taking a sample, we could use this value for $\hat{p}$ and then determine n. Lacking such an estimate, we could take a preliminary sample of size $n \geq 30$ to provide an estimate of p. Then, using Theorem 9.4, we could determine approximately how many observations are needed to provide the desired degree of accuracy. Once again, all fractional values of n are rounded up to the next whole number.

Example 9. How large a sample is required in Example 8 if we want to be 95% confident that our estimate of p is within 0.02?

Solution. Let us treat the 500 people as a preliminary random sample that provides an estimate $\hat{p} = 0.32$. then, by Theorem 9.4,

$$n = \frac{(1.96)^2(0.32)(0.68)}{(0.02)^2} = 2090.$$

Therefore, if we base our estimate of p on a random sample of size 2090 we can be 95% confident that our sample proportion will not differ from the true proportion by more than 0.02.

Occasionally, it will be impractical to obtain an estimate of p to be used in determining the sample size for a specified degree of confidence. If this happens, an upper bound for n is established by noting that $\hat{p}\hat{q} = \hat{p}(1 - \hat{p})$, which must be at most equal to $\frac{1}{4}$, since $\hat{p}$ must lie between zero and 1. This fact may be verified by completing the square. Hence

$$\hat{p}(1 - \hat{p}) = -(\hat{p}^2 - \hat{p}) = \tfrac{1}{4} - (\hat{p}^2 - \hat{p} + \tfrac{1}{4})$$
$$= \tfrac{1}{4} - (\hat{p} - \tfrac{1}{2})^2,$$

which is always less than $\frac{1}{4}$ except when $\hat{p} = \frac{1}{2}$ and then $\hat{p}\hat{q} = \frac{1}{4}$. Therefore, if we substitute $\hat{p} = \frac{1}{2}$ into the formula for n in Theorem 9.4, when, in fact, p actually differs from $\frac{1}{2}$, then n will turn out to be larger than necessary for the specified degree of confidence and as a result our degree of confidence will increase.

THEOREM 9.5

Sample Size for Estimating p. If $\hat{p}$ is used as an estimate of p, we can be at least $(1 - \alpha)100\%$ confident that the error will not exceed a specified amount e when the sample size is

$$n = \frac{z_{\alpha/2}^2}{4e^2}.$$

Example 10. How large a sample is required in Example 8 if we want to be at least 95% confident that our estimate of p is within 0.02?

Solution. Unlike Example 9, we now assume that no preliminary sample has been taken to provide an estimate of p. Consequently, we can be at least 95% confident that our sample proportion will not differ from the population proportion by more than 0.02 if we choose a sample of size

$$n = \frac{(1.96)^2}{(4)(0.02)^2} = 2401.$$

Comparing the results of Examples 9 and 10, we see that information concerning p, provided by a preliminary sample, or perhaps from past experience, enables us to choose a smaller sample while maintaining our required degree of accuracy.

9.6

Estimating the Difference Between Two Proportions

Consider the problem in which we wish to estimate the difference between two binomial parameters p_1 and p_2. For example, we might let p_1 be the proportion of smokers with lung cancer and p_2 the proportion of nonsmokers with lung cancer. Our problem, then, is to estimate the difference between these two proportions. First, we select independent random samples of size n_1 and n_2 from the two binomial populations with means $n_1 p_1$ and $n_2 p_2$ and variances $n_1 p_1 q_1$ and $n_2 p_2 q_2$, respectively, then determine the numbers x_1 and x_2 of people in each sample with lung cancer, and form the proportions $\hat{p}_1 = x_1/n_1$ and $\hat{p}_2 = x_2/n_2$. A point estimator of the difference between the two proportions, $p_1 - p_2$, is given by the statistic $\hat{P}_1 - \hat{P}_2$. Therefore, the difference of the sample proportions, $\hat{p}_1 - \hat{p}_2$, will be used as the point estimate of $p_1 - p_2$.

A confidence interval for $p_1 - p_2$ can be established by considering the sampling distribution of $\hat{P}_1 - \hat{P}_2$. From the preceding section, for n_1 and n_2 sufficiently large, we know that $\hat{P}_1$ and $\hat{P}_2$ are each approximately normally distributed, with means p_1 and p_2 and variances $p_1 q_1/n_1$ and $p_2 q_2/n_2$, respectively. By choosing independent samples from the two populations, the variables $\hat{P}_1$ and $\hat{P}_2$ will be independent, and then by the reproductive property of the normal distribution established in Theorem 8.6 on page 230, we conclude that $\hat{P}_1 - \hat{P}_2$ is approximately normally distributed with mean

$$\mu_{\hat{P}_1 - \hat{P}_2} = p_1 - p_2$$

and variance

$$\sigma^2_{\hat{P}_1 - \hat{P}_2} = \frac{p_1 q_1}{n_1} + \frac{p_2 q_2}{n_2}.$$

Therefore, we can assert that

$$P(-z_{\alpha/2} < Z < z_{\alpha/2}) = 1 - \alpha,$$

where

$$Z = \frac{(\hat{P}_1 - \hat{P}_2) - (p_1 - p_2)}{\sqrt{(p_1 q_1/n_1) + (p_2 q_2/n_2)}}$$

and $z_{\alpha/2}$ is a value of the standard normal curve above which we find an area of $\alpha/2$. Substituting for Z, we write

$$P\left[-z_{\alpha/2} < \frac{(\hat{P}_1 - \hat{P}_2) - (p_1 - p_2)}{\sqrt{(p_1 q_1/n_1) + (p_2 q_2/n_2)}} < z_{\alpha/2}\right] = 1 - \alpha.$$

After performing the usual mathematical manipulations, we replace p_1 and p_2 under the radical sign by their estimates $\hat{p}_1 = x_1/n_1$ and $\hat{p}_2 = x_2/n_2$, provided that n_1 and n_2 are both large, and the following approximate $(1 - \alpha)100\%$ confidence interval for $p_1 - p_2$ is obtained.

Large-Sample Confidence Interval for $p_1 - p_2$. If $\hat{p}_1$ and $\hat{p}_2$ are the proportion of successes in random samples of size n_1 and n_2, respectively, $\hat{q}_1 = 1 - \hat{p}_1$ and $\hat{q}_2 = 1 - \hat{p}_2$, an approximate $(1 - \alpha)100\%$ confidence interval for the difference of two binomial parameters, $p_1 - p_2$, is given by

$$(\hat{p}_1 - \hat{p}_2) - z_{\alpha/2} \sqrt{\frac{\hat{p}_1 \hat{q}_1}{n_1} + \frac{\hat{p}_2 \hat{q}_2}{n_2}} < p_1 - p_2$$

$$< (\hat{p}_1 - \hat{p}_2) + z_{\alpha/2} \sqrt{\frac{\hat{p}_1 \hat{q}_1}{n_1} + \frac{\hat{p}_2 \hat{q}_2}{n_2}},$$

where $z_{\alpha/2}$ is the z value leaving an area of $\alpha/2$ to the right.

Example 11. A poll is taken among the residents of a city and the surrounding county to determine the feasibility of a proposal to construct a civic center. If 2400 of 5000 city residents favor the proposal and 1200 of 2000 county residents favor it, find a 90% confidence interval for the true difference in the fractions favoring the proposal to construct the civic center.

Solution. Let p_1 and p_2 be the true proportions of residents in the city and county, respectively, favoring the proposal. Hence $\hat{p}_1 = 2400/5000 = 0.48$, $\hat{p}_2 = 1200/2000 = 0.60$, and the point estimate of $p_1 - p_2$ is $\hat{p}_1 - \hat{p}_2 = 0.48 - 0.60 = -0.12$. Using Table A.4, we find $z_{0.05} = 1.645$. Therefore, the 90% confidence interval for $p_1 - p_2$ is

$$-0.12 - 1.645 \sqrt{\frac{(0.48)(0.52)}{5000} + \frac{(0.60)(0.40)}{2000}} < p_1 - p_2$$

$$< -0.12 + 1.645 \sqrt{\frac{(0.48)(0.52)}{5000} + \frac{(0.60)(0.40)}{2000}},$$

which simplifies to

$$- 0.1414 < p_1 - p_2 < -0.0986.$$

Since both end points of the interval are negative, we can also conclude that the proportion of county residents favoring the proposal is greater than the proportion of city residents favoring the proposal.

EXERCISES

1. (a) A random sample of 200 voters is selected and 120 are found to support an annexation suit. Find the 96% confidence interval for the fraction of the voting population favoring the suit.
   (b) What can we assert with 96% confidence about the possible size of our error if we estimate the fraction of voters favoring the annexation suit to be 0.6?

2. (a) A random sample of 400 cigarette smokers is selected and 86 are found to have a preference for brand X. Find the 90% confidence interval for the fraction of the population of cigarette smokers who prefer brand X.
   (b) What can we assert with 90% confidence about the possible size of our error if we estimate the fraction of cigarette smokers who prefer brand X to be 0.2?

3. In a random sample of 1000 homes in a certain city, it is found that 628 are heated by natural gas. Find the 98% confidence interval for the fraction of homes in this city that are heated by natural gas.

4. A random sample of 75 college students is selected and 16 are found to have cars on campus. Use a 95% confidence interval to estimate the fraction of students who have cars on campus.

5. A new rocket-launching system is being considered for deployment of small short-range launches. The existing system has $p = 0.8$ as the probability of a successful launch. A sample of 40 experimental launches is made with the new system and 34 are successful.
   (a) Construct a 95% confidence interval for p.
   (b) Would you conclude that the new system is better?

6. How large a sample is needed in Exercise 1 if we wish to be 96% confident that our sample proportion will be within 0.02 of the true fraction of the voting population?

7. How large a sample is needed in Exercise 3 if we wish to be 98% confident that our sample proportion will be within 0.05 of the true proportion of homes in this city that are heated by natural gas?

8. A study is to be made to estimate the percentage of citizens in a town who favor having their water fluoridated. How large a sample is needed if one wishes to be at least 95% confident that our estimate is within 1% of the true percentage?

9. According to Dr. Memory Elvin-Lewis, head of the microbiology department at

Washington University School of Dental Medicine in St. Louis, a couple of cups of either green or oolong tea each day will provide sufficient fluoride to protect your teeth from decay. People who do not like tea and who live in unfluoridated areas should ask their local governments to consider having their water fluoridated. How large a sample is needed to estimate the percentage of citizens in a certain town who favor having their water fluoridated if one wishes to be at least 99% confident that the estimate is within 1% of the true percentage?

10. In a study to estimate the proportion of residents in a certain city and its suburbs who favor the construction of a nuclear power plant, it is found that 52 of 100 urban residents favor the construction while only 34 of 125 suburban residents are in favor. Find a 96% confidence interval for the difference between the proportion of urban and suburban residents who favor construction of the nuclear plant.

11. A cigarette-manufacturing firm claims that its brand A line of cigarettes outsells its brand B line by 8%. If it is found that 42 of 200 smokers prefer brand A and 18 of 150 smokers prefer brand B, compute a 94% confidence interval for the difference between the proportions of sales of the two brands and decide if the 8% difference is a valid claim.

12. A geneticist is interested in the proportion of males and females in the population that have a certain minor blood disorder. In a random sample of 100 males, 24 are found to be afflicted, whereas 13 of 100 females tested appear to have the disorder. Compute a 99% confidence interval for the difference between the proportion of males and females that have this blood disorder.

13. A study is made to determine if a cold climate results in more students being absent from school during a semester than for a warmer climate. Two groups of students are selected at random, one group from Vermont and the other group from Georgia. Of the 300 students from Vermont, 64 were absent at least 1 day during the semester, and of the 400 students from Georgia, 51 were absent 1 or more days. Find a 95% confidence interval for the difference between the fractions of the students who are absent in the two states.

9.7

Estimating the Variance

If a sample of size n is drawn from a normal population with variance σ^2, and the sample variance s^2 is computed, we obtain a value of the statistic S^2. This computed sample variance will be used as a point estimate of σ^2. Hence the statistic S^2 is called an estimator of σ^2.

An interval estimate of σ^2 can be established by using the statistic

$$X^2 = \frac{(n-1)S^2}{\sigma^2},$$

called **chi-square,** whose sampling distribution is known as the **chi-square distribution** with $v = n - 1$ degrees of freedom. As before, v is equal to the divisor in the calculation of s^2. A value of the chi-square statistic is calculated from a random sample by the following theorem.

THEOREM 9.6

Chi-Square Statistic. If s^2 is the variance of a random sample of size n taken from a normal population having the variance σ^2, then

$$\chi^2 = \frac{(n - 1)s^2}{\sigma^2}$$

is a value of a random variable χ^2 having the chi-square distribution with $v = n - 1$ degrees of freedom.

Obviously, the χ^2 values cannot be negative, and therefore the curve of the chi-square distribution cannot by symmetric about $\chi^2 = 0$. The mathematical equation for the curve is rather complicated and fortunately may be omitted. One could easily obtain an experimental sampling distribution of X^2 by selecting several random samples of size n from a normal population and computing the χ^2 value for each sample. The χ^2 curve could then be approximated by drawing a smooth curve over the histogram of the χ^2 values. The curve will have the appearance of those illustrated in Figure 9.7, provided that samples of size 5 or 8 are selected. The curve in Figure 9.7 for $v = 4$ represents the distribution of χ^2 values computed from all possible samples of size 5 from a normal population having the variance σ^2. Similarly, the curve for $v = 7$ represents the distribution of χ^2 values computed from all possible samples of size 8.

The probability that a random sample produces a χ^2 value greater than some specified value is equal to the area under the curve to the right of this

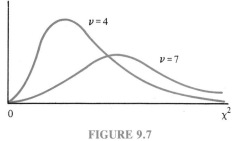

FIGURE 9.7
Chi-square curves for $v = 4$ and $v = 7$.

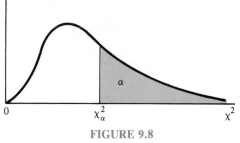

FIGURE 9.8
Tabulated values of the chi-square distribution.

value. It is customary to let χ_α^2 represent the χ^2 value above which we find an area of α. This is illustrated by the shaded region in Figure 9.8.

Table A.6 gives values of χ_α^2 for various values of α and v. The areas, α, are the column headings; the degrees of freedom, v, are given in the left column; and the table entries are the χ^2 values. Hence the χ^2 value with 7 degrees of freedom, leaving an area of 0.05 to the right, is $\chi_{0.05}^2 = 14.067$. Owing to lack of symmetry, we must also use the tables to find $\chi_{0.95}^2 = 2.167$ for $v = 7$.

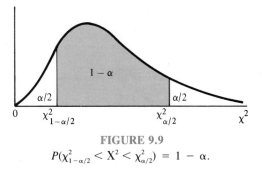

FIGURE 9.9
$P(\chi_{1-\alpha/2}^2 < X^2 < \chi_{\alpha/2}^2) = 1 - \alpha.$

Referring to Figure 9.9, we can assert that

$$P(\chi_{1-\alpha/2}^2 < X^2 < \chi_{\alpha/2}^2) = 1 - \alpha,$$

where $\chi_{1-\alpha/2}^2$ and $\chi_{\alpha/2}^2$ are values of the chi-square distribution with $n - 1$ degrees of freedom, leaving areas under the curve of $1-\alpha/2$ and $\alpha/2$, respectively, to the right. Substituting for X^2, we write

$$P\left(\chi_{1-\alpha/2}^2 < \frac{(n - 1)S^2}{\sigma^2} < \chi_{\alpha/2}^2\right) = 1 - \alpha.$$

Dividing each term in the inequality by $(n - 1)S^2$, and then inverting each term (thereby changing the sense of the inequalities), we obtain

$$P\left[\frac{(n - 1)S^2}{\chi^2_{\alpha/2}} < \sigma^2 < \frac{(n - 1)S^2}{\chi^2_{1-\alpha/2}}\right] = 1 - \alpha.$$

For our particular random sample of size n, the sample variance s^2 is computed, and the following $(1 - \alpha)100\%$ confidence interval for σ^2 is obtained.

Confidence Interval for σ^2. If s^2 is the variance of a random sample of size n from a normal population, a $(1 - \alpha)100\%$ confidence interval for σ^2 is given by

$$\frac{(n - 1)s^2}{\chi^2_{\alpha/2}} < \sigma^2 < \frac{(n - 1)s^2}{\chi^2_{1-\alpha/2}},$$

where $\chi^2_{\alpha/2}$ and $\chi^2_{1-\alpha/2}$ are χ^2 values with $\nu = n - 1$ degrees of freedom, leaving areas of $\alpha/2$ and $1 - \alpha/2$, respectively, to the right.

A $(1 - \alpha)100\%$ confidence interval for σ is obtained by taking the square root of each endpoint of the interval for σ^2.

Example 12. The following are the volumes, in deciliters, of 10 cans of peaches distributed by a certain company: 46.4, 46.1, 45.8, 47.0, 46.1, 45.9, 45.8, 46.9, 45.2, and 46.0. Find a 95% confidence interval for the variance of all such cans of peaches distributed by this company, assuming volume to be a normally distributed variable.

Solution. First we find the sample variance s^2. After subtracting 46 from each observation,

$$s^2 = \frac{n \sum_{i=1}^{n} x_i^2 - \left(\sum_{i=1}^{n} x_i\right)^2}{n(n - 1)} = \frac{(10)(2.72) - (1.2)^2}{(10)(9)} = 0.286.$$

To obtain a 95% confidence interval, we choose $\alpha = 0.05$. Then, using Table A.6 with $\nu = 9$ degrees of freedom, we find $\chi^2_{0.025} = 19.023$ and

$\chi^2_{0.975} = 2.700$. Substituting in the formula

$$\frac{(n-1)s^2}{\chi^2_{\alpha/2}} < \sigma^2 < \frac{(n-1)s^2}{\chi^2_{1-\alpha/2}},$$

we obtain the 95% confidence interval

$$\frac{(9)(0.286)}{19.023} < \sigma^2 < \frac{(9)(0.286)}{2.700},$$

or simply

$$0.135 < \sigma^2 < 0.953.$$

9.8
Estimating the Ratio of Two Variances

A point estimate of the ratio of two population variances σ_1^2/σ_2^2 is given by the ratio s_1^2/s_2^2 of the sample variances. Hence the statistic S_1^2/S_2^2 is the estimator of σ_1^2/σ_2^2.

If σ_1^2 and σ_2^2 are the variances of normal populations, we can establish an interval estimate of σ_1^2/σ_2^2 by using the statistic

$$F = \frac{\sigma_2^2 S_1^2}{\sigma_1^2 S_2^2},$$

whose sampling distribution is called the **F distribution.** Theoretically, we might define the F statistic to be the ratio of two independent chi-square variables, each divided by their degrees of freedom. Hence, if f is a value of the random variable F, we have

$$f = \frac{\chi_1^2/v_1}{\chi_2^2/v_2} = \frac{s_1^2/\sigma_1^2}{s_2^2/\sigma_2^2} = \frac{\sigma_2^2 s_1^2}{\sigma_1^2 s_2^2},$$

where χ_1^2 is a value of a chi-square distribution with $v_1 = n_1 - 1$ degrees of freedom and χ_2^2 is a value of a chi-square distribution with $v_2 = n_2 - 1$ degrees of freedom. We say that f is a value of the F distribution with v_1 and v_2 degrees of freedom.

To obtain an f value, first select a random sample of size n_1 from a normal population having a variance σ_1^2 and compute s_1^2/σ_1^2. An independent sample of size n_2 is then selected from a second normal population with variance σ_2^2 and s_2^2/σ_2^2 is computed. The ratio of the two quantities s_1^2/σ_1^2 and s_2^2/σ_2^2 produces an f value. The distribution of all possible f values where s_1^2/σ_1^2 is the numerator and s_2^2/σ_2^2 is the denominator is called the F distribution, with v_1 and v_2 degrees of freedom. If we consider all possible ratios where s_2^2/σ_2^2 is the numerator and s_1^2/σ_1^2 is the denominator, then we have a distribution of values that also possess an F distribution but with v_2 and v_1 degrees of freedom. The number of degrees of freedom associated with the sample variance in the numerator is always stated first, followed by the number of degrees of freedom associated with the sample variance in the denominator. Thus the curve of the F distribution depends not only on the two parameters v_1 and v_2 but also on the order in which we state them. Once these two values are given, we can identify the curve. Typical f curves are shown in Figure 9.10.

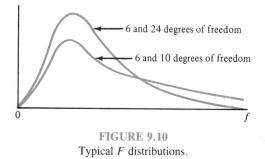

FIGURE 9.10
Typical F distributions.

The foregoing discussion may be summarized in the following theorem.

THEOREM 9.7

F Statistic. If s_1^2 and s_2^2 are the variances of independent random samples of size n_1 and n_2 taken from normal populations with variances σ_1^2 and σ_2^2, respectively, then

$$f = \frac{s_1^2/\sigma_1^2}{s_2^2/\sigma_2^2} = \frac{\sigma_2^2/s_1^2}{\sigma_1^2/s_2^2}$$

is a value of a random variable F having the F distribution with $v_1 = n_1 - 1$ and $v_2 = n_2 - 1$ degrees of freedom.

Let f_α be the f value above which we find an area equal to α. This is illustrated by the shaded region in Figure 9.11. Table A.7 gives values of f_α only for $\alpha = 0.05$ and $\alpha = 0.01$ for various combinations of the degrees of

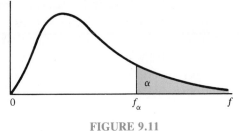

FIGURE 9.11

Tabulated values of the F distribution.

freedom v_1 and v_2. Hence the f value with 6 and 10 degrees of freedom, leaving an area of 0.05 to the right, is $f_{0.05} = 3.22$. By means of the following theorem, Table A.7 can be used to find values of $f_{0.95}$ and $f_{0.99}$.

THEOREM 9.8

Writing $f_\alpha(v_1, v_2)$ for f_α with v_1 and v_2 degrees of freedom, then

$$f_{1-\alpha}(v_1, v_2) = \frac{1}{f_\alpha(v_2, v_1)}.$$

Thus the f value with 6 and 10 degrees of freedom, leaving an area of 0.95 to the right, is

$$f_{0.95}(6, 10) = \frac{1}{f_{0.05}(10, 6)} = \frac{1}{4.06} = 0.246.$$

To establish a confidence interval for σ_1^2/σ_2^2, let us refer to Figure 9.12 and write

$$P[f_{1-\alpha/2}(v_1, v_2) < F < f_{\alpha/2}(v_1, v_2)] = 1 - \alpha,$$

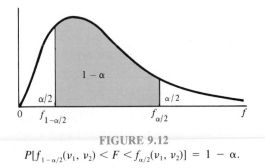

FIGURE 9.12

$P[f_{1-\alpha/2}(v_1, v_2) < F < f_{\alpha/2}(v_1, v_2)] = 1 - \alpha.$

where $f_{1-\alpha/2}(\nu_1, \nu_2)$ and $f_{\alpha/2}(\nu_1, \nu_2)$ are the values of the F distribution with ν_1 and ν_2 degrees of freedom, leaving areas of $1 - \alpha/2$ and $\alpha/2$, respectively, to the right. Substituting for F, we write

$$P\left[f_{1-\alpha/2}(\nu_1, \nu_2) < \frac{\sigma_2^2 S_1^2}{\sigma_1^2 S_2^2} < f_{\alpha/2}(\nu_1, \nu_2)\right] = 1 - \alpha.$$

Multiplying each term in the inequality by S_2^2/S_1^2, and then inverting each term (again changing the sense of the inequalities), we obtain

$$P\left[\frac{S_1^2}{S_2^2}\frac{1}{f_{\alpha/2}(\nu_1, \nu_2)} < \frac{\sigma_1^2}{\sigma_2^2} < \frac{S_1^2}{S_2^2}\frac{1}{f_{1-\alpha/2}(\nu_1, \nu_2)}\right] = 1 - \alpha.$$

The results of Theorem 9.8 enable us to replace $f_{1-\alpha/2}(\nu_1, \nu_2)$ by $1/f_{\alpha/2}(\nu_2, \nu_1)$. Therefore,

$$P\left[\frac{S_1^2}{S_2^2}\frac{1}{f_{\alpha/2}(\nu_1, \nu_2)} < \frac{\sigma_1^2}{\sigma_2^2} < \frac{S_1^2}{S_2^2}f_{\alpha/2}(\nu_2, \nu_1)\right] = 1 - \alpha.$$

For any two independent random samples of size n_1 and n_2 selected from two normal populations, the ratio of the sample variances, s_1^2/s_2^2, is computed, and the following $(1 - \alpha)100\%$ confidence interval for σ_1^2/σ_2^2 is obtained.

Confidence Interval for σ_1^2/σ_2^2. If s_1^2 and s_2^2 are the variances of independent samples of size n_1 and n_2, respectively, from normal populations, then a $(1 - \alpha)100\%$ confidence interval for σ_1^2/σ_2^2 is

$$\frac{s_1^2}{s_2^2}\frac{1}{f_{\alpha/2}(\nu_1, \nu_2)} < \frac{\sigma_1^2}{\sigma_2^2} < \frac{s_1^2}{s_2^2}f_{\alpha/2}(\nu_2, \nu_1),$$

where $f_{\alpha/2}(\nu_1, \nu_2)$ is an f value with $\nu_1 = n_1 - 1$ and $\nu_2 = n_2 - 1$ degrees of freedom leaving an area of $\alpha/2$ to the right, and $f_{\alpha/2}(\nu_2, \nu_1)$ is a similar f value with $\nu_2 = n_2 - 1$ and $\nu_1 = n_1 - 1$ degrees of freedom.

Example 13. A standardized placement test in mathematics was given to 25 boys and 16 girls. The boys made an average grade of 82 with a standard deviation of 8, while the girls made an average grade of 78 with a standard deviation of 7. Find a 98% confidence interval for σ_1^2/σ_2^2 and σ_1/σ_2, where

σ_1^2 and σ_2^2 are the variances of the populations of grades for all boys and girls, respectively, who at some time have taken or will take this test. Assume the populations to be normally distributed.

Solution. We have $n_1 = 25$, $n_2 = 16$, $s_1 = 8$, and $s_2 = 7$. For a 98% confidence interval, $\alpha = 0.02$. Using Table A.7, we find that $f_{0.01}(24, 15) = 3.29$, and $f_{0.01}(15, 24) = 2.89$. Substituting in the formula

$$\frac{s_1^2}{s_2^2} \frac{1}{f_{\alpha/2}(\nu_1, \nu_2)} < \frac{\sigma_1^2}{\sigma_2^2} < \frac{s_1^2}{s_2^2} f_{\alpha/2}(\nu_2, \nu_1),$$

we obtain the 98% confidence interval

$$\frac{64}{49} \left(\frac{1}{3.29} \right) < \frac{\sigma_1^2}{\sigma_2^2} < \frac{64}{49} (2.89),$$

which simplifies to

$$0.397 < \frac{\sigma_1^2}{\sigma_2^2} < 3.775.$$

Upon taking square roots of the confidence limits, a 98% confidence interval for σ_1/σ_2 is

$$0.630 < \frac{\sigma_1}{\sigma_2} < 1.943.$$

EXERCISES

1. (a) Find $\chi_{0.01}^2$ when $\nu = 18$.
   (b) Find $\chi_{0.975}^2$ when $\nu = 29$.
   (c) Find χ_α^2 such that $P(X^2 < \chi_\alpha^2) = 0.99$ when $\nu = 4$.

2. Assuming the sample variance to be a continuous measurement, find the probability that a random sample of 25 observations, from a normal population with variance $\sigma^2 = 6$, will have a variance
   (a) greater than 9.1;
   (b) between 3.462 and 10.745.

3. A manufacturer of car batteries claims that his batteries will last, on the average, 3 years with a variance of 1 year. If 5 of these batteries have lifetimes of 1.9, 2.4, 3.0, 3.5, and 4.2 years, construct a 95% confidence interval for σ^2 and decide if the manufacturer's claim that $\sigma^2 = 1$ is valid. Assume the population of battery lives to be approximately normally distributed.

4. A random sample of 20 students obtained a mean of $\bar{x} = 72$ and a variance of $s^2 = 16$ on a college placement test in mathematics. Assuming the scores to be normally distributed, construct a 98% confidence interval for σ^2.

5. Construct a 95% confidence interval for σ in Exercise 10 on page 263.

6. Construct a 99% confidence interval for σ^2 in Exercise 11 on page 263.

7. Construct a 99% confidence interval for σ in Exercise 12 on page 263.

8. Construct a 90% confidence interval for σ^2 in Exercise 13 on page 264.

9. For an F distribution, find
   (a) $f_{0.05}$ with $v_1 = 7$ and $v_2 = 15$;
   (b) $f_{0.05}$ with $v_1 = 15$ and $v_2 = 7$;
   (c) $f_{0.01}$ with $v_1 = 24$ and $v_2 = 19$;
   (d) $f_{0.095}$ with $v_1 = 19$ and $v_2 = 24$;
   (e) $f_{0.99}$ with $v_1 = 28$ and $v_2 = 12$.

10. If S_1^2 and S_2^2 represent the variances of independent random samples of size $n_1 = 25$ and $n_2 = 31$, taken from normal populations with variances $\sigma_1^2 = 10$ and $\sigma_2^2 = 15$, respectively, find $P(S_1^2/S_2^2 > 1.26)$.

11. If S_1^2 and S_2^2 represent the variances of independent random samples of size $n_1 = 8$ and $n_2 = 12$, taken from normal populations with equal variances, find $P(S_1^2/S_2^2 < 4.89)$.

12. Construct a 98% confidence interval for σ_1/σ_2 in Exercise 17 on page 264. Were we justified in assuming that $\sigma_1 = \sigma_2$?

13. Construct a 90% confidence interval for σ_1^2/σ_2^2 in Exercise 19 on page 264.

14. Construct a 90% confidence interval for σ_1^2/σ_2^2 in Exercise 20 on page 265.

9.9
Bayesian Methods of Estimation

The classical methods of estimation that we have studied so far are based solely on information provided by the random sample. These methods essentially interpret probabilities as relative frequencies. For example, in arriving at a 95% confidence interval for μ, we interpret the statement $P(-1.96 < Z < 1.96) = 0.95$ to mean that 95% of the time in repeated experiments Z will fall between -1.96 and 1.96. Probabilities of this type that can be interpreted in the frequency sense will be referred to as **objective probabilities.** The Bayesian approach to statistical methods of estimation combines sample information with other available prior information that may appear to be pertinent.

Consider the problem of finding a point estimate of the parameter θ for some population. The classical approach would be to take a random sample of size n and substitute the information provided by the sample into the appropriate estimator or decision function. Thus our estimate of p, the proportion of successes in a binomial experiment, would be $\hat{p} = x/n$, and our estimate of the parameter μ for a normal population would be $\bar{x}$. Now, suppose that additional information is given about θ, namely that it is known to vary according to some probability distribution, often called a **prior distribution,** with **prior mean** μ_0 and **prior variance** σ_0^2. The probabilities associated with this prior distribution are called **subjective probabilities,** in that they measure a person's *degree of belief* in the location of the parameter. The person uses his own experience and knowledge as the basis for arriving at the subjective probabilities given by the prior distribution. Bayesian techniques use the prior distribution along with direct sample evidence to compute the **posterior distribution** of the parameter θ. Subsequent inferences concerning the parameter are then based on the posterior distribution. For example, the mean of the posterior distribution may be used as a point estimate of θ.

In this text we restrict our discussion to Bayesian estimates of the parameter p for a binomial population and the parameter μ of a normal population.

Example 14. Let us assume that the prior distribution for the proportion p of defectives produced by a machine is

p	0.1	0.2
$f(p)$	0.6	0.4

Estimate the proportion of defectives being produced by this machine if a random sample of size 2 yields 1 defective.

Solution. If $p = 0.1$, the probability that the random sample of size 2 yields 1 defective is found to be

$$b(1; 2, 0.1) = \binom{2}{1}(0.1)(0.9) = 0.18.$$

Similarly, when $p = 0.2$, we find that

$$b(1; 2, 0.2) = \binom{2}{1}(0.2)(0.8) = 0.32.$$

To find the posterior probabilities of the parameter p, we combine the probabilities 0.18 and 0.32 with the given prior probabilities by using Bayes' rule on page 112. First, let us define the following events:

A: the number of defectives in our sample is 1,

B_1: the proportion of defectives is $p = 0.1$,

B_2: the proportion of defectives is $p = 0.2$.

Now,

$$P(B_1|A) = \frac{P(B_1)P(A|B_1)}{P(B_1)P(A|B_1) + P(B_2)P(A|B_2)}$$

$$= \frac{(0.6)(0.18)}{(0.6)(0.18) + (0.4)(0.32)}$$

$$= 0.46$$

and then by subtraction

$$P(B_2|A) = 1 - P(B_1|A)$$

$$= 1 - 0.46 = 0.54.$$

Hence the posterior distribution for the proportion of defectives, p, when $x = 1$, is

p	0.1	0.2
$f(p|x = 1)$	0.46	0.54

from which we get the posterior mean, denoted by p^*, to be

$$p^* = (0.1)(0.46) + (0.2)(0.54) = 0.154$$

as our point estimate of the true proportion of defectives.

Bayesian methods of estimation concerning the mean μ of a normal population are based on the following theorem.

THEOREM 9.9 _____

Posterior Distribution of the Mean. If $\bar{x}$ is the mean of a random sample of size n from a normal population with known variance σ^2, and the prior distribution of the population mean is a normal distribution with mean μ_0 and variance σ_0^2, then the posterior distribution of the population mean is also a normal distribution with mean μ_1 and variance σ_1^2, where

$$\mu_1 = \frac{n\bar{x}\sigma_0^2 + \mu_0\sigma^2}{n\sigma_0^2 + \sigma^2} \quad and \quad \sigma_1^2 = \frac{\sigma_0^2\sigma^2}{n\sigma_0^2 + \sigma^2}.$$

The Central Limit Theorem allows us to use Theorem 9.9 also when we select random samples of size $n \geq 30$ from nonnormal populations, and when the prior distribution of the mean is approximately normal.

To compute μ_1 and σ_1^2 by the formulas of Theorem 9.9, we have assumed that σ^2 is known. Since this is generally not the case, we shall replace σ^2 by the sample variance s^2, provided that $n \geq 30$. The posterior mean μ_1 is used as a point estimate of the population mean μ, and a $(1 - \alpha)100\%$ **Bayesian interval** for μ can be constructed by computing the interval

$$\mu_1 - z_{\alpha/2}\sigma_1 < \mu < \mu_1 + z_{\alpha/2}\sigma_1,$$

which is centered at the posterior mean and contains $(1 - \alpha)100\%$ of the posterior probability.

Example 15. An electrical firm manufactures light bulbs that have a length of life that is approximately normally distributed with a standard deviation of 100 hours. Prior experience leads us to believe that μ is a value of a normal random variable with a mean $\mu_0 = 800$ hours and a standard deviation $\sigma_0 = 10$ hours. If a random sample of 25 bulbs have an average life of 780 hours, find a 95% Bayesian interval for μ.

Solution. According to Theorem 9.9, the posterior distribution of the mean is also a normal distribution with mean

$$\mu_1 = \frac{(25)(780)(10)^2 + (800)(100)^2}{(25)(10)^2 + (100)^2} = 796$$

and variance

$$\sigma_1^2 = \frac{(10)^2(100)^2}{(25)(10)^2 + (100)^2} = 80.$$

The 95% Bayesian interval for μ is then given by

$$796 - 1.96\sqrt{80} < \mu < 796 + 1.96\sqrt{80}$$

or

$$778.5 < \mu < 813.5.$$

By ignoring the prior information about μ in Example 15, we could proceed as in Section 9.3 and construct the classical 95% confidence interval

$$780 - (1.96)\left(\frac{100}{\sqrt{25}}\right) < \mu < 780 + (1.96)\left(\frac{100}{\sqrt{25}}\right)$$

or

$$740.8 < \mu < 819.2,$$

which is seen to be wider than the corresponding Bayesian interval.

9.10
Decision Theory

In our discussion on the classical approach to point estimation we adopted the criterion that selects the decision function which is most efficient; that is, we choose from all possible unbiased estimators the one with the smallest variance as our "best" estimator. In *decision theory* we also take into account the rewards for making correct decisions and the penalties for making incorrect decisions. This leads to a new criterion, which chooses the decision function $\hat{\Theta}$ that penalizes us the least when the action taken is incorrect. It is convenient now to introduce a **loss function** whose values depend on the true value of the parameter θ and the action $\hat{\theta}$. This is usually written in functional notation as $L(\hat{\Theta}; \theta)$. In many decision-making problems it is desirable to use a loss function of the form

$$L(\hat{\Theta}; \theta) = |\hat{\Theta} - \theta|$$

or perhaps

$$L(\hat{\Theta}; \theta) = (\hat{\Theta} - \theta)^2$$

in arriving at a choice between two or more decision functions.

Since θ is unknown, it must be assumed that it can equal any of several possible values. The set of all possible values that θ can assume is called the **parameter space.** For each possible value of θ in the parameter space, the loss function will vary from sample to sample. We define the **risk function** for the decision function $\hat{\theta}$ to be the expected value of the loss function when the value of the parameter is θ and denote this function by $R(\hat{\Theta}; \theta)$. Hence we have

$$R(\hat{\Theta}; \theta) = E[L(\hat{\Theta}; \theta)].$$

One method of arriving at a choice between $\hat{\theta}_1$ and $\hat{\theta}_2$ as an estimator for θ would be to apply the **minimax criterion.** Essentially, we determine the maximum value of $R(\hat{\Theta}_1; \theta)$ and the maximum value of $R(\hat{\Theta}_2; \theta)$ in the parameter space and then choose the decision function that provided the minimum of these two maximum risks.

Example 16. According to the minimax criterion is $\overline{X}$ or $\tilde{X}$ a better estimator of the mean μ of a normal population with known variance σ^2, based on a random sample of size n when the loss function is of the form $L(\hat{\Theta}, \theta) = (\hat{\Theta} - \theta)^2$?

Solution. The loss function corresponding to $\overline{X}$ is given by

$$L(\overline{X}; \mu) = (\overline{X} - \mu)^2.$$

Hence the risk function is

$$R(\overline{X}; \mu) = E[(\overline{X} - \mu)^2] = \frac{\sigma^2}{n}$$

for every μ in the parameter space. Similarly, one can show that the risk function corresponding to $\tilde{X}$ is given by

$$R(\tilde{X}; \mu) = E[(\tilde{X} - \mu)^2] \simeq \frac{\pi\sigma^2}{2n}$$

for every μ in the parameter space. In view of the fact that $\sigma^2/n < \pi\sigma^2/2n$, the minimax criterion selects $\overline{X}$, rather than $\tilde{X}$, as the better estimator for μ.

In some practical situations we may have additional information concerning the unknown parameter θ. For example, suppose that we wish to estimate the binomial parameter p, the proportion of defectives produced by a machine during a certain day when we know that p varies from day to day. If we can write down $f(p)$, the value of the probability distribution of the binomial parameter at p, then it is possible to determine the expected value of the risk function for each decision function. The expected risk corresponding to the estimator $\hat{P} = X/n$, often referred to as the **Bayes risk,** is written $B(\hat{P}) = E[R(\hat{P}; P)]$, where we are now treating the true proportion of defectives as a random variable. In general, when the unknown parameter is treated as a random variable with probability distribution whose values are given by $f(\theta)$, the Bayes risk in estimating θ by means of the estimator $\hat{\theta}$ is given by

$$B(\hat{\Theta}) = E[R(\hat{\Theta}; \Theta)] = \sum R(\hat{\Theta}; \theta_i) f(\theta_i).$$

The decision function $\hat{\Theta}$ that minimizes $B(\hat{\Theta})$ is called the **Bayes estimator** of θ. We shall make no attempt in this text to derive a Bayes estimator, but instead we shall employ the Bayes risk to establish a criterion for choosing between two estimators.

Bayes Criterion. Suppose that the parameter θ is a value of the random variable θ and $f(\theta)$ is the value of its probability distribution at θ. If $\hat{\Theta}_1$ and $\hat{\Theta}_2$ are two estimators of θ and $B(\hat{\Theta}_1) < B(\hat{\Theta}_2)$, then $\hat{\Theta}_1$ is selected as the better estimator for θ.

The foregoing discussion on decision theory might better be understood if one considers the following two examples.

Example 17. Suppose that a friend has three similar coins except for the fact that the first one has 2 heads, the second one has 2 tails, and the third one is honest. We wish to estimate which coin our friend is flipping on the basis of 2 flips of the coin. Let θ be the number of heads on the coin. consider two decision functions $\hat{\Theta}_1$ and $\hat{\Theta}_2$, where $\hat{\Theta}_1$ is the estimator that assigns to θ the number of heads that occur when the coin is flipped twice and $\hat{\Theta}_2$ is the estimator that assigns the value of 1 to θ no matter what the experiment yields. If the loss function is of the form $L(\hat{\Theta}; \theta) = (\hat{\Theta} - \theta)^2$, which estimator is better according to the minimax procedure?

Solution. For the estimator $\hat{\Theta}_1$, the loss function assumes the values $L(\hat{\theta}_1; \theta) = (\hat{\theta}_1 - \theta)^2$, where $\hat{\theta}_1$ may be 0, 1, or 2, depending on the true value of θ. Clearly, if $\theta = 0$ or 2, both flips will yield all tails or all heads and our decision will be a correct one. Hence $L(0; 0) = 0$ and $L(2; 2) = 0$, from which one may easily conclude that $R(\hat{\Theta}_1; 0) = 0$ and $R(\hat{\Theta}_1; 2) = 0$. However, when $\theta = 1$ we could obtain 0, 1, or 2 heads in the 2 flips with probabilities $\frac{1}{4}$, $\frac{1}{2}$, and $\frac{1}{4}$, respectively. In this case we have $L(0; 1) = 1$, $L(1; 1) = 0$, and $L(2; 1) = 1$, from which we find that

$$R(\hat{\Theta}_1; 1) = 1 \times \tfrac{1}{4} + 0 \times \tfrac{1}{2} + 1 \times \tfrac{1}{4} = \tfrac{1}{2}.$$

For the estimator $\hat{\Theta}_2$, the loss function assumes values given by $L(\hat{\theta}_2; \theta)$ = $(\hat{\theta}_2 - \theta)^2 = (1 - \theta)^2$. Hence $L(1; 0) = 1$, $L(1; 1) = 0$, and $L(1; 2)$ = 1, and the corresponding risks are $R(\hat{\Theta}_2; 0) = 1$, $R(\hat{\Theta}_2; 1) = 0$, and $R(\hat{\Theta}_2; 2) = 1$. Since the maximum risk is $\frac{1}{2}$ for the estimator $\hat{\Theta}_1$ compared to a maximum risk of 1 for $\hat{\Theta}_2$, the minimax criterion selects $\hat{\Theta}_1$ as the better of the two estimators.

Example 18. Let us suppose, referring to Example 17, that our friend flips the honest coin 80% of the time and the other 2 coins each about 10% of the time. Does the Bayes criterion select $\hat{\Theta}_1$ or $\hat{\Theta}_2$ as the better estimator?

Solution. The parameter θ may now be treated as a random variable with the following probability distribution:

θ	0	1	2
$f(\theta)$	0.1	0.8	0.1

For the estimator $\hat{\Theta}_1$ the Bayes risk is

$$B(\hat{\Theta}_1) = R(\hat{\Theta}_1; 0)\, f(0) + R(\hat{\Theta}_1; 1)\, f(1) + R(\hat{\Theta}_1; 2)\, f(2)$$
$$= (0)(0.1) + (\tfrac{1}{2})(0.8) + (0)(0.1) = 0.4.$$

Similarly, for the estimator $\hat{\Theta}_2$, we have.

$$B(\hat{\Theta}_2) = R(\hat{\Theta}_2; 0)\, f(0) + R(\hat{\Theta}_2; 1)\, f(1) + R(\hat{\Theta}_2; 2)\, f(2)$$
$$= (1)(0.1) + (0)(0.8) + (1)(0.1) = 0.2.$$

Since $B(\hat{\Theta}_2) < B(\hat{\Theta}_1)$, the Bayes criterion selects $\hat{\Theta}_2$ as the better estimator for the parameter θ.

EXERCISES

1. Estimate the proportion of defectives being produced by the machine in Example 14 on page 284 if the random sample of size 2 yields 2 defectives.

2. Let us assume that the prior distribution for the proportion p of drinks from a vending machine that overflow is

p	0.05	0.10	0.15
$f(p)$	0.3	0.5	0.2

If 2 of the next 9 drinks from this machine overflow, find
(a) the posterior distribution for the proportion p;
(b) a point estimate of p.

3. The developer of a new condominium complex claims that 3 out of 5 buyers will prefer a two-bedroom unit, while his banker claims that it would be more correct to say that 7 out of 10 buyers will prefer a two-bedroom unit. In previous predictions of this type the banker has been twice as reliable as the developer. If 12 of the next 15 condominiums sold in this complex are two-bedroom units, find
(a) the posterior probabilities associated with the claims of the developer and banker;
(b) a point estimate of the proportion of buyers who prefer a two-bedroom unit.

4. The burn time for the first stage of a rocket is a normal random variable with a standard deviation of 0.8 minute. Assume a normal prior distribution for μ with a mean of 8 minutes and a standard deviation of 0.2 minute. If 10 of these rockets are fired and the first stage has an average burn time of 9 minutes, find a 95% Bayesian interval for μ.

5. The daily profit from a cigarette vending machine placed in a restaurant is a value of a normal random variable with unknown mean μ and variance σ^2. Of course, the mean will vary somewhat from restaurant to restaurant, and the distributor feels that these average daily profits can best be described by a normal distribution with mean $\mu_0 = \$8.00$ and standard deviation $\sigma_0 = \$0.40$. If one of these cigarette machines, placed in a certain restaurant, showed an average daily profit of $\bar{x} = \$6.75$ during the first 30 days with a standard deviation of $s = \$1.20$, find
(a) a Bayesian estimate of the true average daily profit for this restaurant;
(b) a 96% Bayesian interval of μ for this restaurant;
(c) the probability that the average daily profits from the machine in this restaurant is between $6.59 and $7.12.

6. The mathematics department of a large university is designing a placement test to be given to the incoming freshman classes. Members of the department feel that the average grade for this test will vary from one freshman class to another. This variation of the average class grade is expressed subjectively by a normal distribution with mean $\mu_0 = 72$ and variance $\sigma_0^2 = 5.76$.
(a) What prior probability does the department assign to the actual average grade being somewhere between 71.8 and 73.4 for next year's freshman class?
(b) If the test is tried on a random sample of 100 freshman students from the next incoming freshman class resulting in an average grade of 70 with a variance of 64, construct a 95% Bayesian interval for μ.
(c) What posterior probability should the department assign to the event of part (a)?

7. We wish to estimate the binomial parameter p by the decision function $\hat{P}$, the

proportion of successes in a binomial experiment consisting of n trials. Find $R(\hat{P}; p)$ when the loss function is of the form $L(\hat{P}; p) = (\hat{P} - p)^2$.

8. Suppose that an urn contains 3 balls, of which θ are red and the remainder black, where θ can vary from 0 to 3. We wish to estimate θ by selecting two balls in succession without replacement. Let $\hat{\Theta}_1$ be the decision function that assigns to θ the value 0 if neither ball is red, the value 1 if the first ball only is red, the value 2 if the second ball only is red, and the value 3 if both balls are red. Using a loss function of the form $L(\hat{\Theta}_1; \theta) = |\hat{\Theta}_1 - \theta|$, find $R(\hat{\Theta}_1; \theta)$.

9. In Exercise 8, consider the estimator $\hat{\Theta}_2 = X(X + 1)/2$, where X is the number of red balls in our sample. Find $R(\hat{\Theta}_2; \theta)$.

10. Use the minimax criterion to determine whether the estimator $\hat{\Theta}_1$ of Exercise 8 or the estimator $\hat{\Theta}_2$ of Exercise 9 is the better estimator.

11. Use the Bayes criterion to determine whether the estimator $\hat{\Theta}_1$ of Exercise 8 or the estimator $\hat{\Theta}_2$ of Exercise 9 is the better estimator, given the following additional information:

θ	0	1	2	3
$f(\theta)$	0.1	0.5	0.1	0.3

10 Tests of Hypotheses

Often, the problem confronting us is not so much the estimation of a population parameter as discussed in Chapter 9, but rather the formulation of a set of rules that lead to a decision culminating in the acceptance or rejection of some statement or hypothesis about the population. For example, a medical researcher might be required to decide on the basis of experimental evidence whether a certain vaccine is superior to one presently being marketed, an engineer might have to decide on the basis of sample data whether there is a difference between the accuracy of two kinds of gauges, or a sociologist might wish to collect appropriate data to enable her to decide whether the blood type and the eye color of an individual are independent variables. The procedures for establishing a set of rules that lead to the acceptance or rejection of these kinds of statements comprise a major area of statistical inference called **hypothesis testing.**

The testing of a statistical hypothesis is perhaps the most important area of statistical inference. First, let us define precisely what we mean by a statistical hypothesis.

Statistical Hypothesis. A *statistical hypothesis* is an assertion or conjecture concerning one or more populations.

The truth or falsity of a statistical hypothesis is never known with certainty unless we examine the entire population. This, of course, would be impractical in most situations. Instead, we take a random sample from the population of interest and use the information contained in this sample to decide whether the hypothesis is likely to be true or false. Evidence from the sample that is inconsistent with the stated hypothesis leads to a rejection of the hypothesis, whereas evidence supporting the hypothesis leads to its acceptance. We should make it clear at this point that the acceptance of a statistical hypothesis is a result of insufficient evidence to reject it and does not necessarily imply that it is true. For example, in tossing a coin 100 times we might test the hypothesis that the coin is balanced. In terms of population parameters, we are testing the hypothesis that the proportion of heads is $p = 0.5$ if the coin were tossed indefinitely. An outcome of 48 heads would not be surprising if the coin is balanced. Such a result would surely support the hypothesis $p = 0.5$. One might argue that such an occurrence is also consistent with the hypothesis that $p = 0.45$. Thus, in accepting the hypothesis, the only thing we can be reasonably certain about is that the true proportion of heads is not a great deal different from one half. If the 100 trials had resulted in only 35 heads, we would then have evidence to support the rejection of our hypothesis. In view of the fact that the probability of obtaining 35 or fewer heads in 100 tosses of a balanced coin is approximately 0.002, either a very rare event has occurred or we are right in concluding that $p \neq 0.5$.

Although we use the terms ''accept'' and ''reject'' frequently throughout this chapter, it is important to understand that the rejection of a hypothesis is to conclude that it is false, while the acceptance of a hypothesis merely implies that we have no evidence to believe otherwise. Because of this terminology, the statistician or experimenter will often state as his hypothesis that which he hopes to reject. If he is interested in a new cold vaccine, he should assume that it is no better than the vaccine now on the market and then set out to reject this contention. Similarly, to prove that one teaching technique is superior to another, we test the hypothesis that there is no difference in the two techniques.

Hypotheses that were formulated with the hope that they be rejected led to the use of the term **null hypothesis.** Today this term is applied to any hypothesis we wish to test and is denoted by H_0. The rejection of H_0 leads to the acceptance of an **alternative hypothesis,** denoted by H_1. A null hypothesis concerning a population parameter will always be stated so as to specify an exact value of the parameter, whereas the alternative hypothesis allows for the possibility of several values. Hence, if H_0 is the null hypothesis $p = 0.5$ for a binomial population, the alternative hypothesis H_1 might be $p > 0.5$, $p < 0.5$, or $p \neq 0.5$.

10.2
Testing a Statistical Hypothesis

To illustrate the concepts used in testing a statistical hypothesis about a population, consider the following example. A certain type of cold vaccine is known to be only 25% effective after a period of 2 years. In order to determine if a new and somewhat more expensive vaccine is superior in providing protection against the same virus for a longer period of time, 20 people are chosen at random and inoculated. If 9 or more of those receiving the new vaccine surpass the 2-year period without contracting the virus, the new vaccine will be considered superior to the one presently in use. The choice of the number 9 is somewhat arbitrary but appears reasonable in that it represents a modest gain over the 5 people that could be expected to receive protection if the 20 people had been inoculated with the vaccine already in use. We are essentially testing the null hypothesis that the new vaccine is equally effective after a period of 2 years as the one now commonly used against the alternative hypothesis that the new vaccine is in fact superior. This is equivalent to testing the hypothesis that the binomial parameter for the probability of a success on a given trial is $p = \frac{1}{4}$ against the alternative that $p > \frac{1}{4}$. This is usually written as follows:

$$H_0: p = \tfrac{1}{4},$$
$$H_1: p > \tfrac{1}{4}.$$

The statistic on which we base our decision is X, the number of individuals in our test group who receive protection from the new vaccine for a period of at least 2 years. The possible values of X, from 0 to 20, are divided into two groups: those numbers less than 9 and those greater than or equal to 9.

All possible scores above 8.5 constitute the **critical region,** and all possible scores below 8.5 determine the **acceptance region.** The number $x_0 = 8.5$ separating these two regions is called the **critical value.** Thus, if $x > x_0$, we reject H_0 in favor of the alternative hypothesis H_1. If $x < x_0$, we accept H_0.

The decision procedure just described could lead to either of two wrong conclusions. For instance, the new vaccine may be no better than the one now in use and, for this particular randomly selected group of individuals, 9 or more surpass the 2-year period without contracting the virus. We would be committing an error by rejecting H_0 in favor of H_1 when, in fact, H_0 is true. Such an error is called a **type I error.**

DEFINITION

_____

Type I Error. Rejection of the null hypothesis when it is true is called a _type I error_.

_____

A second kind of error is committed if fewer than 9 of the group surpass the 2-year period successfully, and we conclude that the new vaccine is no better when it actually is. In this case we would accept H_0 when it is false. This is called a **type II error.**

DEFINITION

_____

Type II Error. Acceptance of the null hypothesis when it is false is called a _type II error_.

_____

The probability of committing a type I error is also called the **level of significance** of the test and is denoted by α. In our illustration, a type I error will occur when 9 or more individuals surpass a 2-year period without contracting the virus using a new vaccine that is actually equivalent to the one in use. Hence, if X is the number of individuals who remain healthy for at least 2 years,

$$\begin{aligned}
\alpha &= P(\text{type I error}) \\
&= P(X \geq 9 \text{ when } p = \tfrac{1}{4}) \\
&= \sum_{x=9}^{20} b(x; 20, \tfrac{1}{4}) \\
&= 1 - \sum_{x=0}^{8} b(x; 20, \tfrac{1}{4}) \\
&= 1 - 0.9591 \\
&= 0.0409.
\end{aligned}$$

We say that the null hypothesis, $p = \frac{1}{4}$, is being tested at the $\alpha = 0.0409$ level of significance. Sometimes, the level of significance is called the **size** of the critical region. A critical region of size 0.0409 is very small, and therefore it is unlikely that a type I error will be committed. Consequently, it would be most unusual for 9 or more individuals to remain immune to a virus for a 2-year period using a new vaccine that is essentially equivalent to the one now on the market.

The probability of committing a type II error, denoted by β, is impossible to compute unless we have a specific alternative hypothesis. If we test the null hypothesis that $p = \frac{1}{4}$ against the alternative hypothesis that $p = \frac{1}{2}$, then we are able to compute the probability of accepting H_0 when it is false. We simply find the probability of obtaining fewer than 9 in the group that surpass the 2-year period when $p = \frac{1}{2}$. In this case

$$
\begin{aligned}
\beta &= P(\text{type II error}) \\
&= P(X < 9 \text{ when } p = \tfrac{1}{2}) \\
&= \sum_{x=0}^{8} b(x; 20, \tfrac{1}{2}) \\
&= 0.2517.
\end{aligned}
$$

This is a rather high probability, indicating a poor test procedure. It is quite likely that we will reject the new vaccine when, in fact, it is superior to that now in use. Ideally, we like to use a test procedure for which both the type I and type II errors are small.

It is possible that the director of the testing program is willing to make a type II error if the more expensive vaccine is not significantly superior. The only time he wishes to guard against the type II error is when the true value of p is at least 0.7. Letting $p = 0.7$, we find that the test procedure above gives

$$
\begin{aligned}
\beta &= P(\text{type II error}) \\
&= P(X < 9 \text{ when } p = 0.7) \\
&= \sum_{x=0}^{8} b(x; 20, 0.7) \\
&= 0.0051.
\end{aligned}
$$

With such a small probability of committing a type II error it is extremely unlikely that the new vaccine would be rejected when it is 70% effective after a period of 2 years. As the alternative hypothesis approaches unity, the value of β diminishes to zero.

Let us assume that the director of the testing program is unwilling to commit a type II error when the alternative hypothesis $p = \frac{1}{2}$ is true, even though we have found the probability of such an error to be $\beta = 0.2517$. A reduction in β is always possible by increasing the size of the critical region. For example consider what happens to the values of α and β when we change our critical value to 7.5, so that all scores of 8 or more fall in the critical region and those below 8 fall in the acceptance region. Now, in testing $p = \frac{1}{4}$ against the alternative hypothesis that $p = \frac{1}{2}$, we find

$$\alpha = \sum_{x=8}^{20} b(x; 20, \tfrac{1}{4})$$

$$= 1 - \sum_{x=0}^{7} b(x; 20, \tfrac{1}{4})$$

$$= 1 - 0.8982$$

$$= 0.1018,$$

$$\beta = \sum_{x=0}^{7} b(x; 20, \tfrac{1}{2})$$

$$= 0.1316.$$

By adopting a new decision procedure, we have reduced the probability of committing a type II error at the expense of increasing the probability of committing a type I error. For a fixed sample size, a decrease in the probability of one error will always result in an increase in the probability of the other error. Fortunately, the probability of committing both types of errors can be reduced by increasing the sample size. Consider the same problem using a random sample of 100 individuals. If 37 or more of the group surpass the 2-year period, we reject the null hypothesis that $p = \frac{1}{4}$ and accept the alternative hypothesis that $p > \frac{1}{4}$. The critical value is now 36.5. All possible scores above 36.5 constitute the critical region and all possible scores below 36.5 fall in the acceptance region.

To determine the probability of committing a type I error, we shall use the normal-curve approximation with

$$\mu = np = (100)(\tfrac{1}{4}) = 25,$$

$$\sigma = \sqrt{npq} = \sqrt{(100)(\tfrac{1}{4})(\tfrac{3}{4})} = 4.33.$$

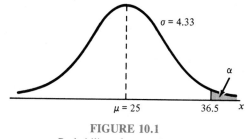

FIGURE 10.1
Probability of a type I error.

Referring to Figure 10.1, we see that

$$\alpha = P(\text{type I error})$$
$$= P(X > 36.5 \text{ when } p = \tfrac{1}{4}).$$

The z value corresponding to $x = 36.5$ is

$$z = \frac{36.5 - 25}{4.33} = 2.66.$$

Therefore,

$$\alpha = P(Z > 2.66)$$
$$= 1 - P(Z < 2.66)$$
$$= 1 - 0.9961$$
$$= 0.0039.$$

If H_0 is false and the true value of H_1 is $p = \tfrac{1}{2}$, we can determine the probability of a type II error using the normal-curve approximation with

$$\mu = np = (100)(\tfrac{1}{2}) = 50,$$
$$\sigma = \sqrt{npq} = \sqrt{(100)(\tfrac{1}{2})(\tfrac{1}{2})} = 5.$$

The probability of falling in the acceptance region when H_1 is true is given by the area of the shaded region in Figure 10.2. Hence

$$\beta = P(\text{type II error})$$
$$= P(X < 36.5 \text{ when } p = \tfrac{1}{2}).$$

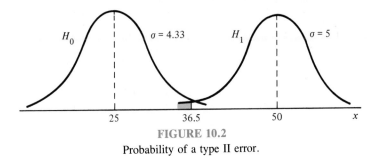

FIGURE 10.2

Probability of a type II error.

The z value corresponding to $x = 36.5$ is

$$z = \frac{36.5 - 50}{5} = -2.7.$$

Therefore,

$$\beta = P(Z < -2.7)$$

$$= 0.0035.$$

Obviously, the type I and type II errors will rarely occur if the experiment consists of 100 individuals.

The concepts discussed above can easily be seen graphically when the population is continuous. Consider the null hypothesis that the average weight of students in a certain college is 68 kilograms against the alternative hypothesis that it is unequal to 68; that is, we wish to test

$$H_0: \mu = 68,$$

$$H_1: \mu \neq 68.$$

The alternative hypothesis allows for the possibility that $\mu < 68$ or $\mu > 68$.

Assume the standard deviation of the population of weights to be $\sigma = 3.6$. Our decision statistic, based on a sample of size $n = 36$, will be $\overline{X}$, the most efficient estimator of μ. From Chapter 8 we know that the sampling distribution of $\overline{X}$ is approximately normally distributed with standard deviation $\sigma_{\overline{X}} = \sigma/\sqrt{n} = 3.6/6 = 0.6$.

A sample mean that falls close to the hypothesized value of 68 would be considered evidence in favor of H_0. On the other hand, a sample mean that is considerably less than or more than 68 would be evidence inconsistent with H_0 and therefore favoring H_1. A critical region, indicated by the shaded area in Figure 10.3, is arbitrarily chosen to be $\overline{x} < 67$ and $\overline{x} > 69$. The acceptance

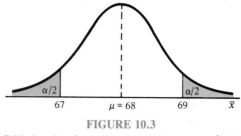

FIGURE 10.3

Critical region for testing $\mu = 68$ versus $\mu \neq 68$.

region will therefore be $67 \leq \bar{x} \leq 69$. Hence, if our sample mean $\bar{x}$ falls inside the critical region, H_0 is rejected; otherwise, we accept H_0.

The probability of committing a type I error, or the level of significance of our test, is equal to the sum of the areas that have been shaded in each tail of the distribution in Figure 10.3. Therefore,

$$\alpha = P(\bar{X} < 67 \text{ when } \mu = 68) + P(\bar{X} > 69 \text{ when } \mu = 68).$$

The z values corresponding to $\bar{x}_1 = 67$ and $\bar{x}_2 = 69$ when H_0 is true are

$$z_1 = \frac{67 - 68}{0.6} = -1.67,$$

$$z_2 = \frac{69 - 68}{0.6} = 1.67.$$

Therefore,

$$\alpha = P(Z < -1.67) + P(Z > 1.67)$$
$$= 2P(Z < -1.67)$$
$$= 0.0950.$$

Thus 9.5% of all samples of size 36 would lead us to reject $\mu = 68$ kilograms when it is true. To reduce α, we have a choice of increasing the sample size or widening the acceptance region. Suppose that we increase the sample size to $n = 64$. Then $\sigma_{\bar{X}} = 3.6/8 = 0.45$. Now

$$z_1 = \frac{67 - 68}{0.45} = -2.22,$$

$$z_2 = \frac{69 - 68}{0.45} = 2.22.$$

Hence

$$\alpha = P(Z < -2.22) + P(Z > 2.22)$$
$$= 2P(Z < -2.22)$$
$$= 0.0264.$$

The reduction in α is not sufficient by itself to guarantee a good testing procedure. We must evaluate β for various alternative hypotheses that we feel should be accepted if true. Therefore, if it is important to reject H_0 when the true mean is some value $\mu \geq 70$ or $\mu \leq 66$, then the probability of committing a type II error should be computed and examined for the alternatives $\mu = 66$ and $\mu = 70$. As a result of symmetry it is only necessary to evaluate the probability of accepting the null hypothesis that $\mu = 68$ when the alternative $\mu = 70$ is true. A type II error will result when the sample mean $\bar{x}$ falls between 67 and 69 when H_1 is true. Therefore, upon referring to Figure 10.4,

$$\beta = P(67 \leq \bar{X} \leq 69 \text{ when } \mu = 70).$$

The z values corresponding to $\bar{x}_1 = 67$ and $\bar{x}_2 = 69$, when H_1 is true, are

$$z_1 = \frac{67 - 70}{0.45} = -6.67,$$

$$z_2 = \frac{69 - 70}{0.45} = -2.22.$$

Therefore,

$$\beta = P(-6.67 < Z < -2.22)$$
$$= P(Z < -2.22) - P(Z < -6.67)$$
$$= 0.0132 - 0.0000$$
$$= 0.0132.$$

If the true value of μ is the alternative $\mu = 66$, the value of β will again be 0.0132. For all possible values of $\mu < 66$ or $\mu > 70$, the value of β will be even smaller when $n = 64$, and consequently there would be little chance of accepting H_0 when it is false.

The probability of committing a type II error increases rapidly when the true value of μ approaches, but is not equal to, the hypothesized value. Of course, this is usually the situation where we do not mind making a type II error. For example, if the alternative hypothesis $\mu = 68.5$ is true, we do not

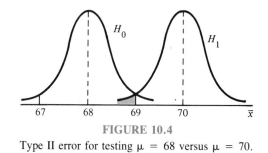

FIGURE 10.4

Type II error for testing $\mu = 68$ versus $\mu = 70$.

mind committing a type II error by concluding that the true answer is $\mu = 68$. The probability of making such an error will be high when $n = 64$. Referring to Figure 10.5, we have

$$\beta = P(67 \leq \overline{X} \leq 69 \text{ when } \mu = 68.5).$$

The z values corresponding to $\overline{x}_1 = 67$ and $\overline{x}_2 = 69$ when $\mu = 68.5$ are

$$z_1 = \frac{67 - 68.5}{0.45} = -3.33,$$

$$z_2 = \frac{69 - 68.5}{0.45} = 1.11.$$

Therefore,

$$\begin{aligned} \beta &= P(-3.33 < Z < 1.11) \\ &= P(Z < 1.11) - P(Z < -3.33) \\ &= 0.8665 - 0.0004 \\ &= 0.8661. \end{aligned}$$

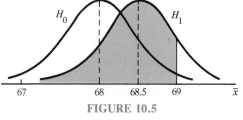

FIGURE 10.5

Type II error for testing $\mu = 68$ versus $\mu = 68.5$.

The preceding examples illustrate the following important properties:

1. The type I error and type II error are related. A decrease in the probability of one results in an increase in the probability of the other.
2. The size of the critical region, and therefore the probability of committing a type I error, can always be reduced by adjusting the critical value(s).
3. An increase in the sample size n will reduce α and β simultaneously.
4. If the null hypothesis is false, β is a maximum when the true value of a parameter is close to the hypothesized value. The greater the distance between the true value and the hypothesized value, the smaller β will be.

10.3

One-Tailed and Two-Tailed Tests

A test of any statistical hypothesis, where the alternative is **one-sided,** such as

$$H_0: \theta = \theta_0,$$
$$H_1: \theta > \theta_0,$$

or perhaps

$$H_0: \theta = \theta_0,$$
$$H_1: \theta < \theta_0,$$

is called a **one-tailed test.** The critical region for the alternative hypothesis $\theta > \theta_0$ lies entirely in the right tail of the distribution, while the critical region for the alternative hypothesis $\theta < \theta_0$ lies entirely in the left tail. In a sense, the inequality symbol points in the direction where the critical region lies. A one-tailed test was used in the vaccine experiment of Section 10.2 to test the hypothesis $p = \frac{1}{4}$ against the one-sided alternative $p > \frac{1}{4}$ for the binomial distribution.

A test of any statistical hypothesis where the alternative is **two-sided,** such as

$$H_0: \theta = \theta_0,$$
$$H_1: \theta \neq \theta_0,$$

is called a **two-tailed test,** since the critical region is split into two equal parts placed in each tail of the distribution of the test statistic. The alternative hypothesis $\theta \neq \theta_0$ states that either $\theta < \theta_0$ or $\theta > \theta_0$. A two-tailed test was used to test the null hypothesis that $\mu = 68$ kilograms against the two-sided alternative $\mu \neq 68$ kilograms for the continuous population of student weights in Section 10.2.

The null hypothesis, H_0, will always be stated using the equality sign so as to specify a single value. In this way the probability of committing a type I error can be controlled. Whether one sets up a one-tailed or a two-tailed test will depend on the conclusion to be drawn if H_0 is rejected. The location of the critical region can be determined only after H_1 has been stated. For example, in testing a new drug, one sets up the hypothesis that it is no better than similar drugs now on the market and tests this against the alternative hypothesis that the new drug is superior. Such an alternative hypothesis will result in a one-tailed test with the critical region in the right tail. However, if we wish to compare a new teaching technique with the conventional classroom procedure, the alternative hypothesis should allow for the new approach to be either inferior or superior to the conventional procedure. Hence the test is two-tailed with the critical region divided equally so as to fall in the extreme left and right tails of the distribution of our statistic.

Example 1. The manufacturer of a certain brand of cigarettes claims that the average nicotine content does not exceed 2.5 milligrams. State the null and alternative hypotheses to be used in testing this claim and determine where the critical region is located.

Solution. The manufacturer's claim should be rejected only if μ is greater than 2.5 milligrams and should be accepted if μ is less than or equal to 2.5 milligrams. Since the null hypothesis always specifies a single value of the parameter, we test

$$H_0: \mu = 2.5,$$
$$H_1: \mu > 2.5.$$

Although we have stated the null hypothesis with an equal sign, it is understood to include any value not specified by the alternative hypothesis. Consequently, the acceptance of H_0 does not imply that μ is exactly equal to 2.5 milligrams but rather that we do not have sufficient evidence favoring H_1. Since we have a one-tailed test, the greater than symbol indicated that the critical region lies entirely in the right tail of the distribution of our statistic $\overline{X}$.

Example 2. A real estate agent claims that 60% of all private residences being built today are 3-bedroom homes. To test this claim, a large sample of new residences are inspected; the proportion of these homes with 3 bedrooms is recorded and used as our test statistic. State the null and alternative hypotheses to be used in this test and determine the location of the critical region.

Solution. If the test statistic is substantially higher or lower than $p = 0.6$, we would reject the agent's claim. Hence we should make the test

$$H_0: p = 0.6,$$

$$H_1: p \neq 0.6.$$

The alternative hypothesis implies a two-tailed test with the critical region divided equally in both tails of the distribution of $\hat{P}$, our test statistic.

In testing hypotheses in which the test statistic is discrete, the critical region may be chosen arbitrarily and its size determined. If the size α is too large, it can be reduced by making an adjustment in the critical value. It may be necessary to increase the sample size to offset the increase that automatically occurs in β. In testing hypotheses in which the test statistic is continuous, it is customary to choose the value of α to be 0.05 or 0.01 and then find the critical region. For example, in a two-tailed test at the 0.05 level of significance, the critical values for a statistic having a standard normal distribution will be $-z_{0.025} = -1.96$ and $z_{0.025} = 1.96$. In terms of z values, the critical region of size 0.05 will be $z < -1.96$ and $z > 1.96$. A test is said to be **significant** if the null hypothesis is rejected at the 0.05 level of significance and is considered to be **highly significant** if the null hypothesis is rejected at the 0.01 level of significance.

In the remaining sections of this chapter we consider several special tests of hypotheses that are frequently used by statisticians. The steps for testing a hypothesis concerning a population parameter θ against some alternative hypothesis may be summarized as follows:

1. State the null hypothesis H_0 that $\theta = \theta_0$.
2. Choose an appropriate alternative hypothesis H_1 from one of the alternatives $\theta < \theta_0$, $\theta > \theta_0$, or $\theta \neq \theta_0$.
3. Choose a significance level of size α.
4. Select the appropriate test statistic and establish the critical region.
5. Compute the value of the test statistic from the sample data.
6. Decision: Reject H_0 if the test statistic has a value in the critical region; otherwise, accept H_0.

1. The proportion of adults living in a small town who are college graduates is estimated to be $p = 0.3$. To test this hypothesis, a random sample of 15 adults is selected. If the number of college graduates in our sample is anywhere from 2 to 7, we shall accept the null hypothesis that $p = 0.3$; otherwise, we shall conclude that $p \neq 0.3$.
   (a) Evaluate α assuming $p = 0.3$.
   (b) Evaluate β for the alternative $p = 0.2$ and $p = 0.4$.
   (c) Is this a good test procedure?

2. Repeat Exercise 1 when 200 adults are selected and the acceptance region is defined to be $48 \leq x \leq 72$, where x is the number of college graduates in our sample.

3. The proportion of families buying milk from company A in a certain city is believed to be $p = 0.6$. If a random sample of 10 families shows that 3 or less buy milk from company A, we shall reject the hypothesis that $p = 0.6$ in favor of the alternative $p < 0.6$.
   (a) Find the probability of committing a type I error if the true proportion is $p = 0.6$.
   (b) Find the probability of committing a type II error for the alternatives $p = 0.3$, $p = 0.4$, and $p = 0.5$.

4. Repeat Exercise 3 when 50 families are selected and the critical region is defined to be $x \leq 24$, where x is the number of families in our sample that buy milk from company A.

5. A random sample of 400 voters in a certain city are asked if they favor an additional 4% gasoline sales tax to provide badly needed revenues for street repairs. If more than 220 but fewer than 260 favor the sales tax, we shall conclude that 60% of the voters are for it.
   (a) Find the probability of committing a type I error if 60% of the voters favor the increased tax.
   (b) What is the probability of committing a type II error using this test procedure if actually only 48% of the voters are in favor of the additional gasoline tax?

6. Suppose, in Exercise 5, we conclude that 60% of the voters favor the gasoline sales tax if more than 214 but fewer than 266 voters in our sample favor it. Show that this new acceptance region results in a smaller value for α at the expense of increasing β.

7. A manufacturer has developed a new fishing line, which he claims has a mean breaking strength of 15 kilograms with a standard deviation of 0.5 kilogram. To test the hypothesis that $\mu = 15$ kilograms against the alternative that $\mu < 15$ kilograms, a random sample of 50 lines will be tested. The critical region is defined to be $\bar{x} < 14.9$.
   (a) Find the probability of committing a type I error when H_0 is true.
   (b) Evaluate β for the alternatives $\mu = 14.8$ and $\mu = 14.9$ kilograms.

8. A soft-drink machine at the Longhorn Steak House is regulated so that the amount of drink dispensed is approximately normally distributed with a mean of 200 milliliters and a standard deviation of 15 milliliters. The machine is checked periodically by taking a sample of 9 drinks and computing the average content. If $\bar{x}$ falls in the interval $191 < \bar{x} < 209$, the machine is thought to be operating satisfactorily; otherwise, we conclude that $\mu \neq 200$ milliliters.
   - (a) Find the probability of committing a type I error when $\mu = 200$ milliliters.
   - (b) Find the probability of committing a type II error when $\mu = 215$ milliliters.

9. Repeat Exercise 8 for samples of size $n = 25$. Use the same critical region.

10. State the null and alternative hypotheses to be used in testing the following claims and determine generally where the critical region is located:
   - (a) The mean snowfall at Lake George during the month of February is 21.8 centimeters.
   - (b) No more than 20% of the faculty at the local university contributed to the annual giving fund.
   - (c) On the average, children attend schools within 6.2 kilometers of their homes in suburban St. Louis.
   - (d) At least 70% of next year's new cars will be in the compact and subcompact category.
   - (e) The proportion of voters favoring the incumbent in the upcoming election is 0.58.
   - (f) The average rib-eye steak at the Longhorn Steak House is at least 340 grams.

11. **OC curve:** If we plot the probabilities of accepting H_0 corresponding to various alternatives for μ (including the value specified by H_0) and connect all the points by a smooth curve, we obtain the *operating characteristic curve* of the test criterion, or simply the *OC curve*. Note that the probability of accepting H_0 when it is true is simply $1 - \alpha$. OC curves are widely used in industrial applications to provide a visual display of the merits of the test criterion. With reference to Exercise 8, find the probabilities of accepting H_0 for the following 9 values of μ and plot the OC curve: 184, 188, 192, 196, 200, 204, 208, 212, and 216.

10.4

Tests Concerning Means

Consider the problem of testing the hypothesis that the mean μ of a population, with known variance σ^2, equals a specified value μ_0 against the two-sided alternative that the mean is not equal to μ_0; that is, we shall test

$$H_0\colon \mu = \mu_0,$$
$$H_1\colon \mu \neq \mu_0.$$

An appropriate statistic on which we base our decision criterion is the random variable $\bar{X}$. From Chapter 8 we already know that the sampling distribution of $\bar{X}$ is approximately normally distributed with mean $\mu_{\bar{X}} = \mu$ and variance $\sigma_{\bar{X}}^2 = \sigma^2/n$, where μ and σ^2 are the mean and variance of the population from which we select random samples of size n. By using a significance level of α, it is possible to find two critical values, $\bar{x}_1$ and $\bar{x}_2$, such that the interval $\bar{x}_1 \leq \bar{x} \leq \bar{x}_2$ defines the acceptance region and the two tails of the distribution, $\bar{x} < \bar{x}_1$ and $\bar{x} > \bar{x}_2$, constitute the critical region.

The critical region can be given in terms of z values by means of the transformation

$$z = \frac{\bar{x} - \mu_0}{\sigma/\sqrt{n}}.$$

Hence, for an α level of significance, the critical values of the random variable Z, corresponding to $\bar{x}_1$ and $\bar{x}_2$, are shown in Figure 10.6 to be

$$-z_{\alpha/2} = \frac{\bar{x}_1 - \mu_0}{\sigma/\sqrt{n}},$$

$$z_{\alpha/2} = \frac{\bar{x}_2 - \mu_0}{\sigma/\sqrt{n}}.$$

From the population we select a random sample of size n and compute the sample mean $\bar{x}$. If $\bar{x}$ falls in the acceptance region, $\bar{x}_1 \leq \bar{x} \leq \bar{x}_2$, then

$$z = \frac{\bar{x} - \mu_0}{\sigma/\sqrt{n}}$$

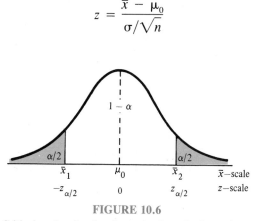

FIGURE 10.6

Critical region for the alternative hypothesis $\mu \neq \mu_0$.

will fall in the region $-z_{\alpha/2} < z < z_{\alpha/2}$ and we conclude that $\mu = \mu_0$; otherwise we reject H_0 and accept the alternative hypothesis that $\mu \neq \mu_0$. The critical region is usually stated in terms of z rather than $\bar{x}$.

The two-tailed test procedure just described is equivalent to finding a $(1 - \alpha)100\%$ confidence interval for μ and accepting H_0 if μ_0 lies in the interval. If μ_0 lies outside the interval, we reject H_0 in favor of the alternative hypothesis H_1. Consequently, when one makes inferences about the mean μ from a population with known variance σ^2, whether it be by the construction of a confidence interval or through the testing of a statistical hypothesis, the same value, $z = (\bar{x} - \mu)/(\sigma/\sqrt{n})$, is employed.

In general, if one uses an appropriate z or t value in Chapter 9 to construct a confidence interval for a population mean μ, or perhaps for the difference $\mu_1 - \mu_2$ of two population means, we can also use that same z or t value to test the hypothesis that $\mu = \mu_0$ or $\mu_1 - \mu_2 = d_0$ against an appropriate alternative. Of course, all the underlying assumptions made in Chapter 9 relative to the use of a given statistic apply to the tests described here. This essentially means that all our samples are selected either from approximately normal populations or are of size $n \geq 30$, in which case we can refer to the Central Limit Theorem to justify using a normal test statistic.

In Table 10.1 we list the values of the statistics used to test specified hypotheses H_0 concerning means and give the corresponding critical regions for one- and two-sided alternative hypotheses H_1. Several of these tests are illustrated in the following examples.

Example 3. A manufacturer of sports equipment has developed a new synthetic fishing line that he claims has a mean breaking strength of 8 kilograms with a standard deviation of 0.5 kilogram. Test the hypothesis that $\mu = 8$ kilograms against the alternative that $\mu \neq 8$ kilograms if a random sample of 50 lines is tested and found to have a mean breaking strength of 7.8 kilograms. Use a 0.01 level of significance.

Solution. Following the six-step procedure outlined in Section 10.3, we have

1. H_0: $\mu = 8$ kilograms.
2. H_1: $\mu \neq 8$ kilograms.
3. $\alpha = 0.01$.
4. Critical region: $z < -2.575$ and $z > 2.575$, where

$$z = \frac{\bar{x} - \mu_0}{\sigma/\sqrt{n}}.$$

TABLE 10.1
Tests Concerning Means

H_0	Value of Test Statistic	H_1	Critical Region
$\mu = \mu_0$	$z = \dfrac{\bar{x} - \mu_0}{\sigma/\sqrt{n}}$; σ known or $n \geq 30$	$\mu < \mu_0$ $\mu > \mu_0$ $\mu \neq \mu_0$	$z < -z_\alpha$ $z > z_\alpha$ $z < -z_{\alpha/2}$ and $z > z_{\alpha/2}$
$\mu = \mu_0$	$t = \dfrac{\bar{x} - \mu_0}{s/\sqrt{n}}$; $v = n - 1$, σ unknown and $n < 30$	$\mu < \mu_0$ $\mu > \mu_0$ $\mu \neq \mu_0$	$t < -t_\alpha$ $t > t_\alpha$ $t < -t_{\alpha/2}$ and $t > t_{\alpha/2}$
$\mu_1 - \mu_2 = d_0$	$z = \dfrac{(\bar{x}_1 - \bar{x}_2) - d_0}{\sqrt{(\sigma_1^2/n_1) + (\sigma_2^2/n_2)}}$; σ_1 and σ_2 known	$\mu_1 - \mu_2 < d_0$ $\mu_1 - \mu_2 > d_0$ $\mu_1 - \mu_2 \neq d_0$	$z < -z_\alpha$ $z > z_\alpha$ $z < -z_{\alpha/2}$ and $z > z_{\alpha/2}$
$\mu_1 - \mu_2 = d_0$	$t = \dfrac{(\bar{x}_1 - \bar{x}_2) - d_0}{s_p\sqrt{(1/n_1) + (1/n_2)}}$; $v = n_1 + n_2 - 2$, $\sigma_1 = \sigma_2$ but unknown, $s_p^2 = \dfrac{(n_1 - 1)s_1^2 + (n_2 - 1)s_2^2}{n_1 + n_2 - 2}$	$\mu_1 - \mu_2 < d_0$ $\mu_1 - \mu_2 > d_0$ $\mu_1 - \mu_2 \neq d_0$	$t < -t_\alpha$ $t > t_\alpha$ $t < -t_{\alpha/2}$ and $t > t_{\alpha/2}$
$\mu_1 - \mu_2 = d_0$	$t' = \dfrac{(\bar{x}_1 - \bar{x}_2) - d_0}{\sqrt{(s_1^2/n_1) + (s_2^2/n_2)}}$; $v = \dfrac{(s_1^2/n_1 + s_2^2/n_2)^2}{\dfrac{(s_1^2/n_1)^2}{n_1 - 1} + \dfrac{(s_2^2/n_2)^2}{n_2 - 1}}$; $\sigma_1 \neq \sigma_2$ and unknown	$\mu_1 - \mu_2 < d_0$ $\mu_1 - \mu_2 > d_0$ $\mu_1 - \mu_2 \neq d_0$	$t' < -t_\alpha$ $t' > t_\alpha$ $t' < -t_{\alpha/2}$ and $t' > t_{\alpha/2}$
$\mu_D = d_0$	$t = \dfrac{\bar{d} - d_0}{s_d/\sqrt{n}}$; $v = n - 1$, paired observations	$\mu_D < d_0$ $\mu_D > d_0$ $\mu_D \neq d_0$	$t < -t_\alpha$ $t > t_\alpha$ $t < -t_{\alpha/2}$ and $t > t_{\alpha/2}$

5. Computations: $\bar{x} = 7.8$ kilograms, $n = 50$, and hence

$$z = \frac{7.8 - 8}{0.5/\sqrt{50}} = -2.83.$$

6. Decision: Reject H_0 and conclude that the average breaking strength is not equal to 8 but is, in fact, less than 8 kilograms.

Example 4. A random sample of 100 recorded deaths in the United States during the past year showed an average life span of 71.8 years, with a standard deviation of 8.9 years. Does this seem to indicate that the average life span today is greater than 70 years? Use a 0.05 level of significance.

Solution

1. H_0: $\mu = 70$ years.
2. H_1: $\mu > 70$ years.
3. $\alpha = 0.05$.
4. Critical region: $z > 1.645$, where

$$z = \frac{\bar{x} - \mu_0}{\sigma/\sqrt{n}}.$$

5. Computations: $\bar{x} = 71.8$ years, $\sigma \simeq s = 8.9$ years, and

$$z = \frac{71.8 - 70}{8.9/\sqrt{100}} = 2.02.$$

6. Decision: Reject H_0 and conclude that the average life span today is greater than 70 years.

Example 5. The average length of time for students to register for fall classes at a certain college has been 50 minutes with a standard deviation of 10 minutes. A new registration procedure using modern computing machines is being tried. If a random sample of 12 students had an average registration time of 42 minutes with a standard deviation of 11.9 minutes under the new system, test the hypothesis that the population mean is now less than 50, using a level of significance of (a) 0.05, and (b) 0.01. Assume the population of times to be normal.

Solution

1. H_0: $\mu = 50$ minutes.
2. H_1: $\mu < 50$ minutes.
3. (a) $\alpha = 0.05$; (b) $\alpha = 0.01$.
4. Critical region: (a) $t < -1.796$; (b) $t < -2.718$, where

$$t = \frac{\bar{x} - \mu_0}{s/\sqrt{n}}$$

with $\nu = 11$ degrees of freedom.

5. Computations: $\bar{x} = 42$ minutes, $s = 11.9$ minutes, and $n = 12$. Hence

$$t = \frac{42 - 50}{11.9/\sqrt{12}} = -2.33.$$

6. Decision: Reject H_0 at the 0.05 level of significance but not at the 0.01 level. This essentially means that the true mean is likely to be less than 50 minutes but does not differ sufficiently to warrant the high cost that would be required to operate a computer.

Example 6. A course in mathematics is taught to 12 students by the conventional classroom procedure. A second group of 10 students was given the same course by means of programed materials. At the end of the semester the same examination was given to each group. The 12 students meeting in the classroom made an average grade of 85 with a standard deviation of 4, while the 10 students using programed materials made an average of 81 with a standard deviation of 5. Test the hypothesis that the two methods of learning are equal using a 0.10 level of significance. Assume the populations to be approximately normal with equal variances.

Solution. Let μ_1 and μ_2 represent the average grades of all students that might take this course by the classroom and programed presentations, respectively. Using the six-step procedure, we have

1. H_0: $\mu_1 = \mu_2$ or $\mu_1 - \mu_2 = 0$.
2. H_1: $\mu_1 \neq \mu_2$ or $\mu_1 - \mu_2 \neq 0$.
3. $\alpha = 0.10$.
4. Critical region: $t < -1.725$ and $t > 1.725$, where

$$t = \frac{(\bar{x}_1 - \bar{x}_2) - d_0}{s_p\sqrt{\dfrac{1}{n_1} + \dfrac{1}{n_2}}}$$

with $v = 20$ degrees of freedom.
5. Computations:

$$\bar{x}_1 = 85, \quad s_1 = 4, \quad n_1 = 12,$$
$$\bar{x}_2 = 81, \quad s_2 = 5, \quad n_2 = 10.$$

Hence

$$s_p = \sqrt{\frac{(11)(16) + (9)(25)}{12 + 10 - 2}} = 4.478,$$

$$t = \frac{(85 - 81) - 0}{4.478\sqrt{\frac{1}{12} + \frac{1}{10}}} = 2.07.$$

6. Decision: Reject H_0 and conclude that the two methods of learning are not equal. Since the computed t value falls in the part of the critical region in the right tail of the distribution, we can conclude that the classroom procedure is superior to the method using programed materials. Note that we arrived at this same conclusion in Example 5 on page 257 by means of a 90% confidence interval.

Example 7. To determine whether membership in a fraternity is beneficial or detrimental to one's grades, the following grade-point averages were collected over a period of 5 years:

	Year				
	1	2	3	4	5
Fraternity	2.0	2.0	2.3	2.1	2.4
Nonfraternity	2.2	1.9	2.5	2.3	2.4

Assuming the populations to be normal, test at the 0.025 level of significance whether membership in a fraternity is detrimental to one's grades.

Solution. Let μ_1 and μ_2 be the average grades of fraternity and nonfraternity students, respectively. We proceed by the six-step rule:

1. H_0: $\mu_1 = \mu_2$ or $\mu_D = \mu_1 - \mu_2 = 0$.
2. H_1: $\mu_1 < | \mu_2$ or $\mu_D = \mu_1 - \mu_2 < 0$.
3. $\alpha = 0.025$.
4. Critical region: $t < -2.776$, where

$$t = \frac{\bar{d} - d_0}{s_d/\sqrt{n}}$$

with $v = 4$ degrees of freedom.

5. Computations:

Fraternity	Nonfraternity	d_i	d_i^2
2.0	2.2	-0.2	0.04
2.0	1.9	0.1	0.01
2.3	2.5	-0.2	0.04
2.1	2.3	-0.2	0.04
2.4	2.4	0.0	0.00
		-0.5	0.13

We find $\bar{d} = -0.5/5 = -0.1$ and

$$s_d^2 = \frac{(5)(0.13) - (-0.5)^2}{(5)(4)} = 0.02.$$

Taking the square root, we have $s_d = 0.14142$. Hence

$$t = \frac{-0.1 - 0}{0.14142/\sqrt{5}} = -1.58.$$

Decision: Accept H_0 and conclude that membership in a fraternity does not significantly affect one's grades.

EXERCISES

1. An electrical firm manufactures light bulbs that have a length of life that is approximately normally distributed with a mean of 800 hours and a standard deviation of 40 hours. Test the hypothesis that $\mu = 800$ hours against the alternative $\mu \neq 800$ hours if a random sample of 30 bulbs has an average life of 788 hours. Use a 0.04 level of significance.

2. In a research report by Richard H. Weindruch of the UCLA Medical School, it is claimed that mice with an average lifespan of 32 months will live to be about 40 months old when 40% of the calories in their food are replaced by vitamins and protein. Is there any reason to believe that $\mu < 40$ if 64 mice that are placed on this diet have an average life of 38 months with a standard deviation of 5.8 months? Use a 0.025 level of significance.

3. The average height of females in the freshman class of a certain college has been 162.5 centimeters with a standard deviation of 6.9 centimeters. Is there reason to believe that there has been a change in the average height if a random sample of 50 females in the present freshman class has an average height of 165.2 centimeters? Use a 0.02 level of significance.

4. It is claimed that an automobile is driven on the average less than 20,000 kilometers per year. To test this claim, a random sample of 100 automobile owners are asked to keep a record of the kilometers they travel. Would you agree with this claim if the random sample showed an average of 23,500 kilometers and a standard deviation of 3900 kilometers? Use a 0.01 level of significance.

5. Test the hypothesis that the average content of containers of a particular lubricant is 10 liters if the contents of a random sample of 10 containers are 10.2, 9.7, 10.1, 10.3, 10.1, 9.8, 9.9, 10.4, 10.3, and 9.8 liters. Use a 0.01 level of significance and assume that the distribution of contents is normal.

6. According to _Dietary Goals for the United States_ (1977), high sodium intake may be related to ulcers, stomach cancer, and migraine headaches. The human requirement for salt is only 220 milligrams per day, which is surpassed in most single servings of ready-to-eat cereals. If a random sample of 20 similar servings of Special K has a mean sodium content of 244 milligrams of sodium and a standard deviation of 24.5 milligrams, does this suggest at the 0.05 level of significance that the average sodium content for single servings of Special K is greater than 220 milligrams? Assume the distribution of sodium contents to be normal.

7. A random sample of 8 cigarettes of a certain brand has an average nicotine content of 4.2 milligrams and a standard deviation of 1.4 milligrams. Is this in line with the manufacturer's claim that the average nicotine content does not exceed 3.5 milligrams? Use a 0.01 level of significance and assume the distribution of nicotine contents to be normal.

8. Last year the employees of the city sanitation department donated an average of $8.00 to the volunteer rescue squad. Test the hypothesis at the 0.01 level of significance that the average contribution this year is still $8.00 if a random sample of 12 employees showed an average donation of $8.90 with a standard deviation of $1.75. Assume that the donations are approximately normally distributed.

9. **Sample size for testing means:** Suppose that we wish to test the null hypothesis that $\mu = \mu_0$ against a specified alternative hypothesis $\mu = \mu_0 + \delta$, where δ may be positive or negative. If the probabilities of type I and type II errors have the preassigned values α and β, and we are sampling from an approximate normal distribution with known variance σ^2, then the required size of the sample is given by

$$n = \frac{(z_\alpha + z_\beta)^2 \sigma^2}{\delta^2}.$$

(a) Suppose that we wish to test the null hypothesis that $\mu = 68$ against the alternative hypothesis $\mu = 69$ for a normal population whose standard deviation is $\sigma = 5$. Find the sample size required if α and β are both to be 0.05.

(b) If the distribution of lifespans in Exercise 2 is approximately normal, how large a sample is required in order that the probability of committing a type II error be 0.1 when the true mean is 35.9 months? Assume that $\sigma = 5.8$ months.

10. A random sample of size $n_1 = 25$, taken from a normal population with a standard deviation $\sigma_1 = 5.2$, has a mean $\bar{x}_1 = 81$. A second random sample of size $n_2 = 36$, taken from a different normal population with a standard deviation $\sigma_2 = 3.4$, has a mean $\bar{x}_2 = 76$. Test the hypothesis at the 0.06 level of significance that $\mu_1 = \mu_2$ against the alternative $\mu_1 \neq \mu_2$.

11. A manufacturer claims that the average tensile strength of thread A exceeds the average tensile strength of thread B by at least 12 kilograms. To test this claim, 50 pieces of each type of thread are tested under similar conditions. Type A thread had an average tensile strength of 86.7 kilograms with a standard deviation of 6.28 kilograms, while type B thread had an average tensile strength of 77.8 kilograms with a standard deviation of 5.61 kilograms. Test the manufacturer's claim using a 0.05 level of significance.

12. A study was made to estimate the difference in salaries of college professors in the private and state colleges of Virginia. A random sample of 100 professors in private colleges showed an average 9-month salary of $26,000 with a standard deviation of $1300. A random sample of 200 professors in state colleges showed an average salary of $26,900 with a standard deviation of $1400. Test the hypothesis that the average salary for professors teaching in state colleges does not exceed the average salary for professors teaching in private colleges by more than $500. Use a 0.02 level of significance.

13. Given two random samples of size $n_1 = 11$ and $n_2 = 14$, from two independent normal populations, with $\bar{x}_1 = 75$, $\bar{x}_2 = 60$, $s_1 = 6.1$, and $s_2 = 5.3$, test the hypothesis at the 0.05 level of significance that $\mu_1 = \mu_2$ against the alternative that $\mu_1 \neq \mu_2$. Assume that the population variances are equal.

14. A study is made to see if increasing the substrate concentration has an appreciable effect on the velocity of a chemical reaction. With the substrate concentration of 1.5 moles per liter, the reaction was run 15 times with an average velocity of 7.5 micromoles per 30 minutes and a standard deviation of 1.5. With a substrate concentration of 2.0 moles per liter, 12 runs were made yielding an average velocity of 8.8 micromoles per 30 minutes and a sample standard deviation of 1.2. Would you say that the increase in substrate concentration increases the mean velocity by more than 0.5 micromole per 30 minutes? Use a 0.01 level of significance and assume the populations to be approximately normally distributed with equal variances.

15. A study was made to determine if the subject matter in a physics course is better understood when a lab constitutes part of the course. Students were allowed to choose between a 3-semester-hour course without labs and a 4-semester-hour

course with labs. In the section with labs 11 students made an average grade of 85 with a standard deviation of 4.7, and in the section without labs 17 students made an average grade of 79 with a standard deviation of 6.1. Would you say that the laboratory course increases the average grade by as much as 8 points? Use a 0.01 level of significance and assume the populations to be approximately normally distributed with equal variances.

16. A large automobile manufacturing company is trying to decide whether to purchase brand A or brand B tires for its new models. To help arrive at a decision, an experiment is conducted using 12 of each brand. The tires are run until they wear out. The results are

$$\text{Brand } A: \quad \bar{x}_1 = 37{,}900 \text{ kilometers, } s_1 = 5100 \text{ kilometers,}$$
$$\text{Brand } B: \quad \bar{x}_2 = 39{,}800 \text{ kilometers, } s_2 = 5900 \text{ kilometers.}$$

Test the hypothesis at the 0.05 level of significance that there is no difference in the two brands of tires. Assume the populations to be approximately normally distributed.

17. The following data represent the running times of films produced by two different motion-picture companies:

	Time (minutes)					
Company 1	102	86	98	109	92	
Company 2	81	165	97	134	92	87	114

Test the hypothesis that the average running time of films produced by company 2 exceeds the average running time of films produced by company 1 by 10 minutes against the one-sided alternative that the difference is more than 10 minutes. Use a 0.1 level of significance and assume the distributions of times to be approximately normal.

18. In Exercise 21 on page 265, test the hypothesis, at the 0.05 level of significance, that the average yields of the two varieties of wheat are equal against the alternative hypothesis that they are unequal.

19. In Exercise 22 on page 265, test the hypothesis, at the 0.01 level of significance, that $\mu_1 = \mu_2$ against the alternative hypothesis that $\mu_1 < \mu_2$.

20. A taxi company is trying to decide whether the use of radial tires instead of regular belted tires improves fuel economy. Twelve cars were equipped with radial tires and driven over a prescribed test course. Without changing drivers, the same cars were then equipped with regular belted tires and driven once again over the test course. The gasoline consumption, in kilometers per liter, was recorded as follows:

| | Kilometers per Liter | |
Car	Radial Tires	Belted Tires
1	4.2	4.1
2	4.7	4.9
3	6.6	6.2
4	7.0	6.9
5	6.7	6.8
6	4.5	4.4
7	5.7	5.7
8	6.0	5.8
9	7.4	6.9
10	4.9	4.7
11	6.1	6.0
12	5.2	4.9

At the 0.025 level of significance, can we conclude that cars equipped with radial tires give better fuel economy than those equipped with belted tires? Assume the populations to be normally distributed.

21. **Sample size for testing differences of means:** Suppose that we wish to test the null hypothesis that $\mu_1 - \mu_2 = d_0$ against the alternative hypothesis that $\mu_1 - \mu_2 = d_0 + \delta$, where δ may be positive or negative. If the probabilities of type I and type II errors have the preassigned values α and β, and we are selecting independent random samples of size $n_1 = n_2 = n$ from approximate normal populations with known variances σ_1^2 and σ_2^2, then the required size of the samples is given by

$$n = \frac{(z_\alpha + z_\beta)^2 (\sigma_1^2 + \sigma_2^2)}{\delta^2}.$$

(a) Suppose that we select independent random samples of size n from two normal populations with known variances $\sigma_1^2 = 80$ and $\sigma_2^2 = 100$ in order to test the null hypothesis that $\mu_1 - \mu_2 = 50$ against the alternative hypothesis that $\mu_1 - \mu_2 = 55$. Find the required size of the samples if $\alpha = 0.05$ and $\beta = 0.01$.

(b) If the distribution of tensile strengths in Exercise 11 are normal, find the size of the samples required in order that the probability of committing a type II error will be 0.05 when the true difference between types A and B is 8 kilograms.

10.5
Tests Concerning Variances

In this section we are concerned with testing hypotheses concerning population variances or standard deviations. In other words, we are interested in testing hypotheses concerning the uniformity of a population or perhaps in comparing the uniformity of one population with that of a second population. We might, therefore, be interested in testing the hypothesis that the variability in the percentage of impurities in a certain kind of fruit preserve does not exceed some specified value, or that the variability in the length of life for some brand of exterior house paint is equal to the variability in the length of life for some competitive brand.

Let us first consider the problem of testing the null hypothesis H_0 that the population variance σ^2 equals a specified value σ_0^2 against one of the usual alternatives $\sigma^2 < \sigma_0^2$, $\sigma^2 > \sigma_0^2$, or $\sigma^2 \neq \sigma_0^2$. The appropriate statistic on which we base our decision is the same chi-square random variable of Theorem 9.6 on page 275 that was used in Chapter 9 to construct a confidence interval for σ^2. Therefore, assuming that the distribution of the population being sampled is at least approximately normal, the **chi-square value for testing $\sigma^2 = \sigma_0^2$** is given by

$$\chi^2 = \frac{(n-1)s^2}{\sigma_0^2},$$

where n is the sample size, s^2 is the sample variance, and σ_0^2 is the value of σ^2 given by the null hypothesis. If H_0 is true, χ^2 is a value of the chi-square distribution with $v = n - 1$ degrees of freedom. Hence, for a two-tailed test at the α level of significance, the critical region is $\chi^2 < \chi_{1-\alpha/2}^2$ and $\chi^2 > \chi_{\alpha/2}^2$. For the one-sided alternative $\sigma^2 < \sigma_0^2$, the critical region is $\chi^2 < \chi_{1-\alpha}^2$, and for the one-sided alternative $\sigma^2 > \sigma_0^2$, the critical region is $\chi^2 > \chi_\alpha^2$.

To test a hypothesis about a population variance, we proceed by the same basic six steps outlined at the end of Section 10.3.

Example 8. A manufacturer of car batteries claims that the life of his batteries have a standard deviation equal to 0.9 year. If a random sample of 10 of these batteries have a standard deviation of 1.2 years, do you think that $\sigma > 0.9$ year? Use a 0.05 level of significance.

Solution

1. H_0: $\sigma^2 = 0.81$.
2. H_1: $\sigma^2 > 0.81$.
3. $\alpha = 0.05$.

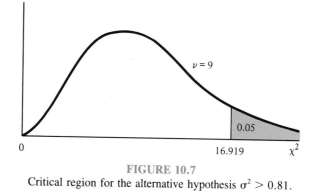

FIGURE 10.7
Critical region for the alternative hypothesis $\sigma^2 > 0.81$.

4. Critical region: From Figure 10.7, we see that the null hypothesis is rejected when $\chi^2 > 16.919$.

5. Computations: $s^2 = 1.44$, $n = 10$, and

$$\chi^2 = \frac{(9)(1.44)}{0.81} = 16.0.$$

6. Decision: Accept H_0 and conclude that there is no reason to doubt that the standard deviation is 0.9 year.

Now let us consider the problem of testing the equality of the variances σ_1^2 and σ_2^2 of two populations. That is, we shall test the null hypothesis H_0 that $\sigma_1^2 = \sigma_2^2$ against one of the usual alternatives $\sigma_1^2 < \sigma_2^2$, $\sigma_1^2 > \sigma_2^2$, or $\sigma_1^2 \neq \sigma_2^2$. For independent random samples of size n_1 and n_2, respectively, from the two populations, the f **value for testing** $\sigma_1^2 = \sigma_2^2$ is the ratio

$$f = \frac{s_1^2}{s_2^2},$$

where s_1^2 and s_2^2 are the variances computed from the two samples. If the two populations are approximately normally distributed and the null hypothesis is true, according to Theorem 9.7 on page 279, the ratio $f = s_1^2/s_2^2$ is a value of the F distribution with $v_1 = n_1 - 1$ and $v_2 = n_2 - 1$ degrees of freedom. Therefore, the critical regions of size α corresponding to the one-sided alternatives $\sigma_1^2 < \sigma_2^2$ and $\sigma_1^2 > \sigma_2^2$ are, respectively, $f < f_{1-\alpha}(v_1, v_2)$ and $f > f_\alpha(v_1, v_2)$. For the two-sided alternative $\sigma_1^2 \neq \sigma_2^2$, the critical region is given by $f < f_{1-\alpha/2}(v_1, v_2)$ and $f > f_{\alpha/2}(v_1, v_2)$.

Example 9. In testing the equality of the two population means in Example 6 on page 313, we assume that the two population variances are equal but unknown. Are we justified in making this assumption? Use a 0.10 level of significance.

Solution. Let σ_1^2 and σ_2^2 be the population variances for the grades of all students that might take this mathematics course by the conventional classroom and programed presentations, respectively. Following the six-step procedure, we have

1. H_0: $\sigma_1^2 = \sigma_2^2$.
2. H_1: $\sigma_1^2 \neq \sigma_2^2$.
3. $\alpha = 0.10$.
4. Critical region:
   From Figure 10.8, we see that

$$f_{0.05}(11,9) = 3.11$$

and, using Theorem 9.8 on page 280,

$$f_{0.95}(11,9) = \frac{1}{f_{0.05}(9,11)} = 0.34.$$

Therefore, the null hypothesis is rejected when $f < 0.34$ or $f > 3.11$.

5. Computations: $s_1^2 = 16$, $s_2^2 = 25$, and hence

$$f = \frac{16}{25} = 0.64.$$

6. Decision: Accept H_0 and conclude that we were justified in assuming the variances equal in Example 6.

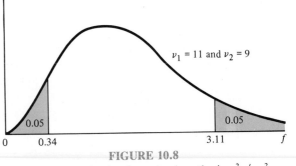

FIGURE 10.8

Critical region for the alternative hypothesis $\sigma_1^2 \neq \sigma_2^2$.

1. The volume of containers of a particular lubricant is known to be normally distributed with a variance of 0.03 liter. Test the hypothesis that $\sigma^2 = 0.03$ against the alternative that $\sigma^2 \neq 0.03$ for the random sample of 10 containers in Exercise 5 on page 316. Use a 0.01 level of significance.

2. Past experience indicates that the time for high school seniors to complete a standardized test is a normal random variable with a standard deviation of 6 minutes. Test the hypothesis that $\sigma = 6$ against the alternative that $\sigma < 6$ if a random sample of 20 high school seniors has a standard deviation $s = 4.51$. Use a 0.05 level of significance.

3. The nicotine content of a certain brand of cigarettes is known to be normally distributed with a variance of 1.3 milligrams. Test the hypothesis that $\sigma^2 = 1.3$ against the alternative that $\sigma^2 \neq 1.3$ if a random sample of 8 of these cigarettes has a standard deviation $s = 1.8$. Use a 0.05 level of significance.

4. Past data indicate that the amount of money contributed by the working residents of a large city to a volunteer rescue squad is a normal random variable with a standard deviation of $1.40. It has been suggested that the contributions to the rescue squad from just the employees of the sanitation department are much more variable. If the contributions of a random sample of 12 employees from the sanitation department had a standard deviation of $1.75, can we conclude at the 0.01 level of significance that the standard deviation of the contributions of all sanitation workers is greater than that of all workers living in this city?

5. A soft-drink dispensing machine is said to be out of control if the variance of the contents exceeds 1.15 deciliters. If a random sample of 25 drinks from this machine has a variance of 2.03 deciliters, does this indicate at the 0.05 level of significance that the machine is out of control? Assume that the contents are approximately normally distributed.

6. **Large-sample test of** $\sigma^2 = \sigma_0^2$: When $n \geq 30$ we can test the null hypothesis that $\sigma^2 = \sigma_0^2$, or $\sigma = \sigma_0$, by computing

$$z = \frac{s - \sigma_0}{\sigma_0/\sqrt{2n}},$$

which is a value of a random variable whose sampling distribution is approximately the standard normal distribution.

(a) With reference to Example 4 on page 312, test at the 0.05 level of significance whether $\sigma = 7.5$ years against the alternative that $\sigma \neq 7.5$ years.

(b) It is suspected that the variance of the distribution of distances in kilometers achieved per 5 liters of fuel by a new automobile model equipped with a diesel engine is less than the variance of the distribution of distances achieved by the same model equipped with a six-cylinder gasoline engine, which is know to be $\sigma^2 = 6.25$. If 72 test runs in the diesel model have a variance of 4.41, can we conclude at the 0.02 level of significance that the variance of the distances achieved by the diesel model is less than that of the gasoline model?

7. A study is conducted to compare the length of time between men and women to assemble a certain product. Past experience indicates that the distribution of times for both men and women are approximately normal but the variance of the times for women is less than that for men. A random sample of times for 11 men and 14 women produced the following data:

Men	Women
$n_1 = 11$	$n_2 = 14$
$s_1 = 6.1$	$s_2 = 5.3$

Test the hypothesis that $\sigma_1^2 = \sigma_2^2$ against the alternative that $\sigma_1^2 > \sigma_2^2$. Use a 0.01 level of significance.

8. With reference to Exercise 16 on page 318, test the hypothesis that $\sigma_1 = \sigma_2$ against the alternative that $\sigma_1 < \sigma_2$. Use a 0.05 level of significance.

9. With reference to Exercise 17 on page 318, test the hypothesis that $\sigma_1^2 = \sigma_2^2$ against the alternative that $\sigma_1^2 \neq \sigma_2^2$, where σ_1^2 and σ_2^2 are the variances for the running times of films produced by company 1 and company 2, respectively. Use a 0.10 level of significance.

10. Two types of instruments for measuring the amount of sulfur monoxide in the atmosphere are being compared in an air-pollution experiment. It is desired to determine whether the two types of instruments yield measurements having the same variability. The following readings were recorded for the two instruments:

	Sulfur Monoxide								
Instrument A	0.96	0.82	0.75	0.61	0.89	0.64	0.81	0.68	0.65
Instrument B	0.87	0.74	0.63	0.55	0.76	0.70	0.69	0.57	0.53

Assuming the populations of measurements to be approximately normally distributed, test the hypothesis that $\sigma_A = \sigma_B$ against the alternative that $\sigma_A \neq \sigma_B$. Use a 0.02 level of significance.

10.6
Tests
Concerning
Proportions

Tests of hypotheses concerning proportions are required in many areas. The politician is certainly interested in knowing what fraction of the voters will favor him in the next election. All manufacturing firms are concerned about the proportion of defectives when a shipment is made. The gambler depends upon a knowledge of the proportion of outcomes that he considers favorable.

We shall consider the problem of testing the hypothesis that the proportion of successes in a binomial experiment equals some specified value. That is, we are testing the null hypothesis H_0 that $p = p_0$, where p is the parameter of the binomial distribution. The alternative hypothesis may be one of the usual one-sided or two-sided alternatives: $p < p_0$, $p > p_0$, or $p \neq p_0$.

The appropriate statistic on which we base our decision criterion is the binomial random variable X, although we could just as well use the statistic $\hat{P} = X/n$. Values of X that are far from the mean $\mu = np_0$ will lead to the rejection of the null hypothesis. To test the hypothesis

$$H_0: p = p_0,$$
$$H_1: p < p_0,$$

the critical region of size α is given by

$$x \leq k'_\alpha,$$

where k'_α is the largest integer for which

$$P(X \leq k'_\alpha \text{ when } p = p_0) = \sum_{x=0}^{k'_\alpha} b(x; n, p_0) \leq \alpha.$$

Similarly, to test the hypothesis

$$H_0: p = p_0,$$
$$H_1: p > p_0,$$

the critical region of size α is given by

$$x \geq k_\alpha,$$

where k_α is the smallest integer for which

$$P(X \geq k_\alpha \text{ when } p = p_0) = \sum_{x=k_\alpha}^{n} b(x; n, p_0) \leq \alpha.$$

Finally, to test the hypothesis

$$H_0: p = p_0,$$
$$H_1: p \neq p_0,$$

the critical region of size α is given by

$$x \leq k'_{\alpha/2} \quad \text{and} \quad x \geq k_{\alpha/2}.$$

Because X is a discrete binomial random variable, the size of these various critical regions should be chosen as close as possible to α without exceeding it.

The steps for testing a null hypothesis about a proportion against various alternatives using the binomial probabilities of Table A.2, are as follows:

───

Testing a Proportion

1. $H_0: p = p_0$.
2. H_1: Alternatives are $p < p_0$, $p > p_0$, or $p \neq p_0$.
3. Choose a level of significance equal to α.
4. Critical region:

$$x \leq k'_\alpha \text{ for the alternative } p < p_0,$$
$$x \geq k_\alpha \text{ for the alternative } p > p_0,$$
$$x \leq k'_{\alpha/2} \text{ and } x \geq k_{\alpha/2} \text{ for the alternative } p \neq p_0.$$

5. Computation: Find x, the number of successes.
6. Decision: Reject H_0 if x falls in the critical region; otherwise, accept H_0.

───

Example 10. A builder claims that heat pumps are installed in 70% of all homes being constructed today in the city of Richmond. Would you agree with this claim if a random survey of new homes in this city shows that 8 out of 15 had heat pumps installed? Use a 0.10 level of significance.

Solution

1. $H_0: p = 0.7$.
2. $H_1: p \neq 0.7$.
3. $\alpha = 0.10$.
4. Critical region: $x \leq 7$ and $x \geq 14$, from Table A.2.
5. Computation: $x = 8$.
6. Decision: Accept H_0 and conclude that there is no reason to doubt the builder's claim.

In Section 7.4 we saw that binomial probabilities were obtainable from the actual binomial formula or from Table A.2 when n is small. for large n approximation procedures are required. When the hypothesized value p_0 is very close to zero or 1, the Poisson distribution, with parameter $\mu = np_0$, may be used. The normal-curve approximation, with parameters $\mu = np_0$ and $\sigma^2 = np_0 q_0$, is usually preferred for large n and is very accurate as long as p_0 is not extremely close to zero or to 1. Using the normal approximation, the **z value for testing $p = p_0$** is given by

$$z = \frac{x - np_0}{\sqrt{np_0 q_0}},$$

which is a value of the standard normal variable Z. Hence for a two-tailed test at the α level of significance, the critical region is $z < -z_{\alpha/2}$ and $z > z_{\alpha/2}$. For the one-sided alternative $p < p_0$, the critical region is $z < -z_\alpha$, and for the alternative $p > p_0$, the critical region is $z > z_\alpha$.

To test a hypothesis about a proportion using the normal-curve approximation, we proceed as follows:

1. $H_0: p = p_0$
2. H_1: Alternatives are $p < p_0$, $p > p_0$, or $p =/ p_0$.
3. Choose a level of significance equal to α.
4. Critical region:

$$z < -z_\alpha \text{ for the alternative } p < p_0,$$

$$z > z_\alpha \text{ for the alternative } p > p_0,$$

$$z < -z_{\alpha/2} \text{ and } z > z_{\alpha/2} \text{ for the alternative } p \neq p_0.$$

5. Computations: Find x from a sample of size n, and then compute

$$z = \frac{x - np_0}{\sqrt{np_0q_0}}.$$

6. Decision: Reject H_0 if z falls in the critical region; otherwise, accept H_0.

Example 11. A commonly prescribed drug on the market for relieving nervous tension is believed to be only 60% effective. Experimental results with a new drug administered to a random sample of 100 adults who were suffering from nervous tension showed that 70 received relief. Is this sufficient evidence to conclude that the new drug is superior to the one commonly prescribed? Use a 0.05 level of significance.

Solution

1. H_0: $p = 0.6$.
2. H_1: $p > 0.6$. |
3. $\alpha = 0.05$.
4. Critical region: $z > 1.645$.
5. Computations: $x = 70$, $n = 100$, $np_0 = (100)(0.6) = 60$, and

$$z = \frac{70 - 60}{\sqrt{(100)(0.6)(0.4)}} = 2.04.$$

6. Decision: Reject H_0 and conclude that the new drug is superior.

10.7

Testing the Difference Between Two Proportions

Situations often arise in which we wish to test the hypothesis that two proportions are equal. For example, we might try to prove that the proportion of doctors who are pediatricians in one state is greater than the proportion of pediatricians in another state. A person may decide to give up smoking only if he is convinced that the proportion of smokers with lung cancer exceeds the proportion of nonsmokers with lung cancer.

In general, we wish to test the null hypothesis

$$H_0: p_1 = p_2 = p$$

against some suitable alternative. The parameters p_1 and p_2 are the two population proportions of the attribute under investigation. The statistic on which
we base our decision criterion is the random variable $\hat{P}_1 - \hat{P}_2$. Independent
samples of size n_1 and n_2 are selected at random from two binomial populations
and the proportion of successes $\hat{P}_1$ and $\hat{P}_2$, for the two samples are computed.
From Section 9.6 we know that

$$z = \frac{\hat{p}_1 - \hat{p}_2}{\sqrt{p_1 q_1/n_1) + (p_2 q_2/n_2)}} = \frac{\hat{p}_1 - \hat{p}_2}{\sqrt{pq[(1/n_1) + (1/n_2)]}}$$

is a value of the standard normal variable Z when H_0 is true and n_1 and n_2 are
large. To compute z, we must estimate the value of p appearing in the radical.
Pooling the data from both samples, the **pooled estimate of the proportion**
_p_ is

$$\hat{p} = \frac{x_1 + x_2}{n_1 + n_2},$$

where x_1 and x_2 are the number of successes in each of the two samples. The
_z_ value for testing $p_1 = p_2$ becomes

$$z = \frac{\hat{p}_1 - \hat{p}_2}{\sqrt{\hat{p}\hat{q}[(1/n_1) + (1/n_2)]}},$$

where $\hat{q} = 1 - \hat{p}$. The critical regions for the appropriate alternative hypotheses are set up as before using the critical points of the standard normal
curve.

To test the hypothesis that two proportions are equal, when the samples
are large, we proceed by the following six steps:

1. H_0: $p_1 = p_2$.
2. H_1: Alternatives are $p_1 < p_2$, $p_1 > p_2$, or $p_1 \neq p_2$.
3. Choose a level of significance equal to α.
4. Critical region:

$$z < -z_\alpha \text{ for the alternative } p_1 < p_2,$$

$$z > z_\alpha \text{ for the alternative } p_1 > p_2,$$

$$z < -z_{\alpha/2} \text{ and } z > z_{\alpha/2} \text{ for the alternative } p_1 \neq p_2.$$

5. Computations: Compute $\hat{p}_1 = x_1/n_1$, $\hat{p}_2 = x_2/n_2$, $\hat{p} = (x_1 + x_2)/(n_1 + n_2)$ and then find

$$z = \frac{\hat{p}_1 - \hat{p}_2}{\sqrt{\hat{p}\hat{q}[(1/n_1) + (1/n_2)]}},$$

. Decision: Reject H_0 if z falls in the critical region; otherwise, accept H_0.

Example 12. A vote is to be taken among the residents of a town and the surrounding country to determine whether a civic center will be constructed. The proposed construction site is within the town limits and for this reason many voters in the county feel that the proposal will pass because of the large proportion of town voters who favor the construction. To determine if there is a significant difference in the proportion of town voters and county voters favoring the proposal, a poll is taken. If 120 or 200 town voters favor the proposal and 240 of 500 county residents favor it, would you agree that the proportion of town voters favoring the proposal is higher than the proportion of county voters? Use a 0.025 level of significance.

Solution. Let p_1 and p_2 be the true proportion of voters in the town and county, respectively, favoring the proposal. We now follow the six-step procedure:

1. H_0: $p_1 = p_2$.
2. H_1: $p_1 > p_2$.
3. $\alpha = 0.025$.
4. Critical region: $z > 1.96$.
5. Computations:

$$\hat{p} = \frac{x_1}{n_1} = \frac{120}{200} = 0.60,$$

$$\hat{p} = \frac{x_2}{n_2} = \frac{240}{500} = 0.48,$$

$$\hat{p} = \frac{x_1 + x_2}{n_1 + n_2} = \frac{120 + 240}{200 + 500} = 0.51.$$

Therefore,

$$z = \frac{0.60 - 0.48}{\sqrt{(0.51)(0.49)[(\frac{1}{200}) + (\frac{1}{500})]}} = 2.9.$$

6. Decision: Reject H_0 and agree that the proportion of town voters favoring the proposal is higher than the proportion of county voters.

1. A manufacturer of cigarettes claims that 20% of the cigarette smokers prefer brand X. To test this claim a random sample of 20 cigarette smokers are selected and asked what brand they prefer. If 6 of the 20 smokers prefer brand X, what conclusion do we draw? Use a 0.05 level of significance.

2. Suppose that in the past 40% of all adults favored capital punishment. Do we have reason to believe that the proportion of adults favoring capital punishment today has increased if, in a random sample of 15 adults, 8 favor capital punishment? Use a 0.05 level of significance.

3. A coin is tossed 20 times resulting in 5 heads. Is this sufficient evidence to reject the hypothesis at the 0.03 level of significance that the coin is balanced in favor of the alternative that heads occur less than 50% of the time?

4. It is believed that at least 60% of the residents in a certain area favor an annexation suit by a neighboring city. What conclusion would you draw if only 110 in a sample of 200 voters favor the suit? Use a 0.04 level of significance.

5. The gas company claims that two-thirds of the houses in a certain city are heated by natural gas. Do we have reason to doubt this claim if, in a random sample of 1000 houses in this city, it is found that 618 are heated by natural gas? Use a 0.02 level of significance.

6. At a certain college it is estimated that fewer than 25% of the students have cars on campus. Does this seem to be a valid estimate if, in a random sample of 90 college students, 28 are found to have cars? Use a 0.05 level of significance.

7. In a study to estimate the proportion of residents in a certain city and its suburbs who favor the construction of a nuclear power plant, it is found that 63 of 100 urban residents favor the construction while only 59 of 125 suburban residents are in favor. Is there a significant difference between the proportion of urban and suburban residents who favor construction of the nuclear plant? Use a 0.04 level of significance.

8. A cigarette-manufacturing firm distributes two brands of cigarettes. If it is found that 56 of 200 smokers prefer brand A and that 29 of 150 smokers prefer brand B, can we conclude at the 0.06 level of significance that brand A outsells brand B?

9. A geneticist is interested in the proportion of males and females in a population that have a certain minor blood disorder. In a random sample of 100 males, 31 are found to be afflicted, whereas only 24 of 100 females tested appear to have the disorder. Can we conclude at the 0.01 level of significance that the proportion of men in the population afflicted with this blood disorder is significantly greater than the proportion of women afflicted?

10. A study is made to determine if a cold climate contributes more to absenteeism from school during a semester than a warmer climate. Two groups of students are selected at random, one group from Maine and the other from Alabama. Of

the 300 students from Maine, 72 were absent at least 1 day during the semester, and of the 400 students from Alabama, 70 were absent 1 or more days. Can we conclude that a colder climate results in a greater number of students being absent from school at least 1 day during the semester? Use a 0.05 level of significance.

11. **z value for testing $p_1 - p_2 = d_o$:** To test the null hypothesis H_0 that $p_1 - p_2 = d_0$, where $d_0 \neq 0$, we base our decision on

$$z = \frac{\hat{p}_1 - \hat{p}_2 - d_0}{\sqrt{\dfrac{\hat{p}_1\hat{q}_1}{n_1} + \dfrac{\hat{p}_2\hat{q}_2}{n_2}}},$$

which is a value of a random variable whose distribution approximates the standard normal distribution as long as n_1 and n_2 are both large.

(a) With reference to Example 12 on page 330, test the hypothesis that the percentage of town voters favoring the construction of a civic center will not exceed the percentage of county voters by more than 3%. Use a 0.025 level of significance.

(b) With reference to Exercise 8, test the hypothesis at the 0.06 level of significance that brand A outsells brand B by 10% against the alternative hypothesis that the difference is less than 10%.

10.8
Goodness-of-Fit Test

Throughout this chapter we have been concerned with the testing of statistical hypotheses about single population parameters such as μ, σ^2, and p. Now we shall consider a test to determine if a population has a specified theoretical distribution. The test is based upon how good a fit we have between the frequency of occurrence of observations in an observed sample and the expected frequencies obtained from the hypothesized distribution.

To illustrate, consider the tossing of a die. We hypothesize that the die is honest, which is equivalent to testing the hypothesis that the distribution of outcomes is uniform. Suppose that the die is tossed 120 times and that each outcome is recorded. Theoretically, if the die is balanced, we would expect each face to occur 20 times. The results are given in Table 10.2. By comparing the observed frequencies with the corresponding expected frequencies, we must decide whether these discrepancies are likely to occur due to sampling fluctuations and the die is balanced, or the die is not honest and the distribution of outcomes is not uniform. It is common practice to refer to each possible outcome of an experiment as a cell. Hence, in our illustration, we have six cells. The appropriate statistic on which we base our decision criterion for an experiment involving k cells is defined in the following theorem.

TABLE 10.2

Observed and Expected Frequencies
of 120 Tosses of a Die

	Faces					
	1	2	3	4	5	6
Observed	20	22	17	18	19	24
Expected	20	20	20	20	20	20

THEOREM 10.1

Goodness-of-Fit Test. A goodness-of-fit test between observed and expected frequencies is based on the quantity

$$\chi^2 = \sum_{i=1}^{k} \frac{(o_i - e_i)^2}{e_i},$$

where χ^2 is a value of the random variable X^2 whose sampling distribution is approximated very closely by the chi-square distribution. The symbols o_i and e_i represent the observed and expected frequencies, respectively, for the ith cell.

If the observed frequencies are close to the corresponding expected frequencies, the χ^2 value will be small, indicating a good fit. If the observed frequencies differ considerably from the expected frequencies, the χ^2 value will be large and the fit is poor. A good fit leads to the acceptance of H_0, whereas a poor fit leads to its rejection. The critical region will, therefore, fall in the right tail of the chi-square distribution. For a level of significance equal to α we find the critical value χ^2_α from Table A.6, and then $\chi^2 > \chi^2_\alpha$ constitutes the critical region. The decision criterion described here should not be used unless each of the expected frequencies is at least equal to 5. This restriction may require the combining of adjacent cells resulting in a reduction of the number of degrees of freedom.

The number of degrees of freedom associated with the chi-square distribution used here depends on two factors: the number of cells in the experiment, and the number of quantities obtained from the observed data that are necessary in the calculation of the expected frequencies. We arrive at this number by the following theorem.

THEOREM 10.2

Degrees of Freedom in a Goodness-of-Fit Test. The number of degrees of freedom in a chi-square goodness-of-fit test is equal to the number of cells minus the number of quantities obtained from the observed data, which are used in the calcualtions of the expected frequencies.

The only quantity provided by the observed data, in computing expected frequencies for the outcome when a die is tossed, is the total frequency. Hence, according to our definition, the computed χ^2 value has $6 - 1 = 5$ degrees of freedom.

From Table 10.2 we find the χ^2 value to be

$$\chi^2 = \frac{(20 - 20)^2}{20} + \frac{(22 - 20)^2}{20} + \frac{(17 - 20)^2}{20} + \frac{(18 - 20)^2}{20}$$

$$+ \frac{(19 - 20)^2}{20} + \frac{(24 - 20)^2}{20}$$

$$= 1.7.$$

Using Table A.6, we find $\chi^2_{0.05} = 11.070$ for $\nu = 5$ degrees of freedom. Since 1.7 is less than the critical value, we fail to reject H_0 and conclude that the distribution is uniform. In other words, the die is balanced.

As a second illustration let us test the hypothesis that the frequency distribution of battery lives given in Table 3.4 on page 50 may be approximated by the normal distribution. The expected frequency for each class (cell), listed in Table 10.3, is obtained from a normal curve having the same mean and standard deviation as our sample. From the data of Table 3.3 we find that the sample of 40 batteries has a mean $\bar{x} = 3.41$ and a standard deviation $s = 0.703$.

TABLE 10.3

Observed and Expected Frequencies of Battery
Lives Assuming Normality

Class Boundaries	o_i	e_i
1.45–1.95	2 ⎫	0.8 ⎫
1.95–2.45	1 ⎬ 7	2.7 ⎬ 10.4
2.45–2.95	4 ⎭	6.9 ⎭
2.95–3.45	15	10.6
3.45–3.95	10	10.2
3.95–4.45	5 ⎫ 8	6.0 ⎫ 8.8
4.45–4.95	3 ⎭	2.8 ⎭

These values will be used for μ and σ in computing z values corresponding to the class boundaries. The z values corresponding to the boundaries of the fourth class, for example, are

$$z_1 = \frac{2.95 - 3.41}{0.703} = -0.65,$$

$$z_2 = \frac{3.45 - 3.41}{0.703} = 0.06.$$

From Table A.4 we find the area between $z_1 = -0.65$ and $z_2 = 0.06$ to be

$$\begin{aligned}
\text{area} &= P(-0.65 < Z < 0.06) \\
&= P(Z < 0.06) - P(Z < -0.65) \\
&= 0.5239 - 0.2578 \\
&= 0.2661.
\end{aligned}$$

Hence the expected frequency for the fourth class is

$$e_4 = (0.2661)(40) = 10.6.$$

The expected frequency for the first class interval is obtained by using the total area under the normal curve to the left of the boundary 1.95. For the last class interval we use the total area to the right of the boundary 4.45. All other expected frequencies are determined by the method described above for the fourth class. Note that we have combined adjacent classes in Table 10.3, where the expected frequencies are less than 5. Consequently, the total number of intervals is reduced from 7 to 4. The χ^2 value is then given by

$$\chi^2 = \frac{(7 - 10.4)^2}{10.4} + \frac{(15 - 10.6)^2}{10.6} + \frac{(10 - 10.2)^2}{10.2} + \frac{(8 - 8.8)^2}{8.8}$$

$$= 3.015.$$

The number of degrees of freedom for this test will be $4 - 3 = 1$, since three quantities—the total frequency, mean, and standard deviation—of the observed data were required to find the expected frequencies. Since the computed χ^2 value is less than $\chi^2_{0.05} = 3.841$ for 1 degree of freedom, we have no reason to reject the null hypothesis and conclude that the normal distribution provides a good fit for the distribution of battery lives.

10.9

Test for Independence

The chi-square test procedure discussed in Section 10.8 can also be used to test the hypothesis of independence of two variables. For example, we may wish to study the relationship between religious affiliation and the pattern of worship for the residents of Illinois. People are chosen at random from the state of Illinois until 1000 people have been classified as to whether they are of the Protestant, Catholic, or Jewish faith and whether or not they worship regularly. The observed frequencies are presented in Table 10.4, which is known as a **contingency table.**

A contingency table containing r rows and c columns is referred to as an $r \times c$ table. The term "$r \times c$" is read "r by c." The row and column totals in Table 10.4 are called **marginal frequencies.** To test the null hypothesis, H_0, of independence between a person's religious faith and his or her pattern of worship, we must first find the expected frequencies for each cell of Table 10.4 under the assumption that H_0 is true.

Let us define the following events:

P: An individual selected from our sample is Protestant.
C: An individual selected from our sample is Catholic.
J: An individual selected from our sample is Jewish.
R: An individual selected from our sample worships regularly.
I: An individual selected from our sample worships infrequently.

Using the marginal frequencies, we can list the following probabilities:

$$P(P) = \frac{336}{1000}, \quad P(C) = \frac{351}{1000},$$

$$P(J) = \frac{313}{1000}, \quad P(R) = \frac{598}{1000},$$

$$P(I) = \frac{402}{1000}.$$

TABLE 10.4
2 × 3 Contingency Table

	Protestant	Catholic	Jewish	Total
Worship regularly	182	213	203	598
Worship infrequently	154	138	110	402
Total	336	351	313	1000

Now, if H_0 is true and the two variables are independent, we should have

$$P(P \cap R) = P(P)P(R) = \left(\frac{336}{1000}\right)\left(\frac{598}{1000}\right),$$

$$P(P \cap I) = P(P)P(I) = \left(\frac{336}{1000}\right)\left(\frac{402}{1000}\right),$$

$$P(C \cap R) = P(C)P(R) = \left(\frac{351}{1000}\right)\left(\frac{598}{1000}\right),$$

$$P(C \cap I) = P(C)P(I) = \left(\frac{351}{1000}\right)\left(\frac{402}{1000}\right),$$

$$P(J \cap R) = P(J)P(R) = \left(\frac{313}{1000}\right)\left(\frac{598}{1000}\right),$$

$$P(J \cap I) = P(J)P(I) = \left(\frac{313}{1000}\right)\left(\frac{402}{1000}\right).$$

The expected frequencies are obtained by multiplying each cell probability by the total number of observations. Thus the expected number of Protestants in our sample who worship regularly will be

$$\left(\frac{336}{1000}\right)\left(\frac{598}{1000}\right)(1000) = \frac{(336)(598)}{1000} = 200.9$$

when H_0 is true. The general formula for obtaining the **expected frequency** of any cell is given by

$$\text{Expected frequency} = \frac{(\text{column total}) \times (\text{row total})}{\text{grand total}}.$$

The expected frequency for each cell is recorded in parentheses beside the actual observed value in Table 10.5. Note that the expected frequencies in any row or column add up to the appropriate marginal total. In our example we need to compute only the two expected frequencies in the top row of Table 10.5 and then find the others by subtraction. By using three marginal totals and the grand total to arrive at the expected frequencies, we have lost 4 degrees of freedom, leaving a total of 2. A simple formula providing the correct number of degrees of freedom is given by $v = (r - 1)(c - 1)$. Hence, for our example, $v = (2 - 1)(3 - 1) = 2$ degrees of freedom.

TABLE 10.5

Observed and Expected Frequencies

	Protestant	Catholic	Jewish	Total
East coast	182 (200.9)	215 (209.9)	203 (187.2)	598
West coast	154 (135.1)	136 (141.1)	110 (125.8)	402
Total	336	351	313	1000

To test the null hypothesis of independence, we use the following decision criterion.

_____

Test for Independence. Calculate

$$\chi^2 = \sum_i \frac{(o_i - e_i)^2}{e_i},$$

where the summation extends over all rc cells in the $r \times c$ contingency table. If $\chi^2 > \chi_\alpha^2$ with $\nu = (r - 1)(c - 1)$ degrees of freedom, reject the null hypothesis of independence at the α level of significance; otherwise, accept the null hypothesis.

_____

Applying this criterion to our example, we find that

$$\chi^2 = \frac{(182 - 200.9)^2}{200.9} + \frac{(215 - 209.9)^2}{209.9} + \frac{(203 - 187.2)^2}{187.2}$$

$$+ \frac{(154 - 135.1)^2}{135.1} + \frac{(136 - 141.1)^2}{141.1} + \frac{(110 - 125.8)^2}{125.8}$$

$$= 8.048.$$

From Table A.6 we find that $\chi_{0.05}^2 = 5.991$ for $\nu = (2 - 1)(3 - 1) = 2$ degrees of freedom. The null hypothesis is rejected at the 0.05 level of significance, and we conclude that religious faith and the pattern of worship are not independent.

10.10

Testing Several Proportions

The chi-square statistic for testing independence is also applicable when testing the hypothesis that k binomial populations have the same parameter p. This is, therefore, an extension of the test presented in Section 10.7 for the difference between two proportions to the differences among k proportions.

Hence we are interested in testing the hypothesis

$$H_0: p_1 = p_2 = \cdots = p_k$$

against the alternative hypothesis that the population proportions are *not all equal,* which is equivalent to testing that the occurrence of a success or failure is independent of the population sampled. To perform this test, we first select independent random samples of size $n_1, n_2, \ldots, n_k$ from the k populations and arrange the data as in the $2 \times k$ contingency table, Table 10.6. The expected cell frequencies are calculated as before and substituted together with the observed frequencies into the chi-square formula for independence, namely

$$\chi^2 = \sum_i \frac{(o_i - e_i)^2}{e_i},$$

with

$$\nu = (2 - 1)(k - 1) = k - 1$$

degrees of freedom. By selecting the appropriate upper tail critical region of the form $\chi^2 > \chi_\alpha^2$, one can now reach a decision concerning H_0.

Observe that the test of independence for the data of Table 10.5 on page 338 is equivalent to testing the hypothesis

$$H_0: p_1 = p_2 = p_3,$$
$$H_1: p_1, p_2, \text{ and } p_3 \text{ are not all equal,}$$

Table 10.6
k Independent Binomial Samples

	Sample			
	1	2	$\cdots$	k
Successes	x_1	x_2	$\cdots$	x_k
Failures	$n_1 - x_1$	$n_2 - x_2$	$\cdots$	$n_k - x_k$

where p_1, p_2, and p_3 are the true proportions of Protestants, Catholics, and Jews who worship frequently.

Example 13. In a shop study, a set of data was collected to determine whether or not the proportion of defectives produced by workers was the same for the day, evening, or night shift worked. The following data were collected:

		Shift	
	Day	Evening	Night
Defectives	45	55	70
Nondefectives	905	890	870

Use a 0.025 level of significance to determine if the proportion of defectives is the same for all three shifts.

Solution. Let p_1, p_2, and p_3 represent the true proportion of defectives for the day, evening, and night shift, respectively. Using the six-step procedure, we have

1. H_0: $p_1 = p_2 = p_3$.
2. H_1: p_1, p_2, and p_3 are not all equal.
3. $\alpha = 0.025$.
4. Critical region: $\chi^2 > 7.378$ for $\nu = 2$ degrees of freedom.
5. Computations: Corresponding to the observed frequencies $o_1 = 45$ and $o_2 = 55$, we find

$$e_1 = \frac{(950)(170)}{2835} = 57.0 \quad \text{and} \quad e_2 = \frac{(945)(170)}{2835} = 56.7.$$

All other expected frequencies are found by subtraction and are displayed in Table 10.7.

TABLE 10.7

Observed and Expected Frequencies

		Shift		
	Day	Evening	Night	Total
Defectives	45 (57.0)	55 (56.7)	70 (56.3)	170
Nondefectives	905 (893.0)	890 (888.3)	870 (883.7)	2665
Total	950	945	940	2835

Now,

$$\chi^2 = \frac{(45 - 57.0)^2}{57.0} + \frac{(55 - 56.7)^2}{56.7} + \frac{(70 - 56.3)^2}{56.3}$$

$$+ \frac{(905 - 893.0)^2}{893.0} + \frac{(890 - 888.3)^2}{888.3} + \frac{(870 - 883.7)^2}{883.7}$$

$$= 6.288.$$

6. Decision: Accept H_0 and conclude that the proportion of defectives produced is about the same for all shifts.

It is important to remember that the statistic on which we base our decision has a distribution that is only approximated by the chi-square distribution. The computed χ^2 values depend on the cell frequencies and consequently are discrete. The continuous chi-square distribution seems to approximate the discrete sampling distribution of X^2 very well, provided that the number of degrees of freedom is greater than 1. In a 2×2 contingency table, where we have only 1 degree of freedom, a correction called **Yates' correction for continuity** is applied. The corrected formula then becomes

$$\chi^2(\text{corrected}) = \sum_i \frac{(|o_i - e_i| - 0.5)^2}{e_i}.$$

If the expected cell frequencies are large, the corrected and uncorrected results are almost the same. When the expected frequencies are between 5 and 10, Yates' correction should be applied. For expected frequencies less than 5 the Fisher–Irwin exact test should be used. A discussion of this test may be found in *Basic Concepts of Probability and Statistics* by Hodges and Lehmann (see the References). The Fisher–Irwin test may be avoided, however, by choosing a larger sample.

EXERCISES

1. A die is tossed 180 times with the following results:

x	1	2	3	4	5	6
f	28	36	36	30	27	23

Is this a balanced die? Use a 0.01 level of significance.

2. In 100 tosses of a coin, 63 heads and 37 tails are observed. Is this a balanced coin? Use a 0.05 level of significance.

3. A machine is supposed to mix peanuts, hazelnuts, cashews, and pecans in the ratio 5:2:2:1. A can containing 500 of these mixed nuts was found to have 269 peanuts, 112 hazelnuts, 74 cashews, and 45 pecans. At the 0.05 level of significance, test the hypothesis that the machine is mixing the nuts in the ratio 5:2:2:1.

4. The grades in a statistics course for a particular semester were as follows:

Grade	A	B	C	D	F
f	14	18	32	20	16

Test the hypothesis, at the 0.05 level of significance, that the distribution of grades is uniform.

5. Three cards are drawn from an ordinary deck of playing cards, with replacement, and the number Y of spades is recorded. After repeating the experiment 64 times, the following outcomes were recorded:

y	0	1	2	3
f	21	31	12	0

Test the hypothesis at the 0.01 level of significance that the recorded data may be fitted by the binomial distribution with values $b(y; 3, \frac{1}{4})$ for $y = 0, 1, 2, 3$.

6. Three marbles are selected from an urn containing 5 red marbles and 3 green marbles. After recording the number X of red marbles, the marbles are replaced in the urn and the experiment repeated 112 times. The results obtained are as follows:

x	0	1	2	3
f	1	31	55	25

Test the hypothesis at the 0.05 level of significance that the recorded data may be fitted by the hypergeometric distribution with values $h(x; 8, 3, 5)$ for $x = 0, 1, 2, 3$.

7. A coin is thrown until a head occurs and the number X of tosses recorded. After repeating the experiment 256 times, we obtained the following results:

x	1	2	3	4	5	6	7	8
f	136	60	34	12	9	1	3	1

Test the hypothesis at the 0.05 level of significance that the observed distribution of X may be fitted by the geometric distribution with values $g(x; \frac{1}{2})$ for $x = 1, 2, 3,\ldots$.

8. Repeat Exercise 5 using a new set of data obtained by actually carrying out the described experiment 64 times.

9. Repeat Exercise 7 using a new set of data obtained by performing the described experiment 256 times.

10. In Exercise 4 on page 64, test the goodness of fit between the observed frequencies in the frequency distribution and the expected normal frequencies, using a 0.05 level of significance.

11. In Exercise 5 on page 64, test the goodness of fit between the observed frequencies in the frequency distribution and the expected normal frequencies, using a 0.01 level of significance.

12. In an experiment to study the dependence of hypertension on smoking habits, the following data were taken on 180 individuals:

	Nonsmokers	Moderate Smokers	Heavy Smokers
Hypertension	21	36	30
No hypertension	48	26	19

Test the hypothesis that the presence or absence of hypertension is independent of smoking habits. Use a 0.05 level of significance.

13. A random sample of 200 married men, all retired, were classified according to education and number of children.

	Number of Children		
Education	0–1	2–3	Over 3
Elementary	14	37	32
Secondary	19	42	17
College	12	17	10

Test the hypothesis, at the 0.05 level of significance, that the size of a family is independent of the level of education attained by the father.

14. A random sample of 30 adults are classified according to sex and the number of hours they watch television during a week.

	Male	Female
Over 25 hours	5	9
Under 25 hours	9	7

Using a 0.01 level of significance, test the hypothesis that a person's sex and time watching television are independent.

15. A random sample of 400 college students are classified according to class status and drinking habits.

	Freshman	Sophomore	Junior	Senior
Heavy drinkers	29	41	33	28
Moderate drinkers	32	29	36	39
Nondrinkers	55	34	27	17

Test the hypothesis that class status and drinking habits are independent. Use a 0.05 level of significance.

16. In a study to estimate the proportion of wives who regularly watch soap operas, it is found that 48 of 200 wives in Denver, 29 of 150 wives in Phoenix, and 35 of 150 wives in Rochester watch at least one soap opera. Use a 0.05 level of significance to test the hypothesis that there is no difference between the true proportions of wives who watch soap operas in these three cities.

17. Three distributors of mixed nuts all advertise that their cans contain up to 60% peanuts. If a can containing 500 mixed nuts is selected at random from each of the three distributors and there are, respectively, 345, 313, and 359 peanuts in each of the cans, can we conclude at the 0.01 level of significance that the mixed nuts of the three distributors contain equal proportions of peanuts?

18. A study was made to determine whether there is a difference between the proportions of parents in the states of Maryland, Virginia, Georgia, and Alabama who favor placing Bibles in the elementary schools. The responses of 100 parents selected at random in each of these states are recorded in the following table:

Preference	Maryland	Virginia	Georgia	Alabama
Yes	84	72	67	81
No	16	28	33	19

Can we conclude that the proportions of parents who favor placing Bibles in the schools is the same for these four states? Use a 0.025 level of significance.

11 Regression and Correlation

There are many statistical investigations in which the main objective is to determine whether a relationship exists between two or more variables. If such a relationship can be expressed by a mathematical formula, we will then be able to use it for the purpose of making predictions. For instance, measurements from meteorological data are used extensively today to predict impact areas for missiles fired under various atmospheric conditions, agronomists predict the yields of farm crops based on the various concentrations of nitrogen, potassium, and phosphorus contained in the fertilizer, admission directors require various tests for entering freshmen in order to predict success in college, and so forth. The reliability of any predictions will, of course, depend on the strength of the relationship between the variables included in the formula.

Scientists, economists, psychologists, and sociologists have always been concerned with the problems of prediction. A mathematical equation that allows us to predict values of one dependent variable from known values of one or more independent variables is called a **regression equation.** This term is derived from the original heredity studies made by Sir Francis Galton (1822–1911) in which he compared the heights of sons to the heights of fathers. Galton showed that the heights of sons of tall fathers over successive generations *regressed* toward the mean height of the population. In other words, sons of unusually tall fathers tend to be shorter than their fathers and sons of unusually short fathers tend to be taller than their fathers. Today the term *regression* is applied to all types of prediction problems and does not necessarily imply a regression toward the population mean.

In this section we consider the problem of estimating or predicting the value of a dependent variable Y on the basis of a known measurement of an independent and frequently controlled variable X. Suppose that we wish to predict a student's grade in freshman chemistry based upon his score on an intelligence test administered prior to his attending college. To make such a prediction we first examine the distribution of chemistry grades corresponding to various intelligence test scores achieved by students in prior years. Denoting an individual's chemistry grade by y and his intelligence test score by x, then the pertinent data of any student in the population can be represented by the coordinates (x, y). A random sample of size n form the population might then be designated by the set $\{(x_i, y_i); i = 1, 2, \ldots, n\}$.

Let us consider the distribution of chemistry grades corresponding to intelligence test scores of 50, 55, 65, and 70. The chemistry grades for a sample of 12 freshmen having these intelligence test scores are presented in Table 11.1. The data of Table 11.1 have been plotted in Figure 11.1 to give a **scatter diagram.** From an inspection of this scatter diagram, it is seen that the points follow closely a straight line, indicating that the two variables are to some extent linearly related. Once a reasonable linear relationship has been ascertained, we usually try to express this mathematically by a straight-line equation called the **linear regression line.** From elementary analytical geometry or high school algebra, we know that the slope-intercept form of a straight line can be written in the form

$$\hat{y} = a + bx,$$

where the constants a and b represent the y intercept and slope, respectively. The symbol $\hat{y}$ is used here to distinguish between the predicted value given

TABLE 11.1
Intelligence Test Scores and
Freshmen Chemistry Grades

Student	Test Score, x	Chemistry Grade y
1	65	85
2	50	74
3	55	76
4	65	90
5	55	85
6	70	87
7	65	94
8	70	98
9	55	81
10	70	91
11	50	76
12	55	74

by the regression line and an actual observed value y for some value of x. Such a regression line has been drawn on the scatter diagram of Figure 11.1. Once the point estimates a and b are determined from the sample data, the linear regression line can then be used to predict the value $\hat{y}$ corresponding to any given value x. Of course, the predicted value of $\hat{y}$ is a point estimate of y and as such is unlikely to hit right on. We hope that it will be close.

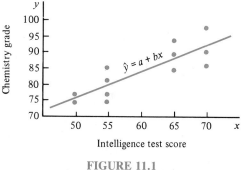

FIGURE 11.1
Scatter diagram with regression line.

Using the regression line in Figure 11.1, we would predict a chemistry grade of 88 for a student whose intelligence test score is 60. However, we would be extremely fortunate if a student with an intelligence test score of 60 made a chemistry grade of exactly 88. In fact, the original data of Table 11.1 show that three students with this intelligence test score received grades of 85,

90, and 94 in freshman chemistry. We must, therefore, interpret the predicted chemistry grade of 88 as an *average* or *expected value* for all students taking the course who have an intelligence test score of 60.

Once we have decided to use a linear regression equation, we face the problem of deriving computational formulas for determining the point estimates a and b from the available sample points. A procedure known as the **method of least squares** will be used. Of all possible lines that one might draw freehand on a scatter diagram, the least-squares procedure selects that particular line for which the sum of the squares of the vertical distances from the observed points to the line is as small as possible. Therefore, if we let e_i represent the vertical deviation from the ith point to the regression line, as indicated in Figure 11.2, the method of least squares yields formulas for calculating a and b so that the sum of the squares of these deviations is a minimum. This sum of the squares of the deviations is often called the **sum of squares of the errors** about the regression line and is denoted by SSE. Thus if we are given a set of paired data $\{(x_i, y_i); i = 1, 2, \ldots, n\}$, we shall find a and b so as to minimize

$$\text{SSE} = \sum_{i=1}^{n} e_i^2 = \sum_{i=1}^{n} (y_i - a - bx_i)^2.$$

The determination of a and b so as to minimize SSE is most easily accomplished by means of differential calculus. We omit the details and state the final formulas.

Estimation of Parameters. Given the sample $\{(x_i, y_i); i = 1, 2, \ldots, n\}$, the least-squares estimates of the parameters in the regression line

$$\hat{y} = a + bx$$

are obtained from the formulas

$$b = \frac{n \sum_{i=1}^{n} x_i y_i - \left(\sum_{i=1}^{n} x_i\right)\left(\sum_{i=1}^{n} y_i\right)}{n \sum_{i=1}^{n} x_i^2 - \left(\sum_{i=1}^{n} x_i\right)^2}$$

and

$$a = \bar{y} - b\bar{x}.$$

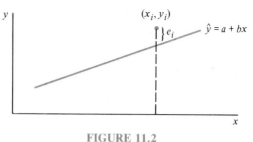

FIGURE 11.2
Least-squares criterion.

Example 1. Find the regression line for the data of Table 11.1.

Solution. We find that

$$\sum_{i=1}^{12} x_i = 725, \qquad \sum_{i=1}^{12} y_i = 1011, \qquad \sum_{i=1}^{12} x_i y_i = 61,685,$$

$$\sum_{i=1}^{12} x_i^2 = 44,475, \qquad \bar{x} = 60.417, \qquad \bar{y} = 84.250.$$

Therefore,

$$b = \frac{(12)(61,685) - (725)(1011)}{(12)(44,475) - (725)^2} = 0.897,$$

$$a = 84.250 - (0.897)(60.417) = 30.056.$$

The regression line is then given by

$$\hat{y} = 30.056 + 0.897x.$$

By substituting any two of the given values of x into this equation, say $x_1 = 50$ and $x_2 = 70$, we obtain the ordinates $\hat{y} = 74.9$ and $\hat{y} = 92.8$. The regression line in Figure 11.2 was drawn by connecting these two points with a straight line.

EXERCISES

1. Consider the following data:

x	1	2	3	4	5	6
y	6	4	3	5	4	2

(a) Find the equation of the regression line.
(b) Graph the line on a scatter diagram.
(c) Find a point estimate of $\mu_{Y|4}$.

2. The grades of a class of 9 students on a midterm report (x) and on the final examination (y) are as follows:

x	77	50	71	72	81	94	96	99	67
y	82	66	78	34	47	85	99	99	68

(a) Find the equation of the regression line.

(b) Estimate the final examination grade of a student who received a grade of 85 on the midterm report but was ill at the time of the final examination.

3. A study was made on the amount of converted sugar in a certain process at various temperatures. The data were coded and recorded as follows:

Temperature, x	Converted Sugar, y
1.0	8.1
1.1	7.8
1.2	8.5
1.3	9.8
1.4	9.5
1.5	8.9
1.6	8.6
1.7	10.2
1.8	9.3
1.9	9.2
2.0	10.5

(a) Estimate the linear regression line.

(b) Estimate the amount of converted sugar produced when the coded temperature is 1.75.

4. The amounts of a chemical compound y, which dissolved in 100 grams of water at various temperatures, x, were recorded as follows:

x °C	y (grams)		
0	8	6	8
15	12	10	14
30	25	21	24
45	31	33	28
60	44	39	42
75	48	51	44

(a) Find the equation of the regression line.

(b) Graph the line on a scatter diagram.

(c) Estimate the amount of chemical that will dissolve in 100 grams of water at 50°C.

5. A mathematics placement test is given to all entering freshmen at a small college. A student who receives a grade below 35 is denied admission to the regular mathematics course and placed in a remedial class. The placement test scores and the final grades for 20 students who took the regular course were recorded as follows:

Placement Test	Course Grade	Placement Test	Course Grade
50	53	90	54
35	41	80	91
35	61	60	48
40	56	60	71
55	68	60	71
65	36	40	47
35	11	55	53
60	70	50	68
90	79	65	57
35	59	50	79

(a) Plot a scatter diagram.
(b) Find the equation of the regression line to predict course grades from placement test scores.
(c) Graph the line on the scatter diagram.
(d) If 60 is the minimum passing grade, below which placement test score should students in the future be denied admission to this course?

6. A study was made by a retail merchant to determine the relation between weekly advertising expenditures and sales. The following data were recorded:

Advertising Costs ($)	Sales ($)
40	385
20	400
25	395
20	365
30	475
50	440
40	490
20	420
50	560
40	525
25	480
50	510

(a) Plot a scatter diagram.
(b) Find the equation of the regression line to predict weekly sales from advertising expenditures.
(c) Estimate the weekly sales when advertising costs are $35.

7. The following data were collected to determine the relationship between high school rank in class and grade-point average at the end of the freshman year in college:

Grade-Point Average, y	Decile Rank, x	Grade-Point Average, y	Decile Rank, x
1.93	3	1.40	8
2.55	2	1.45	4
1.72	1	1.72	8
2.48	1	3.80	1
2.87	1	2.13	5
1.87	3	1.81	6
1.34	4	2.33	1
3.03	1	2.53	1
2.54	2	2.04	2
2.34	2	3.20	2

(a) Find the equation of the regression line to predict the grade-point average of college freshmen from high school rank in class.

(b) Predict the grade-point average for an entering freshman who ranks in the third decile of her graduating class.

8. In a study between the amount of rainfall and the quantity of air pollution removed, the following data were collected:

Daily Rainfall, x (0.01 centimeter)	Particulate Removed, y (micrograms per cubic meter)
4.3	126
4.5	121
5.9	116
5.6	118
6.1	114
5.2	118
3.8	132
2.1	141
7.5	108

(a) Find the equation of the regression line to predict the particulate removed from the amount of daily rainfall.

(b) Estimate the amount of particulate removed when the daily rainfall is $x = 4.8$ units.

11.2
Regression Analysis

In **regression analysis** we assume that the x_i's in the random sample

$$\{(x_i, y_i); i = 1, 2, \ldots, n\}$$

are fixed and not values of random variables. If additional samples of size n were selected using exactly the same values of x, we would expect the values of y to vary. Hence the value y_i in the ordered pair (x_i, y_i) is a value of some random variable Y_i. For convenience we define $Y|x$ to be the random variable Y corresponding to a fixed value x and denote its mean and variance by $\mu_{Y|x}$ and $\sigma^2_{Y|x}$, respectively. Clearly then, if $x = x_i$, the symbol $Y|x_i$ represents the random variable Y_i with mean $\mu_{Y|x_i}$ and variance $\sigma^2_{Y|x_i}$. We are interested in the distributions of the set of random variables $\{Y_i; i = 1, 2, \ldots, n\}$, all of which are assumed to be independent. For the purpose of constructing confidence intervals and making tests of hypotheses, we shall also require Y_1, $Y_2, \ldots, Y_n$ to be normally distributed.

In our prediction problem in Section 11.1, we define $\mu_{Y|x} = E(Y|x)$ to be the mean of the distribution of all chemistry grades for a given or known intelligence test score x, and define $\sigma^2_{Y|x}$ to be the variance of this distribution. We shall assume the variances equal, that is, $\sigma^2_{Y|x} = \sigma^2$, for all x. The parameter $\mu_{Y|x}$ is constant for any fixed x but may vary for different intelligence test scores.

The foregoing discussion is illustrated in Figure 11.3 for three intelligence test scores x_1, x_2, and x_3. The curve connecting the means of all the distributions is called the **regression curve**. If the means $\mu_{Y|x}$ fall on a straight line

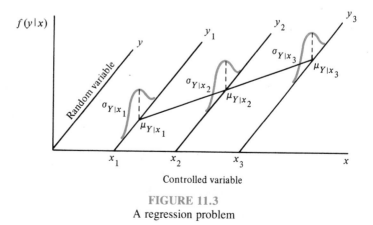

FIGURE 11.3
A regression problem

as is the case in Figure 11.3, the regression is linear and may be represented by the equation

$$\mu_{Y|x} = \alpha + \beta x,$$

where the parameters α and β are called the **regression coefficients.** Denoting their estimates by a and b, respectively, we can then estimate $\mu_{Y|x}$ by $\hat{y}$ from the sample regression line

$$\hat{y} = a + bx.$$

The sample regression line of Example 1 and a hypothetical population regression line have been drawn on the scatter diagram of Figure 11.4 for the data of Table 11.1. The agreement between the sample line and the unknown population line should be good if we have a large number of chemistry grades for each intelligence test score. Once the point estimates a and b of the regression coefficients α and β are evaluated from the sample data, the sample regression line is completely determined.

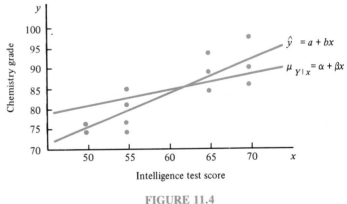

FIGURE 11.4
Scatter diagram with regression lines.

It is clear from Figure 11.4 that for any fixed value of x, each observation (x_i, y_i) in our sample satisfies the relation

$$y_i = \mu_{Y|x_i} + \epsilon_i,$$

where ϵ_i is a random error representing the vertical deviation of the point from the population regression line. From previous assumptions on Y_i, ϵ_i must necessarily be a value of a random variable having a mean of zero and the variance

σ^2. In terms of the sample regression line, we can also write

$$y_i = \hat{y}_i + e_i,$$

where $\hat{y}_i$ is the predicted y value given by the sample regression line when $x = x_i$ and e_i, called the **residual,** is the vertical deviation of the point from the sample regression line as defined in Section 11.1. The difference between ϵ_i and e_i is clearly shown in Figure 11.5.

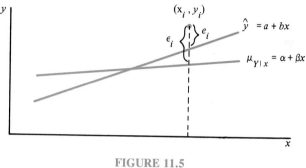

FIGURE 11.5
Comparing ϵ_i with the residual e_i.

An essential part of regression analysis involves the construction of confidence intervals for α and β and testing hypotheses concerning these regression coefficients. First, however, the unknown variance σ^2 must be estimated from the data. An unbiased estimate of σ^2 with $n - 2$ degrees of freedom, denoted by s_e^2, is given by the formula

$$s_e^2 = \frac{\text{SSE}}{n - 2},$$

where

$$\text{SSE} = \sum_{i=1}^{n} (y_i - a - bx_i)^2$$

$$= \sum_{i=1}^{n} (y_i - \hat{y}_i)^2.$$

In the usual sample variance formula we use the divisor $n - 1$ to provide an unbiased estimate of the population variance, since only μ is replaced by the sample mean in our calculations. Here, it is necessary to divide by $n - 2$ in the formula for s_e^2 because 2 degrees of freedom are lost by replacing α and β by a and b in our calculation of the $\hat{y}_i$'s.

Although the preceding formula for SSE appears to have a simple form, the computations involved can be tedious. An equivalent and preferred formula is given by

$$SSE = (n - 1)(s_y^2 - b^2 s_x^2),$$

where

$$s_x^2 = \frac{n \sum_{i=1}^{n} x_i^2 - \left(\sum_{i=1}^{n} x_i \right)^2}{n(n - 1)} \quad \text{and} \quad s_y^2 = \frac{n \sum_{i=1}^{n} y_i^2 - \left(\sum_{i=1}^{n} y_i \right)^2}{n(n - 1)}.$$

Hence we may estimate σ^2 by means of the formula given in the following theorem.

THEOREM 11.1

An unbiased estimate of σ^2 with n − 2 *degrees of freedom is given by the formula*

$$s_e^2 = \frac{n - 1}{n - 2} (s_y^2 - b^2 s_x^2).$$

The calculation of s_e^2 for the data of Table 11.1 is illustrated in the following example.

Example 2. Calculate s_e^2 for the data of Table 11.1.

Solution. In Example 1 we found that

$$\sum_{i=1}^{12} x_i = 725, \qquad \sum_{i=1}^{12} x_i^2 = 44{,}475, \qquad \sum_{i=1}^{12} y_i = 1011.$$

Referring to the data of Table 11.1, we now find $\sum_{i=1}^{12} y_i^2 = 85{,}905$. Therefore,

$$s_x^2 = \frac{(12)(44{,}475) - (725)^2}{(12)(11)} = 61.174,$$

$$s_y^2 = \frac{(12)(85,905) - (1011)^2}{(12)(11)} = 66.205.$$

Recall that $b = 0.897$. Hence

$$s_e^2 = \tfrac{11}{10}[66.205 - (0.805)(61.174)] = 18.656.$$

11.3
Inferences Concerning the Regression Coefficients

It is important to remember that our values of a and b are only estimates of the true parameters α and β based upon a given sample of n observations. The different estimates of α and β that could be computed by drawing several samples of size n may be thought of as the values assumed by the random variables A and B.

Since the values of x remain fixed, the values of A and B depend on the variations in the values of y or, more precisely, on the values of the random variables Y_1, $Y_2,\ldots$, Y_n. Therefore, if we assume Y_1, $Y_2,\ldots$, Y_n to be independent and normally distributed, then it can be shown that the random variable A is also normally distributed with mean

$$\mu_A = \alpha$$

and variance

$$\sigma_A^2 = \left[\frac{\sum\limits_{i=1}^{n} x_i^2}{n(n-1)s_x^2}\right]\sigma^2.$$

Using the Z-transformation of Section 7.2, we can write

$$Z = \frac{A - \alpha}{\sigma\sqrt{\sum\limits_{i=1}^{n} x_i^2} \Big/ s_x \sqrt{n(n-1)}} = \frac{(A - \alpha)s_x \sqrt{n(n-1)}}{\sigma\sqrt{\sum\limits_{i=1}^{n} x_i^2}},$$

which is a random variable having the standard normal distribution.

The standard deviation, σ, is usually unknown and is replaced by its estimator S_e to give

$$T = \frac{(A - \alpha)s_x \sqrt{n(n-1)}}{S_e\sqrt{\sum\limits_{i=1}^{n} x_i^2}},$$

which is a random variable having the t distribution with $n - 2$ degrees of freedom. This quantity can be used to establish the following $(1 - \alpha)100\%$ confidence interval for the parameter α.

Confidence Interval for α. A $(1 - \alpha)100\%$ confidence interval for the parameter α in the regression line $\mu_{Y|x} = \alpha + \beta x$ is

$$a - \frac{t_{\alpha/2} s_e \sqrt{\sum_{i=1}^{n} x_i^2}}{s_x \sqrt{n(n-1)}} < \alpha < a + \frac{t_{\alpha/2} s_e \sqrt{\sum_{i=1}^{n} x_i^2}}{s_x \sqrt{n(n-1)}}.$$

Note that the symbol α is being used here in two totally unrelated ways, first as the level of significance and then as the intercept of the regression line.

Example 3. Find a 95% confidence interval for α in the regression line $\mu_{Y|x} = \alpha + \beta x$, based on the data in Table 11.1.

Solution. In Example 2 we found that $s_x^2 = 61.174$ and $s_e^2 = 18.656$. Therefore, taking square roots we obtain $s_x = 7.8$ and $s = 4.3$. From Example 1 we had $\sum_{i=1}^{n} x_i^2 = 44{,}475$ and $a = 30.056$. Using Table A.5, we find $t_{0.025} = 2.228$ for 10 degrees of freedom. Therefore, a 95% confidence interval for α is given by

$$30.056 - \frac{(2.228)(4.3) \sqrt{44{,}475}}{7.8 \sqrt{(12)(11)}} < \alpha < 30.056 + \frac{(2.228)(4.3) \sqrt{44{,}475}}{7.8 \sqrt{(12)(11)}}$$

which simplifies to

$$7.510 < \alpha < 52.602.$$

To test the null hypothesis H_0 that $\alpha = \alpha_0$ against a suitable alternative, we can use the t distribution with $n - 2$ degrees of freedom to establish a critical region and then base our decision on the value of

$$t = \frac{(a - \alpha_0) s_x \sqrt{n(n-1)}}{s_e \sqrt{\sum_{i=1}^{n} x_i^2}}.$$

Example 4. Using the estimated value $a = 30.056$ in Example 1, test the hypothesis that $\alpha = 35$ at the 0.05 level of significance.

Solution

1. H_0: $\alpha = 35$.
2. H_1: $\alpha \neq 35$.
3. Choose a 0.05 level of significance.
4. Critical region: $t < -2.228$ and $t > 2.228$.
5. Computations:

$$t = \frac{(30.056 - 35)(7.8)\sqrt{(12)(11)}}{4.3\sqrt{44{,}475}} = -0.489.$$

6. Decision: Accept H_0.

The random variable B may also be shown to have a normal distribution with mean

$$\mu_B = \beta$$

and variance

$$\sigma_B^2 = \frac{\sigma^2}{(n-1)s_x^2}.$$

Therefore,

$$Z = \frac{B - \beta}{\sigma/s_x\sqrt{n-1}} = \frac{s_x\sqrt{n-1}\,(B - \beta)}{\sigma}$$

is a random variable having the standard normal distribution. Replacing σ by its estimator S_e we obtain the random variable

$$T = \frac{s_x\sqrt{n-1}\,(B - \beta)}{S_e},$$

which has a t distribution with $n - 2$ degrees of freedom.

The statistic T can be used to construct a $(1 - \alpha)100\%$ confidence interval for the parameter β.

Confidence Interval for β. A $(1 - \alpha)100\%$ confidence interval for the parameter β in the regression line $\mu_{Y|x} = \alpha + \beta x$ is

$$b - \frac{t_{\alpha/2}s_e}{s_x\sqrt{n-1}} < \beta < b + \frac{t_{\alpha/2}s_e}{s_x\sqrt{n-1}}.$$

Example 5. Find a 95% confidence interval for β in the regression line $\mu_{Y|x} = \alpha + \beta x$, based on the data in Table 11.1.

Solution. From Example 3 we note that $s_x = 7.8$, $s_e = 4.3$ and $t_{0.025} = 2.228$ for 10 degrees of freedom. In Example 1 the slope was calculated to be $b = 0.897$. Therefore, a 95% confidence interval for β is given by

$$0.897 - \frac{(2.228)(4.3)}{(7.8)\sqrt{11}} < \beta < 0.897 + \frac{(2.228)(4.3)}{(7.8)\sqrt{11}},$$

which simplifies to

$$0.527 < \beta < 1.267.$$

To test the null hypothesis H_0 that $\beta = \beta_0$, against a suitable alternative, we again use the t distribution with $n - 2$ degrees of freedom to establish a critical region and then base our decision on the value of

$$t = \frac{s_x\sqrt{n-1}\,(b - \beta_0)}{s_e}.$$

The method is illustrated in the following example.

Example 6. Using the estimated value $b = 0.897$ in Example 1, test the hypothesis that $\beta = 0$ at the 0.01 level of significance against the alternative that $\beta > 0$.

Solution

1. H_0: $\beta = 0$.
2. H_1: $\beta > 0$.
3. Choose a 0.01 level of significance.
4. Critical region: $t > 2.764$.

5. Computations:

$$t = \frac{(7.8)\sqrt{11}\,(0.897 - 0)}{4.3} = 5.396.$$

6. Decision: Reject H_0 and conclude that $\beta > 0$.

11.4
Prediction

The equation $\hat{y} = a + bx$ may be used to predict the *mean response* $\mu_{Y|x_0}$ at $x = x_0$, where x_0 is not necessarily one of the prechosen values, or it may be used to predict a single value y_0 of the variable Y_0 when $x = x_0$. We would expect the error of prediction to be higher in the case of a single predicted value than in the case where a mean is predicted. This, then, will affect the width of our confidence intervals for the values being predicted.

Suppose that the experimenter wishes to construct a confidence interval for $\mu_{Y|x_0}$. We shall use the point estimator $\hat{Y}_0 = A + Bx_0$ to estimate $\mu_{Y|x_0} = \alpha + \beta x_0$. It is shown in advanced texts that the sampling distribution of $\hat{Y}_0$ is normal with mean

$$\mu_{\hat{Y}_0} = E(\hat{Y}_0) = \mu_{Y|x_0}$$

and variance

$$\sigma_{\hat{Y}_0}^2 = \left[\frac{1}{n} + \frac{(x_0 - \bar{x})^2}{(n - 1)s_x^2} \right] \sigma^2.$$

In practice we replace σ^2 by s_e^2, a value of the estimator S_e^2. Therefore, the statistic

$$T = \frac{\hat{Y}_0 - \mu_{Y|x_0}}{S_e \sqrt{\dfrac{1}{n} + \dfrac{(x_0 - \bar{x})^2}{(n - 1)s_x^2}}}$$

has a t distribution with $n - 2$ degrees of freedom. A $(1 - \alpha)100\%$ confidence interval for $\mu_{Y|x_0}$ can now be constructed based on the t distribution.

Confidence Interval for $\mu_{Y|x_0}$. A $(1 - \alpha)100\%$ confidence interval for the mean $\mu_{Y|x_0}$ is given by

$$\hat{y}_0 - t_{\alpha/2}s_e \sqrt{\frac{1}{n} + \frac{(x_0 - \bar{x})^2}{(n-1)s_x^2}} < \mu_{Y|x_0} < \hat{y}_0 + t_{\alpha/2}s_e \sqrt{\frac{1}{n} + \frac{(x_0 - \bar{x})^2}{(n-1)s_x^2}}.$$

Example 7. Using the data in Table 11.1, construct a 95% confidence interval for $\mu_{Y|60}$.

Solution. From the regression equation we find for $x = 60$ that

$$\hat{y} = 30.056 + (0.897)(60) = 83.876.$$

Previously, we had $\bar{x} = 60.417$, $s_x^2 = 61.174$, $s_e = 4.3$, and $t_{0.025} = 2.228$ for 10 degrees of freedom. Therefore, a 95% confidence interval for $\mu_{Y|60}$ is given by

$$83.876 - (2.228)(4.3) \sqrt{\frac{1}{12} + \frac{(60 - 60.417)^2}{(11)(61.174)}}$$

$$< \mu_{Y|60} < 83.876$$

$$+ (2.228)(4.3) \sqrt{\frac{1}{12} - \frac{(60 - 60.417)^2}{(11)(61.174)}},$$

or simply

$$81.106 < \mu_{Y|60} < 86.646.$$

To obtain a confidence interval for any single value y_0 of the variable Y_0, it is necessary to estimate the variance of the differences between the ordinates $\hat{y}_0$, obtained from the computed regression lines in repeated sampling when $x = x_0$, and the corresponding true ordinate y_0. We can think of the difference $\hat{y}_0 - y_0$ as a value of the random variable $\hat{Y}_0 - Y_0$, whose sampling distribution can be shown to be normal with mean

$$\mu_{\hat{Y}_0 - Y_0} = E(\hat{Y}_0 - Y_0) = 0$$

and variance

$$\sigma_{\hat{Y}_0 - Y_0}^2 = \left[1 + \frac{1}{n} + \frac{(x_0 - \bar{x})^2}{(n-1)s_x^2}\right]\sigma^2.$$

An estimator for $\sigma_{\hat{Y}_0 - Y_0}^2$ is obtained by substituting S_e^2 for σ^2 in the preceding formula. The statistic

$$T = \frac{\hat{Y}_0 - Y_0}{S_e\sqrt{1 + \frac{1}{n} + \frac{(x_0 - \bar{x})^2}{(n-1)s_x^2}}},$$

which has a t distribution with $n - 2$ degrees of freedom, is then used to construct a confidence interval for y_0 when $x = x_0$.

Confidence Interval for y_0. A $(1 - \alpha)100\%$ confidence interval for the single value y_0 when $x = x_0$ is given by

$$\hat{y}_0 - t_{\alpha/2}s_e\sqrt{1 + \frac{1}{n} + \frac{(x_0 - \bar{x})^2}{(n-1)s_x^2}} < y_0 < \hat{y}_0 + t_{\alpha/2}s_e\sqrt{1 + \frac{1}{n} + \frac{(x_0 - \bar{x})^2}{(n-1)s_x^2}}.$$

Example 8. Using the data of Table 11.1, construct a 95% confidence interval for y_0 when $x = 60$.

Solution. We have $n = 12$, $x_0 = 60$, $\bar{x} = 60.417$, $\hat{y}_0 = 83.876$, $s_x^2 = 61.174$, $s_e = 4.3$, and $t_{0.025} = 2.228$ for 10 degrees of freedom. Therefore, a 95% confidence interval for y_0 when $x = 60$ is given by

$$83.876 - (2.228)(4.3)\sqrt{1 + \frac{1}{12} + \frac{(60 - 60.417)^2}{(11)(61.174)}}$$

$$< y_0 < 83.876$$

$$+ (2.228)(4.3)\sqrt{1 + \frac{1}{12} + \frac{(60 - 60.417)^2}{(11)(61.174)}},$$

which simplifies to

$$73.903 < y_0 < 93.849.$$

EXERCISES

1. With reference to Exercise 1 on page 349,
    (a) evaluate s_e^2;
    (b) test the hypothesis that $\beta = 0$ against the alternative that $\beta \neq 0$ at the 0.05 level of significance.

2. With reference to Exercise 1 on page 349, test the hypothesis, at the 0.01 level of significance, that $\alpha = 6$ against the alternative that $\alpha \neq 6$.

3. With reference to Exercise 2 on page 350,
    (a) evaluate s_x^2, s_y^2, and s_e^2;
    (b) construct a 95% confidence interval for α;
    (c) construct a 95% confidence interval for β.

4. With reference to Exercise 3 on page 350,
    (a) evaluate s_x^2, s_y^2, and s_e^2;
    (b) construct a 95% confidence interval for α;
    (c) construct a 95% confidence interval for β.

5. With reference to Exercise 4 on page 350,
    (a) construct a 99% confidence interval for α;
    (b) construct a 99% confidence interval for β.

6. Construct a 95% confidence interval for α in Exercise 5 on page 351.

7. Construct a 90% confidence interval for β in Exercise 6 on page 351.

8. Consider the following data:

x	0.5	1.5	3.2	4.2	5.1	6.5
y	1.3	3.4	6.7	8.0	10.0	13.2

    (a) Estimate the population regression line $\mu_{Y|x} = \alpha + \beta x$.
    (b) Test the hypothesis that $\alpha = 0$ at the 0.10 level of significance against the alternative that $\alpha \neq 0$.

9. Using the results of Exercise 3(a), construct a 95% confidence interval for $\mu_{Y|80}$ in Exercise 2 on page 350.

10. Using the results of Exercise 4(a), construct a 95% confidence interval for the predicted amount of converted sugar corresponding to $x = 1.6$ in Exercise 3 on page 350.

11. With reference to Exercise 4 on page 350, use the values of s_x^2 and s_e^2 found in Exercise 5 to construct a 99% confidence interval
    (a) for the average amount of chemical compound that will dissolve in 100 grams of water at 50°C;
    (b) for the predicted amount of chemical compound that will dissolve in 100 grams of water at 50°C.

12. With reference to Exercise 5 on page 351, use the values of s_x^2 and s_e^2 found in Exercise 6 to construct a 95% confidence interval
    (a) for the average course grade of students who make a 35 on the placement test;
    (b) for the predicted course grade of a student who made a 35 on the placement test.

13. With reference to Exercise 6 on page 351, use the values of s_x^2 and s_e^2 found in Exercise 7 to construct a 95% confidence interval
    (a) for the average weekly sales when $45 is spent on advertising;
    (b) for the predicted weekly sales when $45 is spent on advertising.

11.5

Test for Linearity of Regression

In Section 11.2 we defined the regression to be linear when all the $\mu_{Y|x}$ fall on a straight line. For any given problem we either assume the regression is linear and proceed with the estimation of parameters as in Section 11.1, or we conclude that the regression is nonlinear and resort to the methods discussed in Sections 11.6 and 11.7. To avoid laborious calculations, a linear regression equation is always preferred over a nonlinear regression curve if the assumption of linearity can be justified. Fortunately, a test for linearity of regression does exist and will now be presented.

Let us select a random sample of n observations using k distinct values of x, say $x_1, x_2, \ldots, x_k$, such that the sample contains n_1 observed values of the random variable Y_1 corresponding to x_1, n_2 observed values of Y_2 corresponding to x_2, $\ldots$, n_k observed values of Y_k corresponding to x_k. Of necessity, $n = \sum_{i=1}^{k} n_i$. We define

$$y_{ij} = j\text{th value of the random variable } Y_i,$$
$$y_{i.} = \text{sum of the values of } Y_i \text{ in our sample.}$$

Hence, if $n_4 = 3$ measurements of Y are made corresponding to $x = x_4$, we could indicate these observations by y_{41}, y_{42}, and y_{43}. Then $y_{4.} = y_{41} + y_{42} + y_{43}$. Now the computed value

$$f = \frac{\chi_1^2/(k-2)}{\chi_2^2/(n-k)},$$

where

$$\chi_1^2 = \sum \frac{y_{i.}^2}{n_i} - \frac{(\sum y_{ij})^2}{n} - b^2(n-1)s_x^2$$

and

$$\chi_2^2 = \sum y_{ij}^2 - \sum \frac{y_{i.}^2}{n_i},$$

is a value of the random variable F, having an F distribution with $k - 2$ and $n - k$ degrees of freedom when the $\mu_{Y|x}$ fall on a straight line and therefore may be used to test the hypothesis H_0 for linearity of regression.

When H_0 is true, $\chi_1^2/(k - 2)$ and $\chi_2^2/(n - k)$ are independent estimates of σ^2. However, if H_0 is false, $\chi_1^2/(k - 2)$ overestimates σ^2. Hence we reject the hypothesis of linearity of regression at the α level of significance when our f value falls in a critical region of size α located in the upper tail of the F distribution.

Example 9. Use the data in Table 11.1 to test the hypothesis that the regression is linear at the 0.05 level of significance.

Solution

1. H_0: the regression is linear.
2. H_1: the regression is nonlinear.
3. Choose a 0.05 level of significance.
4. Critical region: $f > 4.46$.
5. Computations: From Table 11.1 we have

$$x_1 = 50, \quad n_1 = 2, \quad y_{1.} = 150,$$
$$x_2 = 55, \quad n_2 = 4, \quad y_{2.} = 316,$$
$$x_3 = 65, \quad n_3 = 3, \quad y_{3.} = 269,$$
$$x_4 = 70, \quad n_4 = 3, \quad y_{4.} = 276.$$

Therefore,

$$\chi_1^2 = \left(\frac{150^2}{2} + \frac{316^2}{4} + \frac{269^2}{3} + \frac{276^2}{3} \right) - \frac{1011^2}{12}$$
$$- (0.897)^2 (11)(61.174)$$
$$= 8.1506,$$

$$\chi_2^2 = 85{,}905 - \left(\frac{150^2}{2} + \frac{316^2}{4} + \frac{269^2}{3} + \frac{276^2}{3} \right)$$
$$= 178.6667.$$

Hence

$$f = \frac{8.1506/2}{178.6667/8} = 0.182.$$

6. Decision: Accept H_0 and conclude that the data may be fitted by a linear equation.

11.6
Exponential Regression

If a set of data appears to be best represented by a nonlinear regression curve, we must then try to determine the form of the curve and estimate the parameters. Sometimes, a scatter diagram indicates that the means $\mu_{Y|x}$ will probably be best represented by an exponential curve of the form

$$\mu_{Y|x} = \gamma\delta^x,$$

where γ and δ are parameters to be estimated from the data. Denoting these estimates by c and d, respectively, we can estimate $\mu_{Y|x}$ by $\hat{y}$ from the sample regression curve

$$\hat{y} = cd^x.$$

Taking logarithms to the base 10, we obtain the regression curve,

$$\log \hat{y} = \log c + (\log d)x,$$

and each pair of observations in the sample satisfies the relation

$$\log y_i = \log c + (\log d)x_i + e_i$$
$$= a + bx_i + e_i,$$

where $a = \log c$ and $b = \log d$. Therefore, it is possible to find a and b by the formulas in Section 11.1, using the points $(x_i, \log y_i)$, and then determine c and d by taking antilogarithms.

The least-squares procedure for fitting an exponential curve to a set of data is illustrated in the following example.

Example 10. The following data represent the memberships at a local country club during the past 7 years:

x (years)	1	2	3	4	5	6	7
y (membership)	304	341	393	457	548	670	882

Use the method of least squares to estimate a curve of the form $\mu_{Y|x} = \gamma \delta^x$ and predict the membership 5 years from now.

Solution. The logarithms of the y values are, respectively, 2.483, 2.533, 2.594, 2.660, 2.739, 2.826, and 2.945. To compute the estimates a and b, we need the following:

$$\sum_{i=1}^{7} x_i = 28, \qquad \sum_{i=1}^{7} \log y_i = 18.780, \qquad \sum_{i=1}^{7} x_i^2 = 140,$$

$$\sum_{i=1}^{7} x_i \log y_i = 77.237, \qquad \bar{x} = 4, \qquad \overline{\log y} = 2.683.$$

Substituting in the formulas for a and b of Section 11.2, we have

$$b = \frac{(7)(77.237) - (28)(18.780)}{(7)(140) - (28)^2} = 0.076$$

and then

$$a = 2.683 - (0.076)(4) = 2.379.$$

Hence

$$c = 10^{2.379} = 239,$$
$$d = 10^{0.076} = 1.19,$$

and our least-squares regression curve is

$$\hat{y} = (239)(1.19)^x.$$

Based on this rate of growth, we should expect the membership 5 years from now ($x = 12$) to be

$$\hat{y} = 239(1.19)^{12} = 1954.$$

11.7

Multiple Regression

We now consider the problem of estimating or predicting the value of a dependent variable Y on the basis of a set of measurements taken on several independent variables X_1, X_2, ..., X_r. For example, we may wish to estimate the speed of wind as a function of height above the ground, temperature, and pressure. The prediction equation is obtained by using a least-squares procedure on data collected at various heights, temperatures, and pressures, to evaluate the necessary coefficients in the assumed equation.

A random sample of size n from the population might be represented by the set $\{(x_{1i}, x_{2i}, ..., x_{ri}, y_i); i = 1, 2, ..., n\}$. The value y_i is again a value of some random variable Y_i. We assume a theoretical equation of the form

$$\mu_{Y|x_1,x_2, ..., x_r} = \beta_0 + \beta_1 x_1 + \beta_2 x_2 + \cdots + \beta_r x_r,$$

where β_0, β_1, ..., β_r are parameters to be estimated from the data. Denoting these estimates by b_0, b_1 ..., b_r, respectively, we can write the sample regression equation in the form

$$\hat{y} = b_0 + b_1 x_1 + b_2 x_2 + \cdots + b_r x_r.$$

In what follows, we restrict our attention to the case of two independent variables, X_1 and X_2. The results can then be generalized to the case of several independent variables. In dealing with more than two independent variables, a knowledge of matrix theory can facilitate the mathematical manipulations considerably. For a matrix presentation the reader is referred to the text *Probability and Statistics for Engineers and Scientists*, 2nd ed., by Walpole and Myers (see the References).

With only two independent variables the sample regression equation reduces to the form

$$\hat{y} = b_0 + b_1 x_1 + b_2 x_2,$$

and each set of observations satisfies the relation

$$y_i = b_0 + b_1 x_{1i} + b_2 x_{2i} + e_i.$$

The least-squares estimates of b_0, b_1, and b_2 are obtained by solving the simultaneous linear equations

$$n b_0 + b_1 \sum_{i=1}^{n} x_{1i} + b_2 \sum_{i=1}^{n} x_{2i} = \sum_{i=1}^{n} y_i,$$

$$b_0 \sum_{i=1}^{n} x_{1i} + b_1 \sum_{i=1}^{n} x_{1i}^2 + b_2 \sum_{i=1}^{n} x_{1i}x_{2i} = \sum_{i=1}^{n} x_{1i}y_i,$$

$$b_0 \sum_{i=1}^{n} x_{2i} + b_1 \sum_{i=1}^{n} x_{1i}x_{2i} + b_2 \sum_{i=1}^{n} x_{2i}^2 = \sum_{i=1}^{n} x_{2i}y_i.$$

These equations can be solved for b_1 and b_2 by any appropriate method, such as Cramer's rule, and then b_0 can be obtained from the first of the three equations by observing that

$$b_0 = \bar{y} - b_1\bar{x}_1 - b_2\bar{x}_2.$$

Example 11. Suppose in Table 11.1 on page 347 that we are given the number of class periods missed by the 12 students taking the chemistry course. The data are recorded in the following table:

Student	Chemistry Grade, y	Test Score, x_1	Classes Missed, x_2
1	85	65	1
2	74	50	7
3	76	55	5
4	90	65	2
5	85	55	6
6	87	70	3
7	94	65	2
8	98	70	5
9	81	55	4
10	91	70	3
11	76	50	1
12	74	55	4

Fit a regression equation of the form $\mu_{Y|x_1,x_2} = \beta_0 + \beta_1 x_1 + \beta_2 x_2$.

Solution. From the given data, we find that

$$\sum_{i=1}^{12} x_{1i} = 725, \qquad \sum_{i=1}^{12} x_{2i} = 43, \qquad \sum_{i=1}^{12} x_{1i}x_{2i} = 2540,$$

$$\sum_{i=1}^{12} x_{1i}^2 = 44{,}475, \qquad \sum_{i=1}^{12} x_{2i}^2 = 195, \qquad \sum_{i=1}^{12} y_i = 1011,$$

$$\sum_{i=1}^{12} x_{1i}y_i = 61{,}685, \qquad \sum_{i=1}^{12} x_{2i}y_i = 3581.$$

Inserting these values in the simultaneous equations above, we obtain

$$12b_0 + 725b_1 + 43b_2 = 1011,$$
$$725b_0 + 44,475b_1 + 2540b_2 = 61,685,$$
$$43b_0 + 2540b_1 + 195b_2 = 3581.$$

The solution of this set of equations yields the unique estimates $b_0 = 27.547$, $b_1 = 0.922$, and $b_2 = 0.284$. Therefore, our regression equation is

$$\hat{y} = 27.547 + 0.922x_1 + 0.284x_2.$$

If it is desired to fit a regression equation using successive powers of one or more selected independent variables, then the methods in this section can be applied. For example, suppose that we wish to estimate the parameters of the regression equation

$$\mu_{Y|x} = \beta_0 + \beta_1 x + \beta_2 x^2.$$

We are actually fitting a polynomial of the form

$$\mu_{Y|x_1,x_2} = \beta_0 + \beta_1 x_1 + \beta_2 x_2,$$

where we set $x_1 = x$ and $x_2 = x^2$ in the simultaneous equations listed above.

EXERCISES

1. Test for linearity of regression in Exercise 4 on page 350. Use a 0.05 level of significance.

2. Test for linearity of regression in Exercise 5 on page 351. Use a 0.05 level of significance.

3. Observations on the yield of a chemical reaction taken at various temperatures were recorded as follows:

y (%)	x (°C)	y (%)	x (°C)
77.4	150	83.9	250
76.7	150	84.2	250
78.2	150	84.7	250
80.1	200	97.8	300
79.5	200	96.7	300
81.7	200	95.9	300

Test for linearity of regression at the 0.01 level of significance.

4. The following data are the selling prices, y, of a certain make and model of used car x years old:

x (years)	y ($)
1	2350
2	1695
2	1750
3	1395
5	985
5	895

(a) Fit an exponential curve of the form $\mu_{Y|x} = \gamma \delta^x$.
(b) Estimate the selling price of such a car when it is 4 years old.

5. The pressue P of a gas corresponding to various volumes V was recorded as follows:

V, (cm^3)	50	60	70	90	100
P (kg/cm^2)	64.7	51.3	40.5	25.9	7.8

The ideal gas law is given by the equation $PV^{\gamma} = C$, where γ and C are constants.
(a) Find the least-squares estimates of γ and C from the given data.
(b) Estimate P when $V = 80$ cubic centimeters.

6. The following data resulted from 15 experimental runs made on four independent variables and a single response y:

y	x_1	x_2
14.8	11.5	6.3
12.1	14.3	7.4
19.0	9.4	5.9
14.5	15.2	8.7
16.6	8.8	9.1
17.2	9.8	5.6
17.5	11.2	6.8
14.1	10.9	7.4
13.8	14.7	8.2
14.7	15.1	9.2
17.7	8.7	4.7
17.0	8.6	5.5
17.6	9.3	6.6
16.3	10.8	8.7
18.2	11.9	5.4

(a) Fit a regression equation of the form $\mu_{Y|x_1,x_2} = \beta_0 + \beta_1 x_1 + \beta_2 x_2$.
(b) Predict the response y when $x_1 = 10.3$ and $x_2 = 5.8$.

7. The following data were collected to determine a suitable regression equation relating the length of an infant to age and weight at birth:

Infant Length, y(cm)	Age, x_1 (days)	Weight at Birth, x_2(kg)
57.5	78	2.75
52.8	69	2.15
61.3	77	4.41
67.0	88	5.52
53.5	67	3.21
62.7	80	4.32
56.2	74	2.31
68.5	94	4.30
69.2	102	3.71

(a) Fit a regression equation of the form $\mu_{Y|x_1,x_2} = \beta_0 + \beta_1 x_1 + \beta_2 x_2$.
(b) Predict the average length of infants who are 75 days old and weighed 3.15 kg at birth.

8. The following data were collected to determine the relationship between the college entrance aptitude examination score, high school rank in class, and grade-point average at the end of the freshman year.

Grade-Point Average, y	Aptitude Score, x_1	Decile Rank, x_2
1.93	565	3
2.55	525	2
1.72	477	1
2.48	555	1
2.87	502	1
1.87	469	3
1.34	517	4
3.03	555	1
2.54	576	2
2.34	559	2
1.40	574	8
1.45	578	4
1.72	548	8
3.80	656	1
2.13	688	5
1.81	465	6
2.33	661	1
2.53	477	1
2.04	490	2
3.20	524	2

(a) Fit a regression equation of the form $\mu_{Y|x_1, x_2} = \beta_0 + \beta_1 x_1 + \beta_2 x_2$.

(b) Predict the grade-point average for an entering freshman who has an aptitude score of 575 and ranks in the third decile of his graduating class.

9. Given the data

x	0	1	2	3	4	5	6	7	8	9
y	9.1	7.3	3.2	4.6	4.8	2.9	5.7	7.1	8.8	10.2

(a) Fit a regression curve of the form $\mu_{Y|x} = \beta_0 + \beta_1 x + \beta_2 x^2$.

(b) Estimate Y when $x = 2$.

10. (a) Fit a regression curve of the form $\mu_{Y|x} = \beta_0 + \beta_1 x + \beta_2 x^2$ to the data of Exercise 3.

(b) Estimate the yield of the chemical reaction for a temperature of 225°C.

11. An experiment was conducted on a new model of a particular make of automobile to determine the stopping distance at various speeds. The following data were recorded:

Speed, v (kilometers/hr)	35	50	65	80	95	110
Stopping distance, d (meters)	16	26	41	62	88	119

(a) Fit a regression curve of the form $\mu_{D|v} = \beta_0 + \beta_1 v + \beta_2 v^2$.

(b) Estimate the stopping distance when the car is traveling at 70 kilometers per hour.

11.8

Linear Correlation

In this section we shall consider the problem of measuring the relationship between two variables X and Y rather than predicting a value of Y from a knowledge of the independent variable X, as in our study of linear regression. For example, if X represents the amount of money spent yearly on advertising by a retail merchandising firm and Y represents their total yearly sales, we might ask ourselves whether a decrease in the advertising budget is likely to be accompanied by a decrease in the yearly sales. On the other hand, if X represents the age of a used automobile and Y represents the retail book value of the automobile, we would expect large values of X to correspond to small values of Y and small values of X to correspond to large values of Y. **Correlation analysis** attempts to measure the strength of such relationships between two variables by means of a single number called a **correlation coefficient**.

We define a **linear correlation coefficient** to be a measure of the *linear* relationship between the two random variables X and Y, and denote it by r. That is, r measures the extent to which the points cluster about a straight line. Therefore, by constructing a scatter diagram for the n pairs of measurements $\{(x_i, y_i); i = 1, 2, ..., n\}$ in our random sample (see Figure 11.6), we are able to draw certain conclusions concerning r. Should the points follow closely a straight line of positive slope, we have a **high positive correlation** between the two variables. On the other hand, if the points follow closely a straight line of negative slope, we have a **high negative correlation** between the two variables. The correlation between the two variables decreases numerically as the scattering of points from a straight line increases. If the points follow a strictly random pattern as in Figure 11.6c, we have zero correlation and conclude that no linear relationship exists between X and Y.

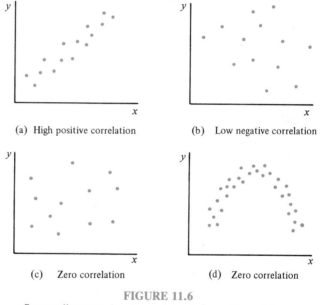

(a) High positive correlation   (b) Low negative correlation

(c)   Zero correlation   (d) Zero correlation

FIGURE 11.6
Scatter diagrams showing various degrees of correlation.

It is important to remember that the correlation coefficient between two variables is a measure of their linear relationship, and a value of $r = 0$ implies a lack of linearity and not a lack of association. Hence, if a strong quadratic relationship exists between X and Y as indicated in Figure 11.6d, we shall still obtain a zero correlation even though there is a strong nonlinear relationship.

The most widely used measure of linear correlation between two variables is called the **Pearson product-moment correlation coefficient** or simply the

sample correlation coefficient. We now present a formula for computing the correlation coefficient in terms of the original measurements that is applicable even when the variables are measured in different units. Thus, if X and Y represent height and weight of an individual, respectively, the following formula will still provide a measure of the linear relationship between these two variables.

Correlation Coefficient. The measure of linear relationship between two variables X and Y is estimated by the *sample correlation coefficient r,* where

$$r = \frac{n \sum_{i=1}^{n} x_i y_i - \left(\sum_{i=1}^{n} x_i \right)\left(\sum_{i=1}^{n} y_i \right)}{\sqrt{\left[n \sum_{i=1}^{n} x_i^2 - \left(\sum_{i=1}^{n} x_i \right)^2 \right]\left[n \sum_{i=1}^{n} y_i^2 - \left(\sum_{i=1}^{n} y_i \right)^2 \right]}} = b \frac{s_x}{s_y}.$$

In Section 11.2 we saw that

$$\text{SSE} = (n - 1)(s_y^2 - b^2 s_x^2).$$

Dividing both sides of this equation by $(n - 1)s_y^2$, we obtain the relation

$$r^2 = 1 - \frac{\text{SSE}}{(n - 1)s_y^2},$$

and since SSE and s_y^2 are always nonnegative, we conclude that r^2 must be between zero and 1. Consequently, r must range from -1 to $+1$. A value of $r = -1$ will occur when SSE $= 0$ and all the sample points lie exactly on a straight line having a negative slope. If the points lie in a straight line having a positive slope, once again SSE $= 0$ and we obtain a value of $r = +1$. Hence a perfect linear relationship exists between the values of X and Y in our sample when $r = \pm 1$. If r is close to $+1$ or -1, the linear relationship between the two variables is strong and we say that we have high correlation. However, if r is close to zero, the linear relationship between X and Y is weak or perhaps nonexistent.

One must be careful in interpreting r beyond what has been stated above. For example, values of r equal 0.3 and 0.6 only mean that we have two positive correlations, one somewhat stronger than the other. It is wrong to conclude that $r = 0.6$ indicates a linear relationship twice as strong as that indicated by the value $r = 0.3$. On the other hand, if we consider r^2, which is usually referred to as the **sample coefficient of determination,** we have a number that expresses the proportion of the total variation in the values of the variable Y that can be accounted for or explained by the linear relationship with the values of the variable X. Thus a correlation of $r = 0.6$ means that 0.36 or 36% of the total variation of the values of Y in our sample is accounted for by a linear relationship with the values of X.

Example 12. compute and interpret the correlation coefficient for the following data:

x (height)	12	10	14	11	12	9
y (weight)	18	17	23	19	20	15

Solution. From the data, we find

$$\sum_{i=1}^{6} x_i = 68, \qquad \sum_{i=1}^{6} y_i = 112, \qquad \sum_{i=1}^{6} x_i y_i = 1292,$$

$$\sum_{i=1}^{6} x_i^2 = 786, \qquad \sum_{i=1}^{6} y_i^2 = 2128.$$

Therefore,

$$r = \frac{(6)(1292) - (68)(112)}{\sqrt{[(6)(786) - (68)^2][(6)(2128) - (112)^2]}}$$

$$= 0.947.$$

A correlation coefficient of 0.947 indicates a very good linear relationship between X and Y. Since $r^2 = 0.90$, we can say that 90% of the variation in the values of Y is accounted for by a linear relationship with X.

The sample correlation coefficient r is a value computed from a random sample of n pairs of measurements. Different random samples of size n from the same population will generally produce different values of r. We might think of r as an estimate of the true linear correlation coefficient for the entire population. Let us denote this **population correlation coefficient** by the symbol ρ. When r is close to zero, we are likely to conclude that $\rho = 0$. However, a sample value of r close to $+1$ or -1 would suggest that $\rho \neq 0$. Our problem, then, is to devise a test procedure that will tell us when r is far enough away from some specified value ρ_0 to justify rejecting the null hypothesis H_0 that $\rho = \rho_0$ in favor of an appropriate alternative. The alternative hypothesis, H_1, may be any one of the usual alternatives: $\rho < \rho_0$, $\rho > \rho_0$, or $\rho \neq \rho_0$.

A test of the null hypothesis $\rho = \rho_0$ is based on the quantity

$$\frac{1}{2} \ln \left(\frac{1 + r}{1 - r} \right),$$

which is a value of a random variable that follows approximately the normal distribution with mean $(\frac{1}{2})\ln[(1 + \rho)/(1 - \rho)]$ and variance $1/(n - 3)$. Thus the test procedure is to compute

$$z = \frac{\sqrt{n - 3}}{2} \left[\ln \left(\frac{1 + r}{1 - r} \right) - \ln \left(\frac{1 + \rho_0}{1 - \rho_0} \right) \right]$$

$$= \frac{\sqrt{n - 3}}{2} \ln \left[\frac{(1 + r)(1 - \rho_0)}{(1 - r)(1 + \rho_0)} \right]$$

and compare to the critical points of the standard normal distribution.

Example 13. For the data of Example 12, test the null hypothesis that there is no linear relationship between the variables. Use a 0.05 level of significance.

Solution

1. H_0: $\rho = 0$.
2. H_1: $\rho \neq 0$.
3. $\alpha = 0.05$.
4. Critical region: $z < -1.96$ and $z > 1.96$.
5. Computations:

$$z = \frac{\sqrt{3}}{2} \ln \left(\frac{1.947}{0.053} \right) = 3.12.$$

6. Decision: Reject the hypothesis of no linear relationship.

11.9

**Multiple and
Partial
Correlation**

The concepts of linear correlation and coefficient of determination introduced
in Section 11.8 provide a measure of the goodness of fit of a least-squares
regression line to a set of paired data. These concepts can also be extended
to the case of multiple variables. Let us assume that the relationship between
the values of the dependent variable Y and the independent variables X_1 and
X_2 can be described by the multiple regression equation

$$\mu_{Y|x_1,x_2} = \beta_0 + \beta_1 x_1 + \beta_2 x_2,$$

which is estimated from the random sample $\{(x_{1i}, x_{2i}, y_i); i = 1, 2, \ldots, n\}$ by
the least-squares sample regression equation

$$\hat{y} = b_0 + b_1 x_1 + b_2 x_2.$$

The **sample coefficient of multiple determination,** denoted by the symbol $R^2_{Y.12}$, indicates the proportion of the total variation among the values of
the variable Y that can be explained by the fitted model. We define $R^2_{Y.12}$ in
exactly the same way that we defined r^2 on page 376. That is,

$$R^2_{Y.12} = 1 - \frac{\text{SSE}}{(n-1)s_y^2},$$

where, as before,

$$\text{SSE} = \sum_{i=1}^{n} (y_i - \hat{y}_i)^2,$$

but now each $\hat{y}_i$ is the predicted value of Y calculated by substituting each of
the coordinates (x_{1i}, x_{2i}), for $i = 1, 2, \ldots, n$ into the multiple regression
equation. Once again, the formula for SSE is extremely tedious to use in
practice and is usually replaced by the equivalent formula

$$\text{SSE} = \sum_{i=1}^{n} y_i^2 - b_0 \sum_{i=1}^{n} y_i - b_1 \sum_{i=1}^{n} x_{1i} y_i - b_2 \sum_{i=1}^{n} x_{2i} y_i.$$

Hence the sample coefficient of multiple determination is given by the following definition.

DEFINITION

Coefficient of Multiple Determination. Given the random sample

$$\{(x_{1i}, x_{2i}, y_i); i = 1, 2, \ldots, n\},$$

the *sample coefficient of multiple determination,* denoted by $R^2_{Y.12}$, is given by

$$R^2_{Y.12} = 1 - \frac{\text{SSE}}{(n-1)s_y^2},$$

where

$$\text{SSE} = \sum_{i=1}^{n} y_i^2 - b_0 \sum_{i=1}^{n} y_i - b_1 \sum_{i=1}^{n} x_{1i} y_i - b_2 \sum_{i=1}^{n} x_{2i} y_i.$$

The **sample multiple correlation coefficient,** denoted by $R_{Y.12}$, is defined to be the positive square root of the coefficient of multiple determination.

Example 14. Compute and interpret the coefficient of multiple determination for the random sample of Example 11 on page 370.

Solution. Using the computations from Example 11 and the fact that

$$\sum_{i=1}^{5} y_i^2 = 85,905 \qquad \text{and} \qquad s_y^2 = 66.205,$$

we find

$$\text{SSE} = 85,905 - (27.547)(1011) - (0.922)(61,685)$$
$$- (0.284)(3581)$$
$$= 164.409$$

and hence

$$R^2_{Y.12} = 1 - \frac{164.409}{(11)(66.205)} = 0.774.$$

The result $R^2_{Y.12} = 0.774$ indicates that the regression plane

$$\hat{y} = 27.547 + 0.922x_1 + 0.284x_2$$

explains 77.4% of the variation in Y.

A coefficient of multiple determination or its square root, the multiple correlation coefficient, provides a measure of association between the observed values of Y and the set of independent variables $X_1, X_2, \ldots, X_r$ contained in the regression equation. Often, an experimenter must decide whether certain variables should be eliminated from the regression equation because they have little or no effect on the predicted values of Y, or perhaps whether additional variables that might conceivably improve the predictive properties of the equation should be included.

An ordinary correlation coefficient or its square, the coefficient of determination, provides a measure of association between Y and some independent variable and could therefore be used to determine the importance of including that particular independent variable in the regression equation. Unfortunately, however, a strong correlation between Y and some additional independent variable, say X_2, may be due entirely to the fact that both Y and X_2 are related to the variable X_1. The true correlation between Y and X_2 can only be observed once the effect of X_1 has been eliminated. We accomplish this by means of the **sample partial correlation coefficient,** denoted by the symbol $r_{Y2.1}$, which measures the correlation between Y and X_2 while X_1 is still considered but held fixed. If we write the ordinary correlation coefficients for Y and X_1, Y and X_2, and X_1 and X_2 as r_{Y1}, r_{Y2}, and r_{12}, respectively, the sample partial correlation coefficient for Y and X_2 with X_1 held fixed is given by the following definition.

DEFINITION

Partial Correlation Coefficient. The measure of linear relationship between the variables Y and X_2, with X_1 fixed, is estimated by the *sample partial correlation coefficient $r_{Y2.1}$*, where

$$r_{Y2.1} = \frac{r_{Y2} - r_{Y1}r_{12}}{\sqrt{(1 - r_{Y1}^2)(1 - r_{12}^2)}}.$$

A similar definition applies to $r_{Y1.2}$, which measures the correlation between Y and X_1 while X_2 is held fixed. We define the square of the sample correlation coefficient to be the **sample coefficient of partial determination,** which represents the ratio of the unexplained variation to the previously unexplained variation. That is, $r_{Y2.1}^2$ gives us the proportion of the variation in the values of Y that was unexplained by a regression line involving only X_1 that can now be explained by including X_2 in the model along with X_1.

Example 15. With reference to Example 11, find and interpret the partial correlation coefficient for Y and X_2 when X_1 is held fixed.

Solution. Substituting the appropriate calculated sums found in Examples 11 and 14 into the definition of the linear correlation coefficient found on page 376, we find that

$$r_{Y1} = \frac{(12)(61,685) - (1011)(725)}{\sqrt{[(12)(85,905) - (1011)^2][(12)(44,475) - (725)^2]}} = 0.862,$$

$$r_{Y2} = \frac{(12)(3581) - (1011)(43)}{\sqrt{[(12)(85,905) - (1011)^2][(12)(195) - (43)^2]}} = -0.242,$$

$$r_{12} = \frac{(12)(2540) - (725)(43)}{\sqrt{[(12)(44,475) - (725)^2][(12)(195) - (43)^2]}} = -0.349.$$

Therefore,

$$r_{Y2.1} = \frac{-0.242 - (0.862)(-0.349)}{\sqrt{[1 - (0.862)^2][1 - (-0.349)^2]}} = 0.124.$$

The value $r_{Y2.1}^2 = 0.015$ indicates that the addition of X_2 to the regression equation results in only a 1.5% reduction in the variation of Y that is unexplained by a regression line using only X_1. Hence the number of class periods missed contributes very little in predicting a student's grade in freshman chemistry.

EXERCISES

1. Compute and interpret the correlation coefficient for the following data:

x	4	5	9	14	18	22	24
y	16	22	11	16	7	3	17

2. Compute and interpret the correlation coefficient for the aptitude scores and grade-point averages in Exercise 8 on page 373.

3. Compute and interpret the correlation coefficient for the following grades of 6 students selected at random:

Mathematics grade	70	92	80	74	65	83
English grade	74	84	63	87	78	90

4. Test the hypothesis that $\rho = -0.8$ in Exercise 1, using a 0.05 level of significance.

5. Test the hypothesis that $\rho = 0$ in Exercise 3, using a 0.05 level of significance.

6. The following data were obtained in a study of the relationship between the weight and chest size of infants at birth:

Weight (kg)	Chest Size (cm)
2.75	29.5
2.15	26.3
4.41	32.2
5.52	36.5
3.21	27.2
4.32	27.7
2.31	28.3
4.30	30.3
3.71	28.7

(a) Calculate r.
(b) Test the null hypothesis that $\rho = 0$ against the alternative that $\rho > 0$ at the 0.01 level of significance.
(c) What percentage of the variation in the infant chest sizes is explained by differences in weight?

7. With reference to Exercise 8 on page 352,
(a) calculate r;
(b) test the null hypothesis that $\rho = -0.5$ against the alternative that $\rho < -0.5$ at the 0.025 level of significance;
(c) determine the percentage of the variation in the amount of particulate removed that is due to changes in the daily amount of rainfall.

8. **Testing $\rho = 0$:** Under the assumption that the random variables X and Y are each normally distributed, a test of the null hypothesis $\rho = 0$ can be based on the equation

$$t = \frac{r\sqrt{n - 2}}{\sqrt{1 - r^2}},$$

which is a value of the t distribution with $\nu = n - 2$ degrees of freedom. Using this test procedure, test the hypothesis at the 0.01 level of significance that $\rho = 0$ against the alternative that $\rho \neq 0$ for the data of Exercise 1.

9. Compute and interpret the coefficient of multiple determination for the data of Exercise 7 on page 373.

10. Compute and interpret the coefficient of multiple determination for the data of Exercise 8 on page 373.

11. With reference to Exercise 7 on page 373, find and interpret
    (a) the partial correlation coefficient for infant length and age when weight at birth is held fixed;
    (b) the partial correlation coefficient for infant length and weight at birth when age is held fixed.

12. With reference to Exercise 8 on page 373, find and interpret
    (a) the partial correlation coefficient for grade-point average and aptitude score when decile rank is held fixed;
    (b) the partial correlation coefficient for grade-point average and decile rank when aptitude score is held fixed.

13. With reference to Example 11 on page 370 and using the calculations carried out in Example 15 on page 381, find the partial correlation coefficient $r_{Y1.2}$ and compare its value with the ordinary correlation coefficient r_{Y1}.

14. An experiment was conducted to determine if the weight, y, of an animal can be predicted after a given period of time on the basis of the initial weight, x_1, and the amount of feed eaten, x_2. The following calculations have been made for a random sample of 10 animals:

$$\sum_{i=1}^{10} x_{1i} = 379, \qquad \sum_{i=1}^{10} x_{2i} = 2417, \qquad \sum_{i=1}^{10} y_i = 825,$$

$$\sum_{i=1}^{10} x_{1i}^2 = 14{,}533, \qquad \sum_{i=1}^{10} x_{2i}^2 = 601{,}365 \qquad \sum_{i=1}^{10} y_i^2 = 70{,}083,$$

$$\sum_{i=1}^{10} x_{1i}y_i = 31{,}726, \qquad \sum_{i=1}^{10} x_{2i}y_i = 204{,}569, \qquad \sum_{i=1}^{10} x_{1i}x_{2i} = 92{,}628.$$

The regression plane is given by $\hat{y} = -22.9915 + 1.3957x_1 + 0.2176x_2$.
    (a) Calculate the multiple correlation coefficient.
    (b) What percentage of the variation in the weights of the animals can be explained by the given regression plane?
    (c) The inclusion of the amount of feed eaten in the regression equation results in what percentage reduction in the variation of the animal weight that is unexplained by a linear regression line using only initial weight?

Analysis of Variance

In Section 10.4 we described an appropriate procedure for testing whether the means of two normal populations with unknown but common variances are equal. It would seem desirable that this technique be extended to cover tests of hypotheses in which there are, say, k population means being compared simultaneously. For instance, we may be interested in testing the null hypothesis that three varieties of wheat produce equal yields on the average, we may wish to determine whether there are significant differences in the behavioral characteristics of rats when placed in four distinct environments, or we may wish to see whether six laboratories give the same results, on the average, in analyzing samples of the same material. Significant contributions in this area of hypothesis testing were advanced by Sir Ronald A. Fisher (1890–1962), one of this century's foremost statisticians.

Consider an experiment in which three varieties of wheat are planted on several plots of equal size and their yields per plot recorded. We are interested in testing the null hypothesis that the three varieties of wheat produce equal yields on the average. To test whether a particular two of the three varieties are significantly different, we use the appropriate tests described in Chapter 10. However, to test for the equality of several means simultaneously, a new technique, called the **analysis of variance,** is required.

The analysis of variance is a method for splitting the total variation of our data into meaningful components that measure different sources of variation. In our experiment we obtain two components, the first measuring the variation due to experimental error and the second measuring variation due to experimental error plus any variation due to the different varieties of wheat. If the null hypothesis is true and the three varieties of wheat produce equal yields on the average, then both components provide independent estimates of the experimental error. Hence we base our test on a comparison of these two components by means of the F distribution.

The yields may vary in part due to different types of soil on which the wheat is planted. This source of variation may be eliminated if the experiment is planned so that all plots of ground have the same soil composition. Failure to eliminate large sources of variation by means of controlled experimentation will result in an inflated estimate of the experimental error and thereby increase the probability of committing a type II error.

Let us now complicate the experiment by considering the yields of three varieties of wheat using four different kinds of fertilizer. We are now interested in testing whether the variation in the yields is caused by differences in the varieties of wheat, differences in the types of fertilizer, or perhaps differences in both. In this case, the analysis of variance provides a means of partitioning the total variation of the yields into three components, the first measuring experimental error only, the second measuring experimental error plus any variation due to the different varieties of wheat, and the third measuring experimental error plus any variation due to the different fertilizers. Hence a comparison of the second component with the first will provide a test of the hypothesis that the varieties of wheat produce equal yields on the average. Similarly, we can test for no differences in the mean yields due to different fertilizers by comparing the third and first components.

The classification of observations on the basis of a single criterion, such as variety of wheat, is called a **one-way classification.** If the observations are classified according to two criteria, such as variety of wheat and type of fertilizer, we have what is called a **two-way classification.** The analysis-of-variance procedures will be considered in detail for each of these classifications in Sections 12.2, 12.5, and 12.6. Methods for the analysis of variance of multiway classifications are not difficult but, with the exception of Section 12.8, are beyond the scope of this text.

12.2
One-Way Classification

Random samples of size n are selected from each of k populations. It will be assumed that the k populations are independent and normally distributed with means $\mu_1, \mu_2, \ldots, \mu_k$ and common variance σ^2. We wish to derive appropriate methods for testing the hypothesis

$$H_0\colon \mu_1 = \mu_2 = \cdots = \mu_k,$$
$$H_1\colon \text{at least two of the means are not equal.}$$

Let x_{ij} denote the jth observation from the ith population and arrange the data as in Table 12.1. Here T_i is the total of all observations in the sample from the ith population, $\bar{x}_{i.}$ the mean of all observations in the sample from the ith population, $T..$ the total of all nk observations, and $\bar{x}..$ the mean of all nk observations. Each observation may be written in the form

$$x_{ij} = \mu_i + \epsilon_{ij},$$

where ϵ_{ij} measures the deviation of the jth observation of the ith sample from the corresponding population mean. An alternative and preferred form of this

TABLE 12.1
k Random Samples

	Population						
	1	2	$\cdots$	i	$\cdots$	k	
	x_{11}	x_{21}	$\cdots$	x_{i1}	$\cdots$	x_{k1}	
	x_{12}	x_{22}	$\cdots$	x_{i2}	$\cdots$	x_{k2}	
	.	.		.		.	
	.	.		.		.	
	.	.		.		.	
	x_{1n}	x_{2n}	$\cdots$	x_{in}	$\cdots$	x_{kn}	
Total	$T_{1.}$	$T_{2.}$	$\cdots$	$T_{i.}$	$\cdots$	$T_{k.}$	$T..$
Mean	$\bar{x}_{1.}$	$\bar{x}_{2.}$	$\cdots$	$\bar{x}_{i.}$	$\cdots$	$\bar{x}_{k.}$	$\bar{x}..$

equation is obtained by substituting $\mu_i = \mu + \alpha_i$, where μ is defined to be the mean of all the μ_i; that is,

$$\mu = \frac{\sum\limits_{i=1}^{k} \mu_i}{k}.$$

Hence we may write

$$x_{ij} = \mu + \alpha_i + \epsilon_{ij},$$

subject to the restriction that $\sum\limits_{i=1}^{k} \alpha_i = \sum\limits_{i=1}^{k} (\mu_i - \mu) = 0$. It is customary to refer to α_i as the **effect** of the ith population.

The null hypothesis that the k population means are equal against the alternative that at least two of the means are unequal may also be stated by the equivalent hypothesis

$$H_0\colon \alpha_1 = \alpha_2 = \cdots = \alpha_k = 0,$$

H_1: At least one of the α_i is not equal to zero.

Our test will be based on a comparison of two independent estimates of the common population variance σ^2. These estimates will be obtained by splitting the total variability of our data into two components.

The variance of all the observations grouped into a single sample of size nk is given by the formula

$$s^2 = \frac{\sum\limits_{i=1}^{k} \sum\limits_{j=1}^{n} (x_{ij} - \bar{x}..)^2}{nk - 1}.$$

The double summation means that we sum all possible terms that are obtained by allowing i to assume values from 1 to k for each value of j from 1 to n. The numerator of s^2, called the **total sum of squares,** measures the total variability of our data. It may be partitioned by means of the following identity.

THEOREM 12.1 _____

One-Way Sum-of-Squares Identity

$$\sum_{i=1}^{k}\sum_{j=1}^{n}(x_{ij} - \bar{x}..)^2 = n\sum_{i=1}^{k}(\bar{x}_{i.} - \bar{x}..)^2 + \sum_{i=1}^{k}\sum_{j=1}^{n}(x_{ij} - \bar{x}_{i.})^2.$$

Proof

$$\sum_{i=1}^{k}\sum_{j=1}^{n}(x_{ij} - \bar{x}..)^2 = \sum_{i=1}^{k}\sum_{j=1}^{n}[(\bar{x}_{i.} - \bar{x}..) + (x_{ij} - \bar{x}_{i.})]^2$$

$$= \sum_{i=1}^{k}\sum_{j=1}^{n}[(\bar{x}_{i.} - \bar{x}..)^2$$

$$+ 2(\bar{x}_{i.} - \bar{x}..)(x_{ij} - \bar{x}_{i.})$$

$$+ (x_{ij} - \bar{x}_{i.})^2]$$

$$= \sum_{i=1}^{k}\sum_{j=1}^{n}(\bar{x}_{i.} - \bar{x}..)^2$$

$$+ 2\sum_{i=1}^{k}\sum_{j=1}^{n}(\bar{x}_{i.} - \bar{x}..)(x_{ij} - \bar{x}_{i.})$$

$$+ \sum_{i=1}^{k}\sum_{j=1}^{n}(x_{ij} - \bar{x}_{i.})^2.$$

The middle term sums to zero, since

$$\sum_{j=1}^{n}(x_{ij} - \bar{x}_{i.}) = \sum_{j=1}^{n}x_{ij} - n\bar{x}_{i.} = \sum_{j=1}^{n}x_{ij} - n\left(\frac{\sum_{j=1}^{n}x_{ij}}{n}\right) = 0.$$

The first sum does not have j as a subscript and therefore may be written

$$\sum_{i=1}^{k}\sum_{j=1}^{n}(\bar{x}_{i.} - \bar{x}..)^2 = n\sum_{i=1}^{k}(\bar{x}_{i.} - \bar{x}..)^2.$$

Hence

$$\sum_{i=1}^{k}\sum_{j=1}^{n}(x_{ij} - \bar{x}..)^2 = n\sum_{i=1}^{k}(\bar{x}_{i.} - \bar{x}..)^2 + \sum_{i=1}^{k}\sum_{j=1}^{n}(x_{ij} - \bar{x}_{i.})^2.$$

It will be convenient in what follows to identify the terms of the sum-of-squares identity by the following notation:

$$SST = \sum_{i=1}^{k} \sum_{j=1}^{n} (x_{ij} - \bar{x}..)^2 = \text{total sum of squares,}$$

$$SSC = n \sum_{i=1}^{k} (\bar{x}_{i.} - \bar{x}..)^2 = \text{sum of squares for column means,}$$

$$SSE = \sum_{i=1}^{k} \sum_{j=1}^{n} (x_{ij} - \bar{x}_{i.})^2 = \text{error sum of squares.}$$

The sum-of-squares identity can then be represented symbolically by the equation

$$SST = SSC + SSE.$$

Many authors refer to the sum of squares for column means as the treatment sum of squares. This terminology is derived from the fact that the k different populations are often classified according to different treatments. Thus the observations x_{ij} ($j = 1, 2, ..., n$) represent the n measurements corresponding to the ith treatment. Today the term "treatment" is used more generally to refer to the various classifications, whether they be different analysts, different fertilizers, different manufacturers, or different regions of the country.

One estimate of σ^2, based on $k - 1$ degrees of freedom, is given by

$$s_1^2 = \frac{SSC}{k - 1}.$$

If H_0 is true, s_1^2 is an unbiased estimate of σ^2. However, if H_1 is true, SSC will tend to have a larger numerical value and s_1^2 will likely overestimate σ^2. A second and independent estimate of σ^2, based on $k(n - 1)$ degrees of freedom, is given by

$$s_2^2 = \frac{SSE}{k(n - 1)}.$$

The estimate s_2^2 is unbiased regardless of the truth or falsity of the null hypothesis. We have already seen that the variance of our grouped data, with

$nk - 1$ degrees of freedom, is

$$s^2 = \frac{\text{SST}}{nk - 1},$$

which is an unbiased estimate of σ^2 when H_0 is true. It is important to note that the sum-of-squares identity has not only partitioned the total variability of the data but also the total number of degrees of freedom; that is,

$$nk - 1 = k - 1 + k(n - 1).$$

When H_0 is true, the ratio

$$f = \frac{s_1^2}{s_2^2}$$

is a value of the random variable F having the F distribution with $k - 1$ and $k(n - 1)$ degrees of freedom. Since s_1^2 overestimates σ^2 when H_0 is false, we have a one-tailed test with the critical region entirely in the right tail of the distribution. The null hypothesis H_0 is rejected at the α level of significance when

$$f > f_\alpha [k - 1, k(n - 1)].$$

The previously defined formulas for SST, SSC, and SSE are not in the best computational form. In actual practice, one computes SST and SSC by the following equivalent formulas and then, making use of the sum-of-squares identity, obtains SSE by subtraction.

Sum-of-Squares Computational Formulas

$$\text{SST} = \sum_{i=1}^{k} \sum_{j=1}^{n} x_{ij}^2 - \frac{T_{..}^2}{nk},$$

$$\text{SSC} = \frac{\sum_{i=1}^{k} T_{i.}^2}{n} - \frac{T_{..}^2}{nk},$$

$$\text{SSE} = \text{SST} - \text{SSC}.$$

TABLE 12.2

Analysis of Variance for the One-Way Classification

Source of Variation	Sum of Squares	Degrees of Freedom	Mean Square	Computed f
Column means	SSC	$k - 1$	$s_1^2 = \dfrac{SSC}{k - 1}$	$\dfrac{s_1^2}{s_2^2}$
Error	SSE	$k(n - 1)$	$s_2^2 = \dfrac{SSE}{k(n - 1)}$	
Total	SST	$nk - 1$		

The computations in an analysis-of-variance problem are usually summarized in tabular form, as shown in Table 12.2. It is customary to refer to the various estimates of σ^2 as the **mean squares.**

Example 1. The data in Table 12.3 represent the number of hours of pain relief provided by 5 different brands of headache tablets administered to 25 subjects. The 25 subjects were randomly divided into 5 groups and each group was treated with a different brand.

TABLE 12.3

Hours of Relief from Headache Tablet

	Tablet					
	A	B	C	D	E	
	5	9	3	2	7	
	4	7	5	3	6	
	8	8	2	4	9	
	6	6	3	1	4	
	3	9	7	4	7	
Total	26	39	20	14	33	132
Mean	5.2	7.8	4.0	2.8	6.6	5.28

Perform the analysis of variance, and test the hypothesis at the 0.05 level of significance that the mean number of hours of relief provided by the tablets is the same for all five brands.

Solution

1. H_0: $\mu_1 = \mu_2 = \mu_3 = \mu_4 = \mu_5$.

2. H_1: at least two of the means are not equal.

3. $\alpha = 0.05$.

4. Critical region: $f > 2.87$.

5. Computations:

$$SST = 5^2 + 4^2 + \cdots + 7^2 - \frac{132^2}{25}$$

$$= 834 - 696.960 = 137.040,$$

$$SSC = \frac{26^2 + 39^2 + \cdots + 33^2}{5} - \frac{132^2}{25}$$

$$= 776.400 - 696.960 = 79.440,$$

$$SSE = 137.040 - 79.440 = 57.600.$$

These results and the remaining computations are exhibited in Table 12.4.

TABLE 12.4

Analysis of Variance for the Data for Table 12.3

Source of Variation	Sum of Squares	Degrees of Freedom	Mean Square	Computed f
Column means	79.440	4	19.860	6.90
Error	57.600	20	2.880	
Total	137.040	24		

6. Decision: Reject the null hypothesis and conclude that the mean number of hours of relief provided by the headache tablets is not the same for all 5 brands.

In experimental work one often loses some of the desired observations. For example, an experiment might be conducted to determine if college students obtain different grades on the average for classes meeting at different times of the day. Because of dropouts during the semester, it is entirely possible to conclude the experiment with unequal numbers of students in the various sections. The previous analysis for equal sample size will still be valid by slightly modifying the sum-of-squares formulas. We now assume the k random samples to be of size $n_1, n_2, \ldots, n_k$, respectively, with $N = \sum_{i=1}^{k} n_i$. The computational formulas for SST, SSC, and SSE are now given by

---

Computational Formulas for Unequal Sample Sizes

$$\text{SST} = \sum_{i=1}^{k} \sum_{j=1}^{n_i} x_{ij}^2 - \frac{T_{..}^2}{N},$$

$$\text{SSC} = \sum_{i=1}^{k} \frac{T_{i.}^2}{n_i} - \frac{T_{..}^2}{N},$$

$$\text{SSE} = \text{SST} - \text{SSC}.$$

The degrees of freedom are then partitioned in the same way: $N - 1$ for SST, $k - 1$ for SSC, and $N - 1 - (k - 1) = N - k$ for SSE.

Example 2. It is suspected that higher-priced automobiles are assembled with greater care than lower-priced automobiles. To investigate whether there is any basis for this feeling, a large luxury model A, a medium-size sedan B, and a subcompact hatchback C were compared for defects when they arrived at the dealer's showroom. All cars were manufactured by the same company. The number of defects for several of the three models are recorded in Table 12.5.

TABLE 12.5

Number of Automobile Defects

| | \multicolumn{3}{c|}{Model} | |
	A	B	C	
	4	5	8	
	7	1	6	
	6	3	8	
	6	5	9	
		3	5	
		4		
Total	23	21	36	80

Test the hypothesis at the 0.05 level of significance that the average number of defects is the same for the three models.

Solution

1. H_0: $\mu_1 = \mu_2 = \mu_3$.
2. H_1: at least two of the means are not equal.
3. $\alpha = 0.05$.
4. Critical region: $f > 3.89$.
5. Computations:

$$\text{SST} = 4^2 + 7^2 + \cdots + 5^2 - \frac{80^2}{15} = 65.333,$$

$$\text{SSC} = \frac{23^2}{4} + \frac{21^2}{6} + \frac{36^2}{5} - \frac{80^2}{15} = 38.283,$$

$$\text{SSE} = 65.333 - 38.283 = 27.050.$$

These results and the remaining computations are exhibited in Table 12.6.

TABLE 12.6

Analysis of Variance for the Data of Table 12.5

Source of Variation	Sum of Squares	Degrees of Freedom	Mean Square	Computed f
Column means	38.283	2	19.142	8.49
Error	27.050	12	2.254	
Total	65.333	14		

6. Decision: Reject H_0 and conclude that the average number of defects for the three models is not the same.

In concluding our discussion on the analysis of variance for the one-way classification, we state the advantages in choosing equal sample sizes as opposed to unequal sample sizes. The first advantage is that the f ratio is insensitive to departures from the assumption of equal variances for the k populations when the samples are of equal size. Second, the choice of equal sample size minimizes the probability of committing a type II error. Finally, the computation of SSC is simplified if the sample sizes are equal.

12.3

Test for the Equality of Several Variances

Although the f ratio obtained from the analysis-of-variance procedure is insensitive to departures from the assumption of equal variances for the k normal populations when the samples are of equal size, we may still prefer to exercise caution and run a preliminary test for homogeneity of variances. Such a test would certainly be advisable in the case of unequal sample sizes if there is a reasonable doubt concerning the homogeneity of the population variances. Suppose, therefore, that we wish to test the null hypothesis

$$H_0: \sigma_1^2 = \sigma_2^2 = \cdots = \sigma_k^2$$

against the alternative

$$H_1: \text{The variances are not all equal.}$$

The test that we shall use, called **Bartlett's test,** is based on a statistic whose sampling distribution provides exact critical values when the sample sizes are equal. These critical values for equal sample sizes can also be used to yield highly accurate approximations to the critical values for unequal sample sizes.

First, we compute the k sample variances $s_1^2, s_2^2, \ldots, s_k^2$ from samples of size $n_1, n_2, \ldots, n_k$, with $\sum_{i=1}^{k} n_i = N$. Second, combine the sample variances to give the pooled estimate

$$s_p^2 = \frac{\sum_{i=1}^{k} (n_i - 1) s_i^2}{N - k}.$$

Now,

$$b = \frac{[(s_1^2)^{n_1-1} (s_2^2)^{n_2-1} \cdots (s_k^2)^{n_k-1}]^{1/(N-k)}}{s_p^2}$$

is a value of a random variable B having the **Bartlett distribution.** For the special case when $n_1 = n_2 = \cdots = n_k = n$, we reject H_0 at the α level of significance if

$$b < b_k(\alpha; n),$$

where $b_k(\alpha; n)$ is the critical value leaving an area of size α in the left tail of the Bartlett distribution. Table A.13 gives the critical values, $b_k(\alpha; n)$, for $\alpha = 0.01$ and 0.05; $k = 2, 3, \ldots, 10$; and selected values of n from 3 to 100.

When the sample sizes are unequal, the null hypothesis is rejected at the α level of significance if

$$b < b_k(\alpha; n_1, n_2, \ldots, n_k),$$

where

$$b_k(\alpha; n_1, n_2, \ldots, n_k) \simeq \frac{[n_1 b_k(\alpha; n_1) + n_2 b_k(\alpha; n_2) + \cdots + n_k b_k(\alpha; n_k)]}{N}.$$

As before, all the $b(\alpha; n_i)$ for sample sizes $n_1, n_2, \ldots, n_k$ are obtained for Table A.13.

Example 3. Use Bartlett's test to test the hypothesis that the variances of the three populations in Example 2 are equal.

Solution

1. H_0: $\sigma_1^2 = \sigma_2^2 = \sigma_3^2$.
2. H_1: the variances are not all equal.
3. $\alpha = 0.05$.
4. Critical region: Referring to Example 2, we have $n_1 = 4$, $n_2 = 6$, $n_3 = 5$, $N = 15$, and $k = 3$. Therefore, we reject when

$$b < b_3(0.05; 4, 6, 5)$$
$$\simeq [(4)(0.4699) + (6)(0.6483) + (5)(0.5762)]/15$$
$$= 0.5767.$$

5. Computations: First compute

$$s_1^2 = 1.583, \qquad s_2^2 = 2.300, \qquad s_3^2 = 2.700,$$

and then

$$s_p^2 = \frac{(3)(1.583) + (5)(2.300) + (4)(2.700)}{12} = 2.254.$$

Now

$$b = \frac{[(1.583)^3(2.300)^5(2.700)^4]^{1/12}}{2.254} = 0.9804.$$

6. Decision: Accept the hypothesis and conclude that the variances of the three populations are equal.

12.4

Multiple-Range Test

The analysis of variance is a powerful procedure for testing the homogeneity of a set of means. However, if we reject the null hypothesis and accept the stated alternative—that the means are not all equal—we still do not know which of the population means are equal and which are different. Several tests are available that separate a set of significantly different means into subsets of homogeneous means. The test that we shall study in this section is called **Duncan's multiple-range test.**

Let us assume that the analysis-of-variance procedure has led to a rejection of the null hypothesis of equal population means. It is also assumed that the k random samples are all of equal size n. The range of any subset of p sample means must exceed a certain value before we consider any of the p population means to be different. This value is called the **least significant range** for the p means, and is denoted by R_p, where

$$R_p = r_p \cdot s_{\bar{x}} = r_p \sqrt{\frac{s^2}{n}}.$$

The sample variance s^2, which is an estimate of the common variance σ^2, is obtained from the error mean square in the analysis of variance. The values of the quantity r_p, called the **least significant studentized-range,** depend on the desired level of significance and the number of degrees of freedom of the error mean square. These values may be obtained from Table A.11 for $p = 2, 3, \ldots, 10$ means.

To illustrate the multiple-range test procedure, let us consider the data in Table 12.3. First we arrange the sample means in increasing order of magnitude:

$\bar{x}_4$	$\bar{x}_3$	$\bar{x}_1$	$\bar{x}_5$	$\bar{x}_2$
2.8	4.0	5.2	6.6	7.8

Next we obtain $s^2 = 2.880$ with 20 degrees of freedom from the error mean square of the analysis of variance in Table 12.4. Let $\alpha = 0.05$. Then the values of r_p are obtained from Table A.11, with $v = 20$ degrees of freedom, for $p = 2, 3, 4,$ and 5. Finally, we obtain R_p by multiplying each r_p by $\sqrt{s^2/n} = \sqrt{2.880/5} = 0.76$. The results of these computations are summarized as follows:

p	2	3	4	5
r_p	2.950	3.097	3.190	3.255
R_p	2.24	2.35	2.42	2.47

Comparing these least significant ranges with the differences in ordered means, we arrive at the following conclusions:

1. Since $\bar{x}_2 - \bar{x}_5 = 1.2 < R_2 = 2.24$, we conclude that $\bar{x}_2$ and $\bar{x}_5$ are not significantly different.
2. Since $\bar{x}_2 - \bar{x}_1 = 2.6 > R_3 = 2.35$, we conclude that $\bar{x}_2$ is significantly larger than $\bar{x}_1$ and therefore $\mu_2 > \mu_1$. It also follows that $\mu_2 > \mu_3$ and $\mu_2 > \mu_4$.
3. Since $\bar{x}_5 - \bar{x}_1 = 1.4 < R_2 = 2.24$, we conclude that $\bar{x}_5$ and $\bar{x}_1$ are not significantly different.
4. Since $\bar{x}_5 - \bar{x}_3 = 2.6 > R_3 = 2.35$, we conclude that $\bar{x}_5$ is significantly larger than $\bar{x}_3$ and therefore $\mu_5 > \mu_3$. Also, $\mu_5 > \mu_4$.
5. Since $\bar{x}_1 - \bar{x}_3 = 1.2 < R_2 = 2.24$, we conclude that $\bar{x}_1$ and $\bar{x}_3$ are not significantly different.
6. Since $\bar{x}_1 - \bar{x}_4 = 2.4 > R_3 = 2.35$, we conclude that $\bar{x}_1$ is significantly larger than $\bar{x}_4$ and therefore $\mu_1 > \mu_4$.
7. Since $\bar{x}_3 - \bar{x}_4 = 1.2 < R_2 = 2.24$, we conclude that $\bar{x}_3$ and $\bar{x}_4$ are not significantly different.

It is customary to summarize the foregoing conclusions by drawing a line under any subset of adjacent means that are not significantly different. Thus we have

$\bar{x}_4$	$\bar{x}_3$	$\bar{x}_1$	$\bar{x}_5$	$\bar{x}_2$
2.8	4.0	5.2	6.6	7.8

One can immediately observe from this manner of presentation that $\mu_2 > \mu_1$, $\mu_2 > \mu_3$, $\mu_2 > \mu_4$, $\mu_5 > \mu_3$, $\mu_5 > \mu_4$, $\mu_1 > \mu_4$, while all other pairs of population means are not considered significantly different.

EXERCISES

1. Show that the computing formula for SSC, in the analysis of variance of the one-way classification, is equivalent to the corresponding term in the identity of Theorem 12.1 on page 389.

2. The following data represent the number of packages of 5 popular brands of cigarettes sold by a supermarket on 8 randomly selected days:

		Brand		
A	B	C	D	E
21	35	45	32	45
35	12	60	53	29
32	27	33	29	31
28	41	36	42	22
14	19	31	40	36
47	23	40	23	29
25	31	43	35	42
38	20	48	42	30

Perform an analysis of variance, at the 0.05 level of significance, and determine whether or not the 5 brands sell, on the average, the same number of cigarettes at this supermarket.

3. Six different machines are being considered for use in manufacturing rubber seals. The machines are being compared with respect to tensile strength of the product. A random sample of 4 seals from each machine is used to determine whether or not the mean tensile strength varies from machine to machine. The following are the tensile-strength measurements in kilograms per square centimeter $\times 10^{-1}$:

		Machine			
1	2	3	4	5	6
17.5	16.4	20.3	14.6	17.5	18.3
16.9	19.2	15.7	16.7	19.2	16.2
15.8	17.7	17.8	20.8	16.5	17.5
18.6	15.4	18.9	18.9	20.5	20.1

Perform the analysis of variance at the 0.05 level of significance and indicate whether or not the treatment means differ significantly.

4. Three sections of the same elementary mathematics course are taught by 3 teachers. The final grades were recorded as follows:

	Teacher	
A	B	C
73	88	68
89	78	79
82	48	56
43	91	91
80	51	71
73	85	71
66	74	87
60	77	41
45	31	59
93	78	68
36	62	53
77	76	79
	96	15
	80	
	56	

Is there a significant difference in the average grades given by the 3 teachers? Use a 0.05 level of significance.

5. In a biological experiment 4 concentrations of a certain chemical are used to enhance the growth of a certain type of plant over a specified period of time. The following growth data, in centimeters, were recorded for the plants that survived:

	Concentration		
1	2	3	4
8.2	7.7	6.9	6.8
8.7	8.4	5.8	7.3
9.4	8.6	7.2	6.3
9.2	8.1	6.8	6.9
	8.0	7.4	7.1
		6.1	

Is there a significant difference in the average growth of these plants for the different concentrations of the chemical? Use a 0.01 level of significance.

6. Test for homogeneity of variances in Exercise 4.

7. Four laboratories are being used to perform chemical analyses. Samples of the same material are sent to the laboratories for analysis as part of the study to

determine whether or not they give, on the average, the same results. The analytical results for the 4 laboratories are as follows:

Laboratory			
A	B	C	D
58.7	62.7	55.9	60.7
61.4	64.5	56.1	60.3
60.9	63.1	57.3	60.9
59.1	59.2	55.2	61.4
58.2	60.3	58.1	62.3

(a) Use Bartlett's test to show that the within-laboratory variances are not significantly different at the 0.05 level of significance.

(b) Perform the analysis of variance and give conclusions concerning the laboratories.

8. Use Duncan's multiple-range test, with a 0.05 level of significance, to analyze the means of the 5 brands of cigarettes in Exercise 2.

9. Use Duncan's multiple-range test in Exercise 7, with a 0.01 level of significance, to determine which laboratories differ, on the average, in their analyses.

10. An investigation was conducted to determine the source of reduction in yield of a certain chemical product. It was known that the loss in yield occurred in the mother liquor; that is, the material removed at the filtration stage. It was felt that different blends of the original material may result in different yield reductions at the mother liquor stage. The following are results of the percentage reduction for 3 batches at each of 4 preselected blends:

Blend			
1	2	3	4
25.6	25.2	20.8	31.6
24.3	28.6	26.7	29.8
27.9	24.7	22.2	34.3

(a) Is there a significant difference in the average percentage reduction in yield for the different blends? Use a 0.05 level of significance.

(b) Use Duncan's multiple-range test to determine which blends differ.

12.5

Two-Way Classification

A set of observations may be classified according to two criteria at once by means of a rectangular array in which the columns represent one criterion of classification and the rows represent a second criterion of classification. For example the rectangular array of observations might be the yields of the three varieties of wheat, discussed in Section 12.1, using four different kinds of fertilizer. The yields are given in Table 12.7. Each treatment combination defines a cell in our array for which we have obtained a single observation.

TABLE 12.7
Yields of Wheat in Kilograms per Plot

Fertilizer Treatment	Variety of Wheat			Total
	v_1	v_2	v_3	
t_1	64	72	74	210
t_2	55	57	47	159
t_3	59	66	58	183
t_4	58	57	53	168
Total	236	252	232	720

In this section we shall derive formulas that will enable us to test whether the variation in our yields is caused by the different varieties of wheat, different kinds of fertilizers, or differences in both.

We shall now generalize and consider a rectangular array consisting of r rows and c columns as in Table 12.8, where x_{ij} denotes an observation in the ith row and jth column. It will be assumed that the x_{ij} are values of independent random variables having normal distributions with means μ_{ij} and the common variance σ^2. In Table 12.8 $T_{i.}$ and $\bar{x}_{i.}$ are the total and mean, respectively, of all observations in the ith row, $T_{.j}$ and $\bar{x}_{.j}$ the total and mean of all observations in the jth column, and $T..$ and $\bar{x}..$ the total and mean of all rc observations.

The average of the population means for the ith row, $\mu_{i.}$, is defined by

$$\mu_{i.} = \frac{\sum_{j=1}^{c} \mu_{ij}}{c}.$$

TABLE 12.8

Two-Way Classification with One Observation per Cell

Row	1	2	$\cdots$	j	$\cdots$	c	Total	Mean
			Column					
1	x_{11}	x_{12}	$\cdots$	x_{1j}	$\cdots$	x_{1c}	$T_{1.}$	$\bar{x}_{1.}$
2	x_{21}	x_{22}	$\cdots$	x_{2j}	$\cdots$	x_{2c}	$T_{2.}$	$\bar{x}_{2.}$
.	.	.		.		.	.	.
.	.	.		.		.	.	.
i	x_{i1}	x_{i2}	$\cdots$	x_{ij}	$\cdots$	x_{ic}	$T_{i.}$	$\bar{x}_{i.}$
.	.	.		.		.	.	.
.	.	.		.		.	.	.
r	x_{r1}	x_{r2}	$\cdots$	x_{rj}	$\cdots$	x_{rc}	$T_{r.}$	$\bar{x}_{r.}$
Total	$T_{.1}$	$T_{.2}$	$\cdots$	$T_{.j}$	$\cdots$	$T_{.c}$	$T_{..}$	
Mean	$\bar{x}_{.1}$	$\bar{x}_{.2}$	$\cdots$	$\bar{x}_{.j}$	$\cdots$	$\bar{x}_{.c}$		$\bar{x}_{..}$

Similarly, the average of the population means for the jth column, $\mu_{.j}$, is defined by

$$\mu_{.j} = \frac{\sum_{i=1}^{r} \mu_{ij}}{r},$$

and the average of the rc population means, μ, is defined by

$$\mu = \frac{\sum_{i=1}^{r} \sum_{j=1}^{c} \mu_{ij}}{rc}.$$

To determine if part of the variation in our observations is due to differences among the rows, we consider the test

$$H_0': \mu_{1.} = \mu_{2.} = \cdots = \mu_{r.} = \mu,$$

H_1': the $\mu_{i.}$ are not all equal.

Also, to determine if part of the variation is due to differences among the columns, we consider the test

$$H_0'': \mu_{.1} = \mu_{.2} = \cdots = \mu_{.c} = \mu,$$

$$H_1'': \text{the } \mu_{.j} \text{ are not all equal.}$$

Each observation may be written in the form

$$x_{ij} = \mu_{ij} + \epsilon_{ij},$$

where ϵ_{ij} measures the deviation of the observed value x_{ij} from the population mean μ_{ij}. The preferred form of this equation is obtained by substituting

$$\mu_{ij} = \mu + \alpha_i + \beta_j,$$

where α_i is the effect of the ith row and β_j is the effect of the jth column. It is assumed that the row and column effects are additive. This is equivalent to stating that $\mu_{ij} - \mu_{ij'} = \mu_{i'j} - \mu_{i'j'}$ or $\mu_{ij} - \mu_{i'j} = \mu_{ij'} - \mu_{i'j'}$, for every value of i, i', j, and j'. That is, the difference between the population means for columns j and j' is the same for every row and the difference between the population means for rows i and i' is the same for every column. Hence we may write

$$x_{ij} = \mu + \alpha_i + \beta_j + \epsilon_{ij}.$$

If we now impose the restrictions that

$$\sum_{i=1}^{r} \alpha_i = 0 \quad \text{and} \quad \sum_{j=1}^{c} \beta_j = 0,$$

then

$$\mu_{i.} = \frac{\sum_{j=1}^{c} (\mu + \alpha_i + \beta_j)}{c} = \mu + \alpha_i,$$

$$\mu_{.j} = \frac{\sum_{i=1}^{r} (\mu + \alpha_i + \beta_j)}{r} = \mu + \beta_j.$$

The null hypothesis that the r row means $\mu_{i.}$ are equal, and therefore equal to μ, is now equivalent to testing the hypothesis

$$H_0': \alpha_1 = \alpha_2 = \cdots = \alpha_r = 0,$$

$$H_1': \text{at least one of the } \alpha_i \text{ is not equal to zero.}$$

Similarly, the null hypothesis that the c column means $\mu_{\cdot j}$ are equal is equivalent to testing the hypothesis

$$H_0'': \beta_1 = \beta_2 = \cdots = \beta_c = 0,$$

H_1'': at least one of the β_j is not equal to zero.

Each of these tests will be based on a comparison of independent estimates of the common population variance σ^2. These estimates will be obtained by splitting the total sum of squares of our data into three components by means of the following identity.

THEOREM 12.2

_____

Two-Way Sum-of-Squares Identity

$$\sum_{i=1}^{r} \sum_{j=1}^{c} (x_{ij} - \bar{x}..)^2 = c \sum_{i=1}^{r} (\bar{x}_{i.} - \bar{x}..)^2 + r \sum_{j=1}^{c} (\bar{x}_{.j} - \bar{x}..)^2$$
$$+ \sum_{i=1}^{r} \sum_{j=1}^{c} (x_{ij} - \bar{x}_{i.} - \bar{x}_{.j} + \bar{x}..)^2.$$

_____

Proof

$$\sum_{i=1}^{r} \sum_{j=1}^{c} (x_{ij} - \bar{x}..)^2 = \sum_{i=1}^{r} \sum_{j=1}^{c} [(\bar{x}_{i.} - \bar{x}..) + (\bar{x}_{.j} - \bar{x}..)$$
$$+ (x_{ij} - \bar{x}_{i.} - \bar{x}_{.j} + \bar{x}..)]^2$$
$$= \sum_{i=1}^{r} \sum_{j=1}^{c} (\bar{x}_{i.} - \bar{x}..)^2 + \sum_{i=1}^{r} \sum_{j=1}^{c} (\bar{x}_{.j} - \bar{x}..)^2$$
$$+ \sum_{i=1}^{r} \sum_{j=1}^{c} (x_{ij} - \bar{x}_{i.} - \bar{x}_{.j} + \bar{x}..)^2$$
$$+ 2 \sum_{i=1}^{r} \sum_{j=1}^{c} (\bar{x}_{i.} - \bar{x}..)(\bar{x}_{.j} - \bar{x}..)$$
$$+ 2 \sum_{i=1}^{r} \sum_{j=1}^{c} (\bar{x}_{i.} - \bar{x}..)(x_{ij} - \bar{x}_{i.} - \bar{x}_{.j} + \bar{x}..)$$
$$+ 2 \sum_{i=1}^{r} \sum_{i=1}^{c} (\bar{x}_{.j} - \bar{x}..)(x_{ij} - \bar{x}_{i.} - \bar{x}_{.j} + \bar{x}..).$$

The cross-product terms are all equal to zero. Hence

$$\sum_{i=1}^{r} \sum_{j=1}^{c} (x_{ij} - \bar{x}..)^2 = c \sum_{i=1}^{r} (\bar{x}_{i.} - \bar{x}..)^2$$

$$+ r \sum_{j=1}^{c} (\bar{x}_{.j} - \bar{x}..)^2$$

$$+ \sum_{i=1}^{r} \sum_{j=1}^{c} (x_{ij} - \bar{x}_{i.} - \bar{x}_{.j} + \bar{x}..)^2$$

The sum-of-squares identity may be represented symbolically by the equation

$$\text{SST} = \text{SSR} + \text{SSC} + \text{SSE},$$

where

$$\text{SST} = \sum_{i=1}^{r} \sum_{j=1}^{c} (x_{ij} - \bar{x}..)^2 = \text{total sum of squares,}$$

$$\text{SSR} = c \sum_{i=1}^{r} (\bar{x}_{i.} - \bar{x}..)^2 = \text{sum of squares for row means,}$$

$$\text{SSC} = r \sum_{j=1}^{c} (\bar{x}_{.j} - \bar{x}..)^2 = \text{sum of squares for column means,}$$

$$\text{SSE} = \sum_{i=1}^{r} \sum_{j=1}^{c} (x_{ij} - \bar{x}_{i.} - \bar{x}_{.j} + \bar{x}..)^2 = \text{error sum of squares.}$$

One estimate of σ^2, based on $r - 1$ degrees of freedom, is given by

$$s_1^2 = \frac{\text{SSR}}{r - 1}.$$

If the row effects $\alpha_1 = \alpha_2 = \cdots = \alpha_r = 0$, s_1^2 is an unbiased estimate of σ^2. However, if the row effects are not all zero, SSR will tend to have an inflated numerical value and s_1^2 is likely to overestimate σ^2. A second estimate of σ^2, based on $c - 1$ degrees of freedom, is given by

$$s_2^2 = \frac{\text{SSC}}{c - 1}.$$

The estimate s_2^2 is an unbiased estimate of σ^2 when the column effects $\beta_1 = \beta_2 = \cdots = \beta_c = 0$. If the column effects are not all zero, SSC will also tend to be inflated and s_2^2 will likely overestimate σ^2. A third estimate of σ^2, based on $(r - 1)(c - 1)$ degrees of freedom and independent of s_1^2 and s_2^2, is given by

$$s_3^2 = \frac{\text{SSE}}{(r - 1)(c - 1)},$$

which is unbiased regardless of the truth or falsity of either null hypothesis.

To test the null hypothesis that the row effects are all equal to zero, we compute the ratio

$$f_1 = \frac{s_1^2}{s_3^2},$$

which is a value of the random variable F_1 having the F distribution with $r - 1$ and $(r - 1)(c - 1)$ degrees of freedom when the null hypothesis is true. The null hypothesis is rejected at the α level of significance when $f_1 > f_\alpha[r - 1, (r - 1)(c - 1)]$.

Similarly, to test the null hypothesis that the column effects are all equal to zero, we compute the ratio

$$f_2 = \frac{s_2^2}{s_3^2},$$

which is a value of the random variable F_2 having the F distribution with $c - 1$ and $(r - 1)(c - 1)$ degrees of freedom when the null hypothesis is true. In this case the null hypothesis is rejected at the α level of significance when $f_2 > f_\alpha[c - 1, (r - 1)(c - 1)]$.

In practice we first compute SST, SSR, and SSC and then, using the sum-of-squares identity, obtain SSE by subtraction. The degrees of freedom associated with SSE are also usually obtained by subtraction; that is,

$$(r - 1)(c - 1) = (rc - 1) - (r - 1) - (c - 1).$$

Preferred computational formulas for the sums of squares are given as follows:

Sum-of-Squares Computational Formulas

$$\text{SST} = \sum_{i=1}^{r} \sum_{j=1}^{c} x_{ij}^2 - \frac{T_{..}^2}{rc},$$

$$\text{SSR} = \frac{\sum_{i=1}^{r} T_{i.}^2}{c} - \frac{T_{..}^2}{rc},$$

$$\text{SSC} = \frac{\sum_{j=1}^{c} T_{.j}^2}{r} - \frac{T_{..}^2}{rc},$$

$$\text{SSE} = \text{SST} - \text{SSR} - \text{SSC}.$$

The computations in an analysis-of-variance problem, for a two-way classification with a single observation per cell, may be summarized as in Table 12.9.

TABLE 12.9
Analysis of Variance for the Two-Way Classification with a Single Observation per Cell

Source of Variation	Sum of Squares	Degrees of Freedom	Mean Square	Computed f
Row means	SSR	$r - 1$	$s_1^2 = \dfrac{\text{SSR}}{r - 1}$	$f_1 = \dfrac{s_1^2}{s_3^2}$
Column means	SSC	$c - 1$	$s_2^2 = \dfrac{\text{SSC}}{c - 1}$	$f_2 = \dfrac{s_2^2}{s_3^2}$
Error	SSE	$(r - 1)(c - 1)$	$s_3^2 = \dfrac{\text{SSE}}{(r - 1)(c - 1)}$	
Total	SST	$rc - 1$		

Example 4. For the data in Table 12.7, test the hypothesis H_0', at the 0.05 level of significance, that there is no difference in the average yield of wheat when different kinds of fertilizer are used. Also test the hypothesis H_0'' that there is no difference in the average yield of the three varieties of wheat.

Solution

1. (a) H_0': $\alpha_1 = \alpha_2 = \alpha_3 = \alpha_4 = 0$ (row effects are zero),
   (b) H_0'': $\beta_1 = \beta_2 = \beta_3 = 0$ (column effects are zero).
2. (a) H_1': at least one of the α_i is not equal to zero,
   (b) H_1'': at least one of the β_j is not equal to zero.
3. $\alpha = 0.05$.
4. Critical regions: (a) $f_1 > 4.76$, (b) $f_2 > 5.14$.
5. Computations:

$$\text{SST} = 64^2 + 55^2 + \cdots + 53^2 - \frac{720^2}{12} = 662,$$

$$\text{SSR} = \frac{210^2 + 159^2 + 183^2 + 168^2}{3} - \frac{720^2}{12} = 498,$$

$$\text{SSC} = \frac{236^2 + 252^2 + 232^2}{4} - \frac{720^2}{12} = 56,$$

$$\text{SSE} = 662 - 498 - 56 = 108.$$

These results and the remaining computations are exhibited in Table 12.10.

TABLE 12.10
Analysis of Variance for the Data of Table 12.7

Source of Variation	Sum of Squares	Degrees of Freedom	Mean Square	Computed f
Row means	498	3	166	9.22
Column means	56	2	28	1.56
Error	108	6	18	
Total	662	11		

6. Decisions:
   (a) Reject H_0' and conclude that there is a difference in the average yield of wheat when different kinds of fertilizer are used.
   (b) Accept H_0'' and conclude that there is no difference in the average yield of the three varieties of wheat.

12.6
Two-Way Classification with Interaction

In Section 12.5 it was assumed that the row and column effects were additive. If we refer to Table 12.7, this implies that if variety v_2 produces on the average 5 more kilograms of wheat per plot than variety v_1 when fertilizer treatment t_1 is used, then v_2 will still produce on the average 5 more kilograms than v_1 when t_2, t_3, or t_4 is used. Similarly, if v_1 produces an average of 3 more kilograms per plot using fertilizer treatment t_2 rather than t_4, then v_2 or v_3 will also produce an average of 3 more kilograms per plot using fertilizer treatment t_2 instead of t_4.

In many experiments the assumption of additivity does not hold and the analysis of Section 12.5 leads to erroneous conclusions. Suppose, for instance, that variety v_2 produces on the average 5 more kilograms of wheat per plot than variety v_1 when fertilizer treatment t_1 is used but produces an average of 2 kilograms per plot less than v_1 when fertilizer treatment t_2 is used. The varieties of wheat and the kinds of fertilizer are now said to **interact.**

An inspection of Table 12.7 suggests the presence of interaction. This apparent interaction may be real or it may be due to experimental error. The analysis of Example 4 was based on the assumption that the apparent interaction was due entirely to experimental error. If the total variability of our data was in part the result of an interaction effect, this source of variation

TABLE 12.11

Yields of Wheat for Three
Replications

Fertilizer Treatment	Variety of Wheat		
	v_1	v_2	v_3
t_1	64	72	74
	66	81	51
	70	64	65
t_2	65	57	47
	63	43	58
	58	52	67
t_3	59	66	58
	68	71	39
	65	59	42
t_4	58	57	53
	41	61	59
	46	53	38

remained a part of the error sum of squares, causing the error mean square to overestimate σ^2, and thereby increased the probability of committing a type II error.

To test for differences among the row and column means when the interaction is a significant factor, we must obtain an unbiased and independent estimate of σ^2. This is best accomplished by considering the variation of repeated measurements obtained under similar conditions. Suppose that we have reason to believe that the varieties of wheat and kinds of fertilizer in Table 12.7 interact. We repeat the experiment twice, using 36 one-acre plots, rather than 12, and record the results in Table 12.11. It is customary to say that the experiment has been **replicated** three times.

To present general formulas for the analysis of variance using repeated observations, we shall consider the case of n replications and then demonstrate the use of these formulas by applying them to the data of Table 12.11. As before, we consider a rectangular array consisting of r rows and c columns. Thus we still have rc cells, but now each cell contains n observations. We denote the kth observation in the ith row and jth column by x_{ijk}; the rcn observations are shown in Table 12.12.

The observations in the ijth cell constitute a random sample of size n from a population that is assumed to be normally distributed with mean μ_{ij} and variance σ^2. All rc populations are assumed to have the same variance σ^2. Let us define the following useful symbols, some of which are used in Table 12.12.

$$T_{ij} = \text{sum of the observations in the } ij\text{th cell,}$$

$$T_{i..} = \text{sum of the observations in the } i\text{th row,}$$

$$T_{.j.} = \text{sum of the observations in the } j\text{th column,}$$

$$T... = \text{sum of all } rcn \text{ observations,}$$

$$\bar{x}_{ij.} = \text{mean of the observations in the } ij\text{th cell,}$$

$$\bar{x}_{i..} = \text{mean of the observations in the } i\text{th row,}$$

$$\bar{x}_{.j.} = \text{mean of the observations in the } j\text{th column,}$$

$$\bar{x}... = \text{mean of all } rcn \text{ observations.}$$

Each observation in Table 12.12 may be written in the form

$$x_{ijk} = \mu_{ij} + \epsilon_{ijk},$$

where ϵ_{ijk} measures the deviations of the observed x_{ijk} values in the ijth cell from the population mean μ_{ij}. If we let $(\alpha\beta)_{ij}$ denote the interaction effect of

TABLE 12.12

Two-Way Classification with Several Observations per Cell

Row	Column				Total	Mean
	1	2	$\cdots$	c		
1	x_{111} x_{112} . . . x_{11n}	x_{121} x_{122} . . x_{12n}	$\cdots$	x_{1c1} x_{1c2} . . . x_{1cn}	$T_{1..}$	$\bar{x}_{1..}$
2	x_{211} x_{212} . . . x_{21n}	x_{221} x_{222} . . x_{22n}	$\cdots$	x_{2c1} x_{2c2} . . . x_{2cn}	$T_{2..}$	$\bar{x}_{2..}$
. . .	. . .	. . .	$\cdots$	. . .	. . .	. . .
r	x_{r11} x_{r12} . . . x_{r1n}	x_{r21} x_{r22} . . x_{r2n}	$\cdots$	x_{rc1} x_{rc2} . . . x_{rcn}	$T_{r..}$	$\bar{x}_{r..}$
Total Mean	$T_{.1.}$ $\bar{x}_{.1.}$	$T_{.2.}$ $\bar{x}_{.2.}$	$\cdots$ $\cdots$	$T_{.c.}$ $\bar{x}_{.c.}$	$T...$	$\bar{x}...$

the ith row and the jth column, α_i the effect of the ith row, β_j the effect of the jth column, and μ the overall mean, we can write

$$\mu_{ij} = \mu + \alpha_i + \beta_j + (\alpha\beta)_{ij},$$

and then

$$x_{ijk} = \mu + \alpha_i + \beta_j + (\alpha\beta)_{ij} + \epsilon_{ijk},$$

on which we impose the restrictions

$$\sum_{i=1}^{r} \alpha_i = 0, \qquad \sum_{j=1}^{c} \beta_j = 0, \qquad \sum_{i=1}^{r} (\alpha\beta)_{ij} = 0, \qquad \sum_{j=1}^{c} (\alpha\beta)_{ij} = 0.$$

The three hypotheses to be tested are as follows:

1. H_0': $\alpha_1 = \alpha_2 = \cdots = \alpha_r = 0$,
   H_1': at least one of the α_i is not equal to zero.
2. H_0'': $\beta_1 = \beta_2 = \cdots = \beta_c = 0$,
   H_1'': at least one of the β_j is not equal to zero.
3. H_0''': $(\alpha\beta)_{11} = (\alpha\beta)_{12} = \cdots = (\alpha\beta)_{rc} = 0$,
   H_1''': at least one of the $(\alpha\beta)_{ij}$ is not equal to zero.

Each of these tests will be based on a comparison of independent estimates of σ^2 provided by the splitting of the total sum of squares of our data into four components by means of the identity

$$SST = SSR + SSC + SS(RC) + SSE,$$

where

$$SST = \sum_{i=1}^{r} \sum_{j=1}^{c} \sum_{k=1}^{n} (x_{ijk} - \bar{x}...)^2 = \text{total sum of squares},$$

$$SSR = cn \sum_{i=1}^{r} (\bar{x}_{i..} - \bar{x}...)^2 = \text{sum of squares for row means},$$

$$SSC = rn \sum_{j=1}^{c} (\bar{x}_{.j.} - \bar{x}...)^2 = \text{sum of squares for column means},$$

$$SS(RC) = n \sum_{i=1}^{r} \sum_{j=1}^{c} (\bar{x}_{ij.} - \bar{x}_{i..} - \bar{x}_{.j.} + \bar{x}...)^2 = \text{sum of squares for}$$

interaction of rows and columns,

$$SSE = \sum_{i=1}^{r} \sum_{j=1}^{c} \sum_{k=1}^{n} (x_{ijk} - \bar{x}_{ij.})^2 = \text{error sum of squares}.$$

The degrees of freedom are partitioned according to the identity

$$rcn - 1 = (r - 1) + (c - 1) + (r - 1)(c - 1) + rc(n - 1).$$

Dividing each of the sum of squares on the right side of the sum of squares identity by their corresponding number of degrees of freedom, we obtain the four estimates

$$s_1^2 = \frac{SSR}{r - 1}, \quad s_2^2 = \frac{SSC}{c - 1}, \quad s_3^2 = \frac{SS(RC)}{(r - 1)(c - 1)}, \quad s_4^2 = \frac{SSE}{rc(n - 1)},$$

of σ^2, which are all unbiased when H_0', H_0'', and H_0''' are true.

To test the hypothesis H_0' that the row effects are all equal to zero, we compute the ratio

$$f_1 = \frac{s_1^2}{s_4^2},$$

which is a value of the random variable F_1 having the F distribution with $r - 1$ and $rc(n - 1)$ degrees of freedom when H_0' is true. The null hypothesis is rejected at the α level of significance when $f_1 > f_\alpha[r - 1, rc(n - 1)]$. Similarly, to test the hypothesis H_0'' that the column effects are all equal to zero, we compute the ratio

$$f_2 = \frac{s_2^2}{s_4^2},$$

which is a value of the random variable F_2 having the F distribution with $(r - 1)(c - 1)$ and $rc(n - 1)$ degrees of freedom when H_0'' is true. This hypothesis is rejected at the α level of significance when $f_2 > f_\alpha[c - 1, rc(n - 1)]$. Finally, to test the hypotheses H_0''' that the interaction effects are all equal to zero, we compute the ratio

$$f_3 = \frac{s_3^2}{s_4^2},$$

which is a value of the random variable F_3 having the F distribution with $(r - 1)(c - 1)$ and $rc(n - 1)$ degrees of freedom.

The presence of interaction in an experiment can sometimes disguise or mask significant differences among the row or column effects, perhaps resulting in the acceptance of a hypothesis that should be rejected. For this reason any test that leads to the acceptance of the hypothesis that the row or column effects are zero must be considered invalid when interaction is significant.

The computations in an analysis-of-variance problem, for a two-way classification with several observations per cell, are usually summarized as in Table 12.13.

TABLE 12.13

Analysis of Variance for the Two-Way Classification with Interaction

Source of Variation	Sum of Squares	Degrees of Freedom	Mean Square	Computed f
Row means	SSR	$r - 1$	$s_1^2 = \dfrac{SSR}{r - 1}$	$f_1 = \dfrac{s_1^2}{s_4^2}$
Column means	SSC	$c - 1$	$s_2^2 = \dfrac{SSC}{c - 1}$	$f_2 = \dfrac{s_2^2}{s_4^2}$
Interaction	SS(RC)	$(r - 1)(c - 1)$	$s_3^2 = \dfrac{SS(RC)}{(r - 1)(c - 1)}$	$f_3 = \dfrac{s_3^2}{s_4^2}$
Error	SSE	$rc(n - 1)$	$s_4^2 = \dfrac{SSE}{rc(n - 1)}$	
Total	SST	$rcn - 1$		

The sums of squares are usually obtained by means of the following computational formulas:

Sum-of-Squares Computational Formulas

$$SST = \sum_{i=1}^{r} \sum_{j=1}^{c} \sum_{k=1}^{n} x_{ijk}^2 - \frac{T_{...}^2}{rcn},$$

$$SSR = \frac{\sum_{i=1}^{r} T_{i..}^2}{cn} - \frac{T_{...}^2}{rcn},$$

$$SSC = \frac{\sum_{j=1}^{c} T_{.j.}^2}{rn} - \frac{T_{...}^2}{rcn},$$

$$SS(RC) = \frac{\sum_{i=1}^{r} \sum_{j=1}^{c} T_{ij.}^2}{n} - \frac{\sum_{i=1}^{r} T_{i..}^2}{cn} - \frac{\sum_{j=1}^{c} T_{.j.}^2}{rn} + \frac{T_{...}^2}{rcn},$$

$$SSE = SST - SSR - SSC - SS(RC).$$

Example 5. For the data in Table 12.11, use a 0.05 level of significance to test the following hypotheses: (a) H_0': there is no difference in the average yield of wheat when different kinds of fertilizer are used; (b) H_0'': there is no difference in the average yield of the three varieties of wheat; and (c) H_0''': there is no interaction between the different kinds of fertilizer and the different varieties of wheat.

Solution

1. (a) H_0': $\alpha_1 = \alpha_2 = \alpha_3 = \alpha_4 = 0$,
   (b) H_0'': $\beta_1 = \beta_2 = \beta_3 = 0$,
   (c) H_0''': $(\alpha\beta)_{11} = (\alpha\beta)_{12} = \cdots = (\alpha\beta)_{43} = 0$.
2. (a) H_1': at least one of the α_i is not equal to zero,
   (b) H_1'': at least one of the β_j is not equal to zero,
   (c) H_1''': at least one of the $(\alpha\beta)_{ij}$ is not equal to zero.
3. $\alpha = 0.05$.
4. Critical regions: (a) $f_1 > 3.01$, (b) $f_2 > 3.40$,
   (c) $f_3 > 2.51$.
5. Computations: From Table 12.11 we first construct the following table of totals:

	v_1	v_2	v_3	Total
t_1	200	217	190	607
t_2	186	152	172	510
t_3	192	196	139	527
t_4	145	171	150	466
Total	723	736	651	2110

Now

$$SST = 64^2 + 66^2 + \cdots + 38^2 - \frac{2110^2}{36}$$

$$= 127{,}448 - 123{,}669 = 3779,$$

$$SSR = \frac{607^2 + 510^2 + 527^2 + 466^2}{9} - \frac{2110^2}{36}$$

$$= 124{,}826 - 123{,}669 = 1157,$$

$$SSE = \frac{723^2 + 736^2 + 651^2}{12} - \frac{2110^2}{36}$$

$$= 124{,}019 - 123{,}669 = 350,$$

TABLE 12.14
Analysis of Variance for the Data of Table 12.11

Source of Variation	Sum of Squares	Degrees of Freedom	Mean Square	Computed f
Row means	1157	3	385.667	6.17
Column means	350	2	175.000	2.80
Interaction	771	6	128.500	2.05
Error	1501	24	62.542	
Total	3779	35		

$$SS(RC) = \frac{200^2 + 186^2 + \cdots + 150^2}{3}$$

$$- 124{,}826 - 124{,}019 + 123{,}669$$

$$= 771,$$

$$SSE = 3779 - 1157 - 350 - 771 = 1501.$$

These results, with the remaining computations, are given in Table 12.14.

6. Decisions:
   (a) Reject H_0' and conclude that a difference in the average yields of wheat exists when different kinds of fertilizer are used.
   (b) Accept H_0'' and conclude that there is no difference in the average yield of the three varieties of wheat.
   (c) Accept H_0''' and conclude that there is no interaction between the different kinds of fertilizer and the different varieties of wheat.

EXERCISES

1. Referring to the proof of Theorem 12.2 on page 406, show that the cross-product term

$$\sum_{i=1}^{r} \sum_{j=1}^{c} (x_{ij} - \bar{x}_{i.} - \bar{x}_{.j} + \bar{x}_{..})(\bar{x}_{.j} - \bar{x}_{..}) = 0.$$

2. Show that the computing formula for SSR, in the analysis of variance of the two-way classification with a single observation per cell, is equivalent to the corresponding term in the identity of Theorem 12.2 on page 406.

3. The following data represent the final grades obtained by 5 students in mathematics, English, French, and biology:

	Subject			
Student	Mathematics	English	French	Biology
1	68	57	73	61
2	83	94	91	86
3	72	81	63	59
4	55	73	77	66
5	92	68	75	87

Use a 0.05 level of significance to test the hypothesis that
(a) the courses are of equal difficulty;
(b) the students have equal ability.

4. Use the multiple-range test, with a 0.05 level of significance, to analyze the mean grades obtained by the 5 students in Exercise 3.

5. An experiment is conducted in which 4 treatments are to be compared using 5 subjects. The following data are generated:

	Subject				
Treatment	1	2	3	4	5
1	12.8	10.6	11.7	10.7	11.0
2	11.7	14.2	11.8	9.9	13.8
3	11.5	14.7	13.6	10.7	15.9
4	12.6	16.5	15.4	9.6	17.1

Perform the analysis of variance, separating out the treatment, subject, and error sums of squares. Use a 0.05 level of significance to test the hypothesis that there is no difference between the treatment means.

6. A study is made to determine the force required to pull apart pieces of glued plastic. Three types of plastic were tested using 4 different levels of humidity. The results, in kilograms, are given as follows:

	Humidity			
Plastic Type	30%	50%	70%	90%
A	39.0	33.1	33.8	33.0
B	36.9	27.2	29.7	28.5
C	27.4	29.2	26.7	30.9

Use a 0.05 level of significance to test the hypothesis that there is no difference in the mean force required to pull the glued plastic apart

(a) when different types of plastic are used;

(b) for different humidity conditions.

7. The following are the number of words per minute typed by 4 secretaries on 4 different typewriter models:

Secretary	Typewriter Model			
	Royal	IBM	Underwood	Olivetti
Kim	78	62	71	77
Doug	57	49	62	60
Rhonda	69	78	72	83
Kevin	71	66	59	67

Use a 0.05 level of significance to test the hypothesis that

(a) the secretaries type with equal speed;

(b) different typewriters have no effect on typing speed.

8. The following data represent the results of 4 quizzes obtained by 5 students in mathematics, English, French, and biology:

Student	Mathematics		English		French		Biology	
1	88	63	51	58	73	81	87	81
	79	80	72	65	77	77	92	76
2	79	96	85	95	82	36	80	93
	56	68	67	88	80	68	62	67
3	67	66	74	47	91	95	77	70
	51	89	59	82	59	92	84	73
4	35	60	76	49	43	52	55	49
	64	70	26	76	42	32	53	56
5	99	77	84	94	95	81	83	76
	87	95	83	76	98	96	87	80

Use a 0.05 level of significance to test the hypotheses that

(a) the courses are of equal difficulty;

(b) the students have equal ability;

(c) the students and subjects do not interact.

9. Three varieties of potatoes are being compared for yield. The experiment was conducted using 9 uniform plots at each of 4 different locations. Each variety of potato was planted at each location on 3 plots selected at random. The yields, in 100 kilograms per plot, were as follows:

Location	Variety of Potatoes		
	A	B	C
1	15	20	22
	19	24	17
	12	18	14
2	17	24	26
	10	18	19
	13	22	21
3	9	12	10
	12	15	5
	6	10	8
4	14	21	19
	8	16	15
	11	14	12

Use a 0.05 level of significance to test the hypotheses that
(a) there is no difference in the yielding capabilities of the 3 varieties of potatotes;
(b) different locations have no effect on the yields;
(c) the locations and varieties of potatoes do not interact.

10. In an experiment conducted to determine which of 3 missile systems is preferable, the propellant burning rate for 24 static firings was measured. Four propellant types were used. The experiment yielded duplicate observations of burning rates at each combination of the treatments. The data, after coding, were recorded as follows:

Missile System	Propellant Type			
	b_1	b_2	b_3	b_4
a_1	34.0	30.1	29.8	29.0
	32.7	32.8	26.7	28.9
a_2	32.0	30.2	28.7	27.6
	33.2	29.8	28.1	27.8
a_3	28.4	27.3	29.7	28.8
	29.3	28.9	27.3	29.1

Use a 0.05 level of significance to test the following hypotheses:
(a) H_0': there is no difference in the mean propellant burning rates when different missile systems are used.
(b) H_0'': there is no difference in the mean propellant burning rates of the 4 propellant types.
(c) H_0''': there is no interaction between the different missile systems and the different propellant types.

11. Three strains of rats were studied under 2 environmental conditions for their performance in a maze test. The error scores for the 48 rats were recorded as follows:

Environment	Strain					
	Bright		Mixed		Dull	
Free	28	12	33	83	101	94
	22	23	36	14	33	56
	25	10	41	76	122	83
	36	86	22	58	35	23
Restricted	72	32	60	89	136	120
	48	93	35	126	38	153
	25	31	83	110	64	128
	91	19	99	118	87	140

Use a 0.01 level of significance to test the hypotheses that
(a) there is no difference in error scores for different environments;
(b) there is no difference in error scores for different strains;
(c) the environments and strains of rats do not interact.

12.7

Design of Experiments

The analysis-of-variance technique has been described as a process for splitting the total variation of a set of experimental data into meaningful components that measure different sources of variation. The precise steps in carrying out the analysis will depend on the experimental design used to generate the data. In most cases the scientist or statistician may solve his problem by planning his experiment to fit one of several cataloged designs.

The simplest of all the experimental designs is called the **completely randomized design.** This design may be defined as one in which the treatments are randomly arranged over the whole of the experimental material. If the treatments in question represent three different varieties of wheat planted at random on several plots of equal size, the design is said to be completely randomized and the data are recorded and analyzed using the methods of Section 12.2 for the one-way classification. When the treatments represent all possible combinations of two criteria, such as the 12 possible combinations using 3 varieties of wheat and 4 kinds of fertilizer, and the experimental material is large enough for one or more replications of the 12 treatments, then the design is analyzed by the methods described in Section 12.5 or in Section 12.6 for the two-way classification. The completely randomized design is very easy to lay out and the analysis is simple to perform. It should be used only when the number of treatments is small and the experimental material is homogeneous. If, in the above illustration, the plots are not uniform in their soil composition, it is possible to select a more efficient design.

Perhaps the next simplest experimental design is called the **randomized block design.** The experimental material is divided into groups or blocks such that the plots or units making up a particular block are homogeneous. Each block constitutes a **replication** of the treatments. The treatments are assigned at random to the units in each block. A typical layout for a randomized block design using 3 treatments in 4 blocks is as follows:

Block 1	Block 2	Block 3	Block 4
t_2	t_1	t_3	t_2
t_1	t_3	t_2	t_1
t_3	t_2	t_1	t_3

The t's denote the random assignment to blocks of each of the 3 treatments. If the treatments represent a single criterion of classification, the data can be recorded as in the following 4×3 array

	Treatment		
Block	t_1	t_2	t_3
1	x_{11}	x_{12}	x_{13}
2	x_{21}	x_{22}	x_{23}
3	x_{31}	x_{32}	x_{33}
4	x_{41}	x_{42}	x_{43}

where x_{11} represents the observed data obtained from block 1 using treatment t_1, x_{12} represents the data obtained from block 1 using treatment t_2, ..., and x_{43} represents the data obtained from block 4 using treatment t_3. The analysis-of-variance procedure is now exactly the same as for the two-way classification discussed in Section 12.5, where rows and columns represent blocks and treatments, respectively.

Analysis-of-variance procedures for multiway classifications are required to analyze a randomized block design where the treatments are all possible combinations of two or more criteria of classification. The chief disadvantage of the randomized block design is that it is not suitable for large numbers of treatments, owing to the difficulty in obtaining the homogeneous blocks. This disadvantage may be overcome by choosing a design from the catalog of **incomplete block designs.** These designs allow one to investigate differences among v treatments arranged in b blocks containing k experimental units, where $k < v$. When the treatments in a randomized block design are all possible combinations of two criteria of classification, it is sometimes possible to choose a design for which the number of experimental units can be substantially reduced. Such a design will be discussed in detail in Section 12.8.

We have considered here only a few of the simplest and often used experimental designs. It would take a separate book to give a complete and detailed account of the numerous designs that are now available. Lacking knowledge in this field of specialization, the experimenter would be wise to consult with a competent statistician prior to conducting his investigation. Only a well-planned experiment can guarantee reliable conclusions.

12.8
Latin Squares

The randomized block design is very effective in reducing the experimental error by removing one source of variation. Another design that is particularly useful in controlling two sources of variation, and at the same time reduces the required number of treatment combinations, is called the **Latin square.** Suppose that we are interested in the yields of four varieties of wheat using four different fertilizers over a period of 4 years. The total number of treatment combinations for a completely randomized design would be 64. By selecting the same number of categories for all three criteria of classification, we may select a Latin square design and perform the analysis of variance using the results of only 16 treatment combinations. A typical Latin square, selected at random from all possible 4 × 4 squares, might be the following:

	Column			
Row	1	2	3	4
1	A	B	C	D
2	D	A	B	C
3	C	D	A	B
4	B	C	D	A

The four letters A, B, C, and D represent the 4 varieties of wheat which are referred to as the treatments. The rows and columns, represented by the 4 fertilizers and the 4 years, respectively, are the two sources of variation that we wish to control. We now see that each treatment occurs exactly once in each row and each column. With such a balanced arrangement the analysis of variance enables one to separate the variation due to the different fertilizers and different years from the error sum of squares and thereby obtain a more accurate test for differences in the yielding capabilities of the 4 varieties of wheat. When there is interaction present between any of the sources of variation, the f values in the analysis of variance are no longer valid. In this case, the Latin square design would be inappropriate.

We shall now generalize and consider an $r \times r$ Latin square where x_{ijk} denotes an observation in the ith row and jth column corresponding to the kth letter. Note, that once i and j are specified for a particular Latin square, we automatically know the letter given by k. For example, when $i = 2$ and $j = 3$ in the 4×4 Latin square above, we have $k = B$. Hence k is a function of i and j. If α_i and β_j are the effects of the ith row and jth column, τ_k the effect of the kth treatment, μ the grand mean, and ϵ_{ijk} the random error, then we can write

$$x_{ijk} = \mu + \alpha_i + \beta_j + \tau_k + \epsilon_{ijk}$$

on which we impose the restrictions

$$\sum_i \alpha_i = \sum_j \beta_j = \sum_k \tau_k = 0.$$

As before, the x_{ijk} are assumed to be values of independent random variables having normal distributions with means

$$\mu_{ijk} = \mu + \alpha_i + \beta_j + \tau_k$$

and common variance σ^2.

Tests of hypotheses concerning row and column effects are carried out using the same computational steps that were given for the two-way classification with a single observation per cell. To test the hypothesis that the treatment effects are all equal to zero; that is, $\tau_A = \tau_B = \tau_C = \cdots = 0$, we must remove from the error sum of squares an additional sum of squares with $r - 1$ degrees of freedom, designated SSTr, which measures the variability associated with the different treatments. Thus the sum-of-squares identity is written as

$$SST = SSR + SSC + SSTr + SSE$$

and the degrees of freedom are partitioned according to the identity

$$r^2 - 1 = (r - 1) + (r - 1) + (r - 1) + (r - 1)(r - 2).$$

Since the subscript k of the observation x_{ijk} is a function of i and j, it will be to our advantage in writing the computational formulas for the sums of squares to introduce the following notation:

$T_{i..}$ = sum of the observations in the ith row,

$T_{.j.}$ = sum of the observations in the jth column,

$T_{..k}$ = sum of the observations for treatment k,

$T_{...}$ = sum of all r^2 observations.

The sums of squares are now easily computed for the following formulas:

Sum-of-Squares Computational Formulas

$$SST = \sum_i \sum_j \sum_k x_{ijk}^2 - \frac{T_{...}^2}{r^2},$$

$$SSR = \frac{\sum_i T_{i..}^2}{r} - \frac{T_{...}^2}{r^2},$$

$$SSC = \frac{\sum_j T_{.j.}^2}{r} - \frac{T_{...}^2}{r^2},$$

$$SSTr = \frac{\sum_k T_{..k}^2}{r} - \frac{T_{...}^2}{r^2},$$

$$SSE = SST - SSR - SSC - SSTr.$$

All three tests of hypotheses in a Latin square design are carried out by computing the appropriate f values as indicated in Table 12.15.

TABLE 12.15
Analysis of Variance for an $r \times r$ Latin Square

Source of Variation	Sum of Squares	Degrees of Freedom	Mean Square	Computed f
Row means	SSR	$r - 1$	$s_1^2 = \dfrac{SSR}{r - 1}$	$f_1 = \dfrac{s_1^2}{s_4^2}$
Column means	SSC	$r - 1$	$s_2^2 = \dfrac{SSC}{r - 1}$	$f_2 = \dfrac{s_2^2}{s_4^2}$
Treatment means	SSTr	$r - 1$	$s_3^2 = \dfrac{SSTr}{r - 1}$	$f_3 = \dfrac{s_3^2}{s_4^2}$
Error	SSE	$(r - 1)(r - 2)$	$s_4^2 = \dfrac{SSE}{(r - 1)(r - 2)}$	
Total	SST	$r^2 - 1$		

Example 6. To illustrate the analysis of a Latin square design, let us return to the experiment in which the letters A, B, C, and D represent 4 varieties of wheat, the rows represent 4 different fertilizers, and the columns account for 4 different years. The data in Table 12.16 are the yields for the four varieties of wheat measured in kilograms per plot. It is assumed that the various sources of variation do not interact. Using a 0.05 level of signifi-

TABLE 12.16
Yields of Wheat in Kilograms per Plot

Fertilizer Treatment	Year			
	1978	1979	1980	1981
t_1	A 70	B 75	C 68	D 81
t_2	D 66	A 59	B 55	C 63
t_3	C 59	D 66	A 39	B 42
t_4	B 41	C 57	D 39	A 55

cance, test the hypotheses that (a) H_0': there is no difference in the average yields of wheat when different kinds of fertilizer are used; (b) H_0'': there is no difference in the average yield of wheat due to different years; and (c) H_0''': there is no difference in the average yield of the four varieties of wheat.

Solution

1. (a) H_0': $\alpha_1 = \alpha_2 = \alpha_3 = \alpha_4 = 0$,
   (b) H_0'': $\beta_1 = \beta_2 = \beta_3 = \beta_4 = 0$,
   (c) H_0''': $\tau_A = \tau_B = \tau_C = \tau_D = 0$.
2. (a) H_1': at least one of the α_i is not equal to zero,
   (b) H_1'': at least one of the β_j is not equal to zero,
   (c) H_1''': at least one of the τ_k is not equal to zero.
3. $\alpha = 0.05$.
4. Critical regions: (a) $f_1 > 4.76$, (b) $f_2 > 4.76$, (c) $f_3 > 4.76$.
5. Computations: From Table 12.16, we find the row, column, and treatment totals to be

$$T_{1..} = 294, \quad T_{2..} = 243, \quad T_{3..} = 206, \quad T_{4..} = 192,$$

$$T_{.1.} = 236, \quad T_{.2.} = 257, \quad T_{.3.} = 201, \quad T_{.4.} = 241,$$

$$T_{..A} = 223, \quad T_{..B} = 213, \quad T_{..C} = 247, \quad T_{..D} = 252.$$

Hence

$$\text{SST} = 70^2 + 75^2 + \cdots + 55^2 - \frac{935^2}{16} = 2500,$$

$$\text{SSR} = \frac{294^2 + 243^2 + 206^2 + 192^2}{4} - \frac{935^2}{16} = 1557,$$

$$\text{SSC} = \frac{236^2 + 257^2 + 201^2 + 241^2}{4} - \frac{935^2}{16} = 418,$$

$$\text{SSTr} = \frac{223^2 + 213^2 + 247^2 + 252^2}{4} - \frac{935^2}{16} = 264,$$

$$\text{SSE} = 2500 - 1557 - 418 - 264 = 261.$$

These results, along with the remaining computations, are given in Table 12.17.

TABLE 12.17
Analysis of Variance for the Data of Table 12.16

Source of Variation	Sum of Squares	Degrees of Freedom	Mean Square	Computed f
Row means	1557	3	519.000	11.93
Column means	418	3	139.333	3.20
Treatment means	264	3	88.000	2.02
Error	261	6	43.500	
Total	2500	15		

6. Decisions:
   (a) Reject H_0' and conclude that a difference in the average yields of wheat exists when different kinds of fertilizer are used.
   (b) Accept H_0'' and conclude that there is no difference in the average yields of wheat due to different years.
   (c) Accept H_0''' and conclude that there is no difference in the average yields of the four varieties of wheat.

EXERCISES

1. Three varieties of potatoes are being compared for yield. The experiment was conducted by assigning each variety at random to 3 equal-size plots at each of 4 different locations. The following yields for varieties A, B, and C, in 100 kilograms per plot, were recorded:

Location 1	Location 2	Location 3	Location 4
B 13	C 21	C 9	A 11
A 18	A 20	B 12	C 10
C 12	B 23	A 14	B 17

Perform a two-way analysis of variance to test the hypothesis that there is no difference in the yielding capabilities of the 3 varieties of potatoes. Use a 0.05 level of significance. Was it necessary to plant each variety of potato at each location to reach a valid conclusion concerning the 3 varieties?

2. Four different machines, M_1, M_2, M_3, and M_4 are to be considered in the assembling of a particular product. It is decided that 6 different operators are to be used in a randomized block experiment to compare the machines. The machines are assigned in a random order to each operator. The operation of the machines requires a certain amount of physical dexterity and it is known that there is a difference among the operators in the speed with which they operate the machines. The following times, in seconds, were recorded for assembling the given product:

Operator 1	M_2	M_4	M_3	M_1
	39.8	41.3	40.2	42.5

Operator 2	M_3	M_1	M_2	M_4
	40.5	39.3	40.1	42.2

Operator 3	M_2	M_1	M_4	M_3
	40.5	39.6	43.5	41.3

Operator 4	M_4	M_2	M_1	M_3
	44.2	42.3	39.9	43.4

Operator 5	M_1	M_3	M_2	M_4
	42.9	44.9	42.5	45.9

Operator 6	M_2	M_4	M_3	M_1
	43.1	42.3	45.1	43.6

(a) Test the hypothesis H_0', at the 0.05 level of significance, that the machines perform at the same mean rate of speed.

(b) Test the hypothesis H_0'', at the 0.05 level of significance, that the operators perform at the same mean rate of speed.

3. The following data are the percents of foreign additives measured by 5 analysts for 3 similar brands of strawberry jam, A, B, and C:

Analyst 1		Analyst 2		Analyst 3		Analyst 4		Analyst 5	
B	2.7	C	7.5	B	2.8	A	1.7	C	8.1
C	3.6	A	1.6	A	2.7	B	0.9	A	2.0
A	3.8	B	5.2	C	6.4	C	0.6	B	4.8

Perform the analysis of variance and test the hypothesis, at the 0.05 level of significance, that

(a) there is no difference in the percents of foreign additives due to different analysts;

(b) the percent of foreign additives is the same for all three brands of jam.

4. The mathematics department of a large university wishes to evaluate the teaching capabilities of 4 professors. In order to eliminate any effects due to different mathematics courses and different times of the day, it was decided to conduct an experiment using a Latin square design in which the letters A, B, C, and D represent the 4 different professors. Each professor taught one section of each of 4 different courses scheduled at each of 4 different times during the day. The data in the following table show the grades assigned by these professors to 16 students of approximately equal ability:

Time Period	Course			
	Algebra	Geometry	Statistics	Calculus
1	A 84	B 79	C 63	D 97
2	B 91	C 82	D 80	A 93
3	C 59	D 70	A 77	B 80
4	D 75	A 91	B 75	C 68

Use a 0.05 level of significance to test the hypothesis that

(a) there is no difference in the grades due to different time periods;

(b) the courses are of equal difficulty;

(c) different professors have no effect on the grades.

5. A manufacturing firm wants to investigate the effects of 5 color additives on the setting time of a new concrete mix. Variations in the setting times can be expected from day to day changes in temperature and humidity and also from the different workers who prepare the test molds. To eliminate these extraneous sources of variation a 5 × 5 Latin square design was used in which the letters A, B, C, D, and E represent the 5 additives. The setting times, in hours, for the 25 molds are shown in the following table:

Worker	Day				
	1	2	3	4	5
1	D 10.7	E 10.3	B 11.2	A 10.9	C 10.5
2	E 11.3	C 10.5	D 12.0	B 11.5	A 10.3
3	A 11.8	B 10.9	C 10.5	D 11.3	E 7.5
4	B 14.1	A 11.6	E 11.0	C 11.7	D 11.5
5	C 14.5	D 11.5	A 11.5	E 12.7	B 10.9

At the 0.05 level of significance, can we say that the color additives have any effect on the setting time of the concrete mix?

6. Construct a 5 × 5 **magic square** containing the numbers 1 through 25 by completing the following array:

		1	8	
	5	7		
4	6			
10				3
11			2	9

(a) Show that the numbers in every row, column, and diagonal sum to 65.
(b) If each number in the magic square is divided by 5 and replaced by its remainder, show that we obtain a Latin square in which the digits 0, 1, 2, 3, and 4 appear exactly once in every row and every column.

13 Nonparametric Statistics

Most of the hypothesis-testing procedures discussed so far in this book are based on the assumption that the random samples are selected from normal populations. Fortunately, most of these tests are still reasonably reliable for slight departures from normality, particularly when the sample size is large. Traditionally, these testing procedures have been referred to as parametric methods. *In this chapter we consider a number of alternative test procedures, called* nonparametric *or* distribution-free methods, *that assume no knowledge whatsoever about the distributions of the underlying populations, except perhaps that they are continuous.*

13.1
Nonparametric Tests

Nonparametric tests have gained a certain appeal in recent years for several reasons. First, the computations involved are usually very quick and easy to carry out. Second, the data need not be quantitative measurements but could be in the form of qualitative responses such as "defective" versus "nondefective," "yes" versus "no," and so forth, or frequently are values of an ordinal scale to which we assign ranks. On an ordinal scale the subjects are ranked according to a specified order, and a nonparametric test analyzes the various ranks. For example, two judges might rank five brands of premium beer by assigning a rank of 1 to the brand believed to have the best overall quality, a rank of 2 to the second best, and so forth. A nonparametric test could then be used to determine whether there is any agreement between the two judges. A third and perhaps the most important advantage in using nonparametric tests is that they are encumbered with less restrictive assumptions than their parametric counterparts.

We should also point out that there are a number of disadvantages associated with nonparametric tests. Primarily, they do not utilize all the information provided by the sample. As a result of this wastefulness, a nonparametric test will be slightly less efficient than the corresponding parametric procedure when both methods are applicable. Consequently, a nonparametric test will require a larger sample size than will the corresponding parametric test in order to achieve the same probability of committing a type II error.

In summary, if a parametric and a nonparametric test are both applicable to the same set of data, one should probably avoid the "quick and easy" nonparametric test and carry out the more efficient parametric technique. However, recognizing the fact that the assumptions of normality often cannot be justified, and also the fact that we do not always have quantitative measurements, it is fortunate that statisticians have provided us with a number of useful nonparametric procedures.

13.2
Sign Test

The procedures discussed in Section 10.4 for testing the null hypothesis that $\mu = \mu_0$ are valid only if the population is approximately normal or if the sample is large. However, if $n < 30$ and the population is decidedly nonnormal, we must resort to a nonparametric test. Perhaps the easiest and quickest to perform is a test called the **sign test.** In testing the null hypothesis H_0 that $\mu = \mu_0$ against an appropriate alternative on the basis of a random sample of

434

size n, we replace each sample value exceeding μ_0 with a *plus* sign and each sample value less than μ_0 with a *minus* sign. If the null hypothesis is true and the population is symmetric, the sum of the plus signs should be approximately equal to the sum of the minus signs. When one sign appears more frequently than it should, based on chance alone, we reject the hypothesis that the population mean μ is equal to μ_0.

The sign test is applicable only in situations where μ_0 cannot equal any of the observations. Although it is theoretically impossible to obtain a sample observation exactly equal to μ_0 when the population is continuous, nevertheless in practice a sample value equal to μ_0 often will occur from a lack of precision in recording the data. When sample values equal to μ_0 are observed, they must be excluded from the analysis and the sample size is correspondingly reduced.

An appropriate test statistic for the sign test is the random variable X representing the number of plus signs in our random sample. If the null hypothesis that $\mu = \mu_0$ is true, the probability that a sample value results in either a plus or a minus sign is equal to $\frac{1}{2}$. Consequently, the test statistic X has a binomial probability distribution with the parameter $p = \frac{1}{2}$ when H_0 is true from which levels of significance for both one-sided and two-sided alternatives can be computed. For example, in testing

$$H_0: \mu = \mu_0,$$
$$H_1: \mu < \mu_0,$$

we shall reject H_0 in favor of H_1 only if the proportion of plus signs is sufficiently less than $\frac{1}{2}$, that is, when the value of x of our random variable is small. Hence the largest critical region of size not exceeding α is established by forming the inequality

$$x \leq k'_\alpha$$

where k'_α, as defined on page 325, is the largest interger for which

$$P(X \leq k'_\alpha \text{ when } \mu = \mu_0) = \sum_{x=0}^{k'_\alpha} b(x; n, \tfrac{1}{2}) \leq \alpha.$$

Because X is a discrete random variable, the size of the critical region is likely to be less than α, but as close as possible to α without exceeding it. This is usually a problem only for small sample sizes. For example, when $n = 15$ and $\alpha = 0.05$, we find from Table A.2 that

$$P(X \le 3 \text{ when } \mu = \mu_0) = \sum_{x=0}^{3} b(x; 15, \tfrac{1}{2}) = 0.0176$$

and

$$P(X \le 4 \text{ when } \mu = \mu_0) = \sum_{x=0}^{4} b(x; 15, \tfrac{1}{2}) = 0.0592,$$

so that the critical region of size not exceeding $\alpha = 0.05$ is given by $x \le 3$.

To test the hypothesis

$$H_0: \mu = \mu_0,$$
$$H_1: \mu > \mu_0,$$

we reject H_0 in favor of H_1 only if the proportion of plus signs is sufficiently greater than $\tfrac{1}{2}$, that is, when x is large. Hence the largest critical region of size not exceeding α is established by forming the inequality

$$x \ge k_\alpha,$$

where k_α is computed as on page 326 by the binomial probability distribution. Finally, to test the hypothesis

$$H_0: \mu = \mu_0,$$
$$H_1: \mu \ne \mu_0,$$

we reject H_0 in favor of H_1 when the proportion of plus signs is significantly less than or greater than $\tfrac{1}{2}$. This, of course, is equivalent to x being sufficiently small or sufficiently large. Therefore, the largest critical region of size not exceeding α is given by

$$x \le k'_{\alpha/2} \quad \text{and} \quad x \ge k_{\alpha/2}.$$

Since the values of k' and k are found from a table of binomial probabilities with $p = \tfrac{1}{2}$ when the sample size is small, one could then use the normal-curve approximation to the binomial distribution whenever $n > 10$. Suppose, for example, that we wish to test the hypothesis

$$H_0: \mu = \mu_0,$$
$$H_1: \mu < \mu_0,$$

at the $\alpha = 0.05$ level of significance for a random sample of size $n = 20$ that yields $x = 6$ plus signs. Using the normal-curve approximation, the critical region is given by $z < -1.645$. It follows that

$$\mu = np = (20)(0.5) = 10,$$
$$\sigma = \sqrt{npq} = \sqrt{(20)(0.5)(0.5)} = 2.236,$$

and hence

$$z = \frac{6 - 10}{2.236} = -1.79,$$

which leads to the rejection of the null hypothesis.

Example 1. The following data represent the number of hours that a rechargeable hedge trimmer operates before a recharge is required: 1.5, 2.2, 0.9, 1.3, 2.0, 1.6, 1.8, 1.5, 2.0, 1.2, and 1.7. Use the sign test to test the hypothesis at the 0.05 level of significance that this particular trimmer operates, on the average, 1.8 hours before requiring a recharge.

Solution. Using the six-step procedure, we have

1. H_0: $\mu = 1.8$.
2. H_1: $\mu \neq 1.8$.
3. $\alpha = 0.05$.
4. Critical region: $x \leq k'_{0.025}$ and $x \geq k_{0.025}$, where x is the number of plus signs.
5. Computations: Replacing each value by the symbol " $+$ " if it exceeds 1.8, by the symbol " $-$ " if it is less than 1.8, and discarding the one measurement that equals 1.8, we obtain the sequence

$$- \ + \ - \ - \ + \ - \ - \ + \ - \ -$$

for which $n = 10$ and $x = 3$.
6. Decision: From Table A.2 we find that $k'_{0.025} = 1$ and $k_{0.025} = 9$. Since $x = 3$ falls in the acceptance region, we accept the null hypothesis and conclude that the average operating time is not significantly different from 1.8 hours.

The sign test for testing $\mu = \mu_0$ on the basis of a random sample from a single population can also be used when n pairs of observations are selected from two nonnormal populations defined over a continuous sample space. In testing the null hypothesis H_0 that $\mu_1 = \mu_2$, or $\mu_D = 0$, we simply replace each difference d_i of the paired observations with a plus or minus sign depending on whether d_i is positive or negative and then proceed as before. One can also use the sign test to test the null hypothesis $\mu_1 - \mu_2 = d_0$ for paired observations. Hence we adjust each d_i by subtracting d_0 and then once again replace each adjusted difference with a plus or minus sign depending on whether the adjusted difference is positive or negative. Throughout this section we have assumed that the populations are symmetric. However, even if populations are skewed, we can carry out the same test procedure, but the hypotheses refer to the population medians rather than the means.

Example 2. A taxi company is trying to decide whether the use of radial tires instead of regular belted tires improves fuel economy. Twelve cars were equipped with radial tires and driven over a prescribed test course. Without changing drivers, the same cars were then equipped with the regular belted tires and driven once again over the test course. The gasoline consumption, in kilometers per liter, was recorded as follows:

Car	Radial Tires	Belted Tires
1	4.2	4.1
2	4.7	4.9
3	6.6	6.2
4	7.0	6.9
5	6.7	6.8
6	4.5	4.4
7	5.7	5.7
8	6.0	5.8
9	7.4	6.9
10	4.9	4.9
11	6.1	6.0
12	5.2	4.9

Can we conclude at the 0.05 level of significance that cars equipped with radial tires give better fuel economy than those equipped with regular belted tires? Use the normal approximation to the binomial distribution.

Solution. Let μ_1 and μ_2 represent the mean kilometers per liter for cars equipped with radial and belted tires, respectively.

1. H_0: $\mu_1 - \mu_2 = 0$.
2. H_1: $\mu_1 - \mu_2 > 0$.
3. $\alpha = 0.05$.
4. Critical region: $z > 1.645$.
5. Computations: Examination of the data indicates 8 plus signs, 2 minus signs, and 2 zeros. Therefore, after throwing out the zeros, $n = 10$ and $x = 8$. Hence

$$\mu = np = (10)(0.5) = 5,$$
$$\sigma = \sqrt{npq} = \sqrt{(10)(0.5)(0.5)} = 1.581,$$

from which we obtain

$$z = \frac{8 - 5}{1.581} = 1.90.$$

6. Decision: Reject H_0 and conclude that, on the average, radial tires do improve fuel economy.

Not only is the sign test one of our simplest nonparametric procedures to apply, it has the additional advantage of being applicable to dichotomous data that cannot be recorded on a numerical scale but can be represented by positive and negative responses. For example, the sign test is applicable in experiments where a qualitative response such as "hit" or "miss" is recorded, and in sensory-type experiments where a plus or minus sign is recorded, depending on whether the taste tester correctly or incorrectly identifies the desired ingredient.

13.3
Wilcoxon Signed-Rank Test

The sign test utilizes only the plus and minus signs of the differences between the observations and μ_0 in the one-sample case, or the plus and minus signs of the differences between the pairs of observations in the paired-sample case, but it does not take into consideration the magnitudes of these differences. A test utilizing both direction and magnitude was proposed in 1945 by Frank

Wilcoxon and is now commonly referred to as the **Wilcoxon signed-rank test,** or in the case of paired observations, it is also often called the **Wilcoxon test for paired observations.**

To test the hypothesis that $\mu = \mu_0$ for a continuous symmetrical population or $\mu_1 = \mu_2$ for two continuous symmetrical populations by the Wilcoxon signed-rank test, first discard all differences equal to zero and then rank the remaining d_i's without regard to sign. A rank of 1 is assigned to the smallest d_i in absolute value, a rank of 2 to the next smallest, and so on. When the absolute value of two or more differences is the same, assign to each the average of the ranks that would have been assigned if the differences were distinguishable. For example, if the 5th and 6th smallest differences are equal in absolute value, each would be assigned a rank of 5.5. If the hypothesis $\mu = \mu_0$ or $\mu_1 = \mu_2$ is true, the total of the ranks corresponding to the positive differences should be almost equal to the total of the ranks corresponding to the negative differences. Let us represent these totals by w_+ and w_-, respectively. We shall designate the smaller of the w_+ and w_- by w.

In selecting repeated samples, we would expect w_+ and w_-, and therefore w, to vary. Thus we may think of w_+, w_-, and w as values of the corresponding random variables W_+, W_-, and W. The null hypothesis $\mu = \mu_0$ (or $\mu_1 = \mu_2$) can be rejected in favor of the alternative $\mu < \mu_0$ (or $\mu_1 < \mu_2$) only if w_+ is small and w_- is large. Likewise, the alternative $\mu > \mu_0$ (or $\mu_1 > \mu_2$) can be accepted only if w_+ is large and w_- is small. For a two-sided alternative we may reject H_0 in favor of H_1 if either w_+ or w_- and hence if w is sufficiently small. Therefore, no matter what the alternative hypothesis may be, we reject the null hypothesis when the value of the appropriate statistic W_+, W_-, or W is sufficiently small. The various test procedures are summarized in Table 13.1.

It is not difficult to show that whenever $n < 5$ and the level of significance does not exceed 0.05 for a one-tailed test or 0.10 for a two-tailed test, the

TABLE 13.1

Wilcoxon Signed-Rank Test

To test H_0	Versus H_1	Compute
$\mu = \mu_0$	$\mu < \mu_0$	w_+
	$\mu > \mu_0$	w_-
	$\mu \neq \mu_0$	w
$\mu_1 = \mu_2$	$\mu_1 < \mu_2$	w_+
	$\mu_1 > \mu_2$	w_-
	$\mu_1 \neq \mu_2$	w

value of w will always lead to the acceptance of the null hypothesis. However, when $5 \leq n \leq 30$, Table A.8 gives approximate critical values of W_+ and W_- for levels of significance equal to 0.01, 0.025, and 0.05 for a one-tailed test, and critical values of W for levels of significance equal to 0.02, 0.05, and 0.10 for a two-tailed test. For example, when $n = 12$ Table A.8 shows that a value of $w_+ \leq 17$ is required for the one-sided alternative $\mu < \mu_0$ to be significant at the 0.05 level.

Example 3. Rework Example 1 on page 437 by using the Wilcoxon signed-rank test.

Solution. Following the six-step procedure, we have

1. H_0: $\mu = 1.8$.
2. H_1: $\mu \neq 1.8$.
3. $\alpha = 0.05$.
4. Critical region: Since $n = 10$, after discarding the one measurement that equals 1.8, Table A.8 shows the critical region to be $w \leq 8$.
5. Computations: Subtracting 1.8 from each measurement and then ranking the differences without regard to sign, we have

d_i	−0.3	0.4	−0.9	−0.5	0.2	−0.2	−0.3	0.2	−0.6	−0.1
Ranks	5.5	7	10	8	3	3	5.5	3	9	1

Now $w_+ = 13$ and $w_- = 42$, so that $w = 13$, the smaller of w_+ and w_-.

6. Decision: Accept H_0 as before and conclude that the average operating time is not significantly different from 1.8 hours.

When $n > 15$, the sampling distribution of W_+ (or W_-) approaches the normal distribution with mean

$$\mu_{W_+} = \frac{n(n + 1)}{4}$$

and variance

$$\sigma^2_{W_+} = \frac{n(n + 1)(2n + 1)}{24}.$$

In this case the statistic

$$Z = \frac{(W_+ - \mu_{W_+})}{\sigma_{W_+}}$$

can be used to determine the critical region for our test.

The Wilcoxon signed-rank test may also be used to test the null hypothesis that $\mu_1 - \mu_2 = \mu_D = d_0$. In this case the populations need not be symmetric. We simply apply the same procedure as above after each d_i is adjusted by subtracting d_0.

Example 4. It is claimed that a college senior can increase his score in the major field area of the graduate record examination by at least 50 points if he is provided sample problems in advance. To test this claim, 20 college seniors were divided into 10 pairs such that each matched pair had almost the same overall quality grade-point average for their first 3 years in college. Sample problems and answers were provided at random to 1 member of each pair 1 week prior to the examination. The following examination scores were recorded:

	Pair									
	1	2	3	4	5	6	7	8	9	10
With sample problems	531	621	663	579	451	660	591	719	543	575
Without sample problems	509	540	688	502	424	683	568	748	530	524

Test the null hypothesis at the 0.05 level of significance that sample problems increase the scores by 50 points against the alternative hypothesis that the increase is less than 50 points.

Solution Let μ_1 and μ_2 represent the mean score of all students taking the test in question with and without sample problems, respectively. We follow the six-step procedure:

1. H_0: $\mu_1 - \mu_2 = 50$.
2. H_1: $\mu_1 - \mu_2 | < 50$.
3. $\alpha = 0.05$.
4. Critical region: Since $n = 10$, Table A.8 shows the critical region to be $w_+ \leq 11$.

5. Computations:

	Pair									
	1	2	3	4	5	6	7	8	9	10
d_i	22	81	−25	77	27	−23	23	−29	13	51
$d_i - d_0$	−28	31	−75	27	−23	−73	−27	−79	−37	1
Ranks	5	6	9	3.5	2	8	3.5	10	7	1

Now we find $w_+ = 6 + 3.5 + 1 = 10.5$.

6. Decision: Reject H_0 and conclude that the sample problems do not, on the average, increase one's graduate record score by as much as 50 points.

EXERCISES

1. On 12 visits to a doctor's office a patient had to wait 17, 32, 25, 15, 28, 25, 20, 12, 35, 20, 26, and 24 minutes before being seen by the doctor. Use the sign test with $\alpha = 0.05$ to test the doctor's claim that, on the average, her patients do not wait more than 20 minutes before being admitted to the examination room.

2. The following data represent the number of hours of flight training received by 18 student pilots from a certain instructor prior to their first solo flight: 9, 12, 13, 12, 10, 11, 18, 16, 13, 14, 11, 15, 12, 9, 13, 14, 11, and 14. Use the sign test with $\alpha = 0.02$ to test the instructor's claim that, on the average, his students solo after 12 hours of flight training.

3. A food inspector examined 15 jars of a certain brand of jam to determine the percent of foreign impurities. The following data were recorded: 2.4, 2.3, 1.7, 1.7, 2.3, 1.2, 1.1, 3.6, 3.1, 1.0, 4.2, 1.6, 2.5, 2.4, and 2.3. Using the normal approximation to the binomial distribution, perform a sign test at the 0.01 level of significance to test the null hypothesis that the average percent of impurities in this brand of jam is 2.5% against the alternative that the average percent of impurities is not 2.5%.

4. An international electronics firm is considering an expense-free vacation trip for its senior executives and their families. In order to determine a preference between a week in Hawaii or a week in Spain, a random sample of 18 executives were asked their preference. Using the normal approximation to the binomial distribution, perform a sign test at the 0.05 level of significance to test the null hypothesis that the two locations are equally preferred against the alternative that the preferences differ if 4 of the 18 stated that they preferred Spain.

5. A paint supplier claims that a new additive will reduce the drying time of his acrylic paint. To test his claim, 12 panels of wood are painted, one half of each panel with paint containing the regular additive and the other half with paint containing the new additive. The drying times, in hours, were recorded as follows:

	Drying Time (hours)	
Panel	New Additive	Regular Additive
1	6.4	6.6
2	5.8	5.8
3	7.4	7.8
4	5.5	5.7
5	6.3	6.0
6	7.8	8.4
7	8.6	8.8
8	8.2	8.4
9	7.0	7.3
10	4.9	5.8
11	5.9	5.8
12	6.5	6.5

Use the sign test at the 0.05 level to test the hypothesis that the new additive is no better than the regular additive in reducing the drying time of this kind of paint.

6. It is claimed that a new diet will reduce a person's weight by 4.5 kilograms on the average in a period of 2 weeks. The weights of 10 women who followed this diet were recorded before and after a 2-week period, yielding the following data:

Woman	Weight Before	Weight After
1	58.5	60.0
2	60.3	54.9
3	61.7	58.1
4	69.0	62.1
5	64.0	58.5
6	62.6	59.9
7	56.7	54.4
8	63.6	60.2
9	68.2	62.3
10	59.4	58.7

Use the sign test at the 0.05 level of significance to test the hypothesis that the diet reduces a person's weight by 4.5 kilograms on the average, against the alternative hypothesis that the mean difference in weight is less than 4.5 kilograms.

7. Two types of instruments for measuring the amount of sulfur monoxide in the atmosphere are being compared in an air-pollution experiment. The following readings were recorded daily for a period of 2 weeks:

Day	Sulfur Monoxide	
	Instrument A	Instrument B
1	0.96	0.87
2	0.82	0.74
3	0.75	0.63
4	0.61	0.55
5	0.89	0.76
6	0.64	0.70
7	0.81	0.69
8	0.68	0.57
9	0.65	0.53
10	0.84	0.88
11	0.59	0.51
12	0.94	0.79
13	0.91	0.84
14	0.77	0.63

Using the normal approximation to the binomial distribution, perform a sign test to determine whether the different instruments lead to different results. Use a 0.01 level of significance.

8. Analyze the data of Exercise 1 using the Wilcoxon signed-rank test.

9. Analyze the data of Exercise 2 using the Wilcoxon signed-rank test.

10. The weights of five people before they stopped smoking and 5 weeks after they stopped smoking, in kilograms, are as follows:

	Individual				
	1	2	3	4	5
Before	66	80	69	52	75
After	71	82	68	56	73

Use the Wilcoxon signed-rank test to test the hypothesis, at the 0.05 level of significance, that giving up smoking has no effect on a person's weight against the alternative that one's weight increases if he quits smoking.

11. Analyze the data of Exercise 5 using the Wilcoxon signed-rank test.

12. Rework Exercise 6 using the Wilcoxon signed-rank test.

13.4

Wilcoxon Rank-Sum Test

In this section we consider a very simple nonparametric procedure proposed by Wilcoxon for the comparison of the means of two continuous nonnormal populations when independent samples are selected from the populations. This nonparametric alternative to the two-sample t test discussed in Section 10.4 is called the **Wilcoxon rank-sum test** or the **Wilcoxon two-sample test.**

We shall test the null hypothesis H_0 that $\mu_1 = \mu_2$ against some suitable alternative. First we select a random sample from each of the populations. Let n_1 be the number of observations in the smaller sample and n_2 the number of observations in the larger sample. When the samples are of equal size, n_1 and n_2 may be randomly assigned. Arrange the $n_1 + n_2$ observations of the combined samples in ascending order and substitute a rank of 1, 2, ..., $n_1 + n_2$ for each observation. In the case of ties (identical observations) we replace the observations by the mean of the ranks that the observations would have if they were distinguishable. For example, if the seventh and eighth observations are identical, we would assign a rank of 7.5 to each of the two observations.

The sum of the ranks corresponding to the n_1 observations in the smaller sample is denoted by w_1. Similarly, the value w_2 represents the sum of the n_2 ranks corresponding to the larger sample. The total $w_1 + w_2$ depends only on the number of observations in the two samples and is in no way affected by the results of the experiment. Hence if $n_1 = 3$ and $n_2 = 4$, then $w_1 + w_2 = 1 + 2 + \cdots + 7 = 28$, regardless of the numerical values of the observations. In general,

$$w_1 + w_2 = \frac{(n_1 + n_2)(n_1 + n_2 + 1)}{2},$$

the arithmetic sum of the integers 1, 2, ..., $n_1 + n_2$. Once we have determined w_1, it may be easier to find w_2 by the formula

$$w_2 = \frac{(n_1 + n_2)(n_1 + n_2 + 1)}{2} - w_1.$$

In choosing repeated samples of size n_1 and n_2, we would expect w_1, and therefore w_2, to vary. Thus we may think of w_1 and w_2 as values of the random variables W_1 and W_2, respectively. The null hypothesis $\mu_1 = \mu_2$ will be rejected in favor of the alternative $\mu_1 < \mu_2$ only if w_1 is small and w_2 is large. Similarly, the alternative $\mu_1 > \mu_2$ can be accepted only if w_1 is large and w_2 is small. For a two-tailed test we may reject H_0 in favor of H_1 if w_1 is small and w_2 is large or if w_1 is large and w_2 is small. In other words, the alternative $\mu_1 < \mu_2$ is accepted if w_1 is sufficiently small, the alternative $\mu_1 < \mu_2$ is accepted if w_2 is sufficiently small, and the alternative $\mu_1 \neq \mu_2$ is accepted if the minimum of w_1 and w_2 is sufficiently small. In actual practice we usually base our decision on the value

$$u_1 = w_1 - \frac{n_1(n_1 + 1)}{2}$$

or

$$u_2 = w_2 - \frac{n_2(n_2 + 1)}{2}$$

of the related statistic U_1 or U_2, or on the value of the statistic U, the minimum of U_1 and U_2. These statistics simplify the construction of tables of critical values since both U_1 and U_2 have symmetric sampling distributions and assume values in the interval from 0 to $n_1 n_2$ such that $u_1 + u_2 = n_1 n_2$.

From the formulas for u_1 and u_2 we see that u_1 will be small when w_1 is small and u_2 will be small when w_2 is small. Consequently, the null hypothesis will be rejected whenever the appropriate statistic U_1, U_2, or U assumes a value less than or equal to the desired critical value given in Table A.9. The various test procedures are summarized in Table 13.2.

TABLE 13.2
Wilcoxon Rank-Sum Test

To Test H_0	Versus H_1	Compute
	$\mu_1 < \mu_2$	u_1
$\mu_1 = \mu_2$	$\mu_1 > \mu_2$	u_2
	$\mu_1 \neq \mu_2$	u

Table A.9 gives critical values of U_1 and U_2 for levels of significance equal to 0.001, 0.01, 0.025, and 0.05 for a one-tailed test, and critical values of U for levels of significance equal to 0.002, 0.02, 0.05, and 0.10 for a two-tailed test. If the observed value of u_1, u_2, or u is less than or equal to the tabled critical value, the null hypothesis is rejected at the level of significance indicated by the table. Suppose, for example, that we wish to test the null hypothesis that $\mu_1 = \mu_2$ against the one-sided alternative that $\mu_1 < \mu_2$ at the 0.05 level of significance for random samples of size $n_1 = 3$ and $n_2 = 5$ that yield the value $w_1 = 8$. It follows that

$$ u_1 = 8 - \left[\frac{(3)(4)}{2} \right] = 2. $$

Our one-tailed test is based on the statistic U_1. Using Table A.9, we reject the null hypothesis of equal means when $u_1 \leq 1$. Since $u_1 = 2$ falls in the acceptance region, the null hypothesis cannot be rejected.

Example 5. The nicotine content of two brands of cigarettes, measured in milligrams, was found to be as follows:

Brand A	2.1	4.0	6.3	5.4	4.8	3.7	6.1	3.3	
Brand B	4.1	0.6	3.1	22.5	4.0	6.2	1.6	2.2	1.9 5.4

Test the hypothesis, at the 0.05 level of significance, that the average nicotine contents of the two brands are equal against the alternative that they are unequal.

**Solution.** We proceed by the six-step rule with $n_1 = 8$ and $n_2 = 10$.

1. H_0: $\mu_1 = \mu_2$.
2. H_1: $\mu_1 \neq \mu_2$.
3. $\alpha = 0.05$.
4. Critical region: $u \leq 17$ (from Table A.9).
5. Computations: The observations are arranged in ascending order and ranks from 1 to 18 assigned.

Original Data	Ranks
0.6	1
1.6	2
1.9	3
2.1	4
2.2	5
2.5	6
3.1	7
3.3	8
3.7	9
4.0	10.5
4.0	10.5
4.1	12
4.8	13
5.4	14.5
5.4	14.5
6.1	16
6.2	17
6.3	18

The ranks of the observations belonging to the smaller sample are underscored. Now,

$$w_1 = 4 + 8 + 9 + 10.5 + 13 + 14.5 + 16 + 18 = 93$$

and

$$w_2 = \left[\frac{(18)(19)}{2} \right] - 93 = 78.$$

Therefore,

$$u_1 = 93 - \left[\frac{(8)(9)}{2} \right] = 57,$$

$$u_2 = 78 - \left[\frac{(10)(11)}{2} \right] = 23,$$

so that $u = 23$.

6. Decision: Accept H_0 and conclude that there is no difference in the average nicotine contents of the two brands of cigarettes.

When n_1 and n_2 increase in size, the sampling distribution of U_1 (or U_2) approaches the normal distribution with mean

$$\mu_{U_1} = \frac{n_1 n_2}{2}$$

and variance

$$\sigma^2_{U_1} = \frac{n_1 n_2 (n_1 + n_2 + 1)}{12}.$$

Consequently, when n_2 is greater than 20 and n_1 is at least 10, one could use the statistic

$$Z = \frac{U_1 - \mu_{U_1}}{\sigma_{U_1}}$$

for our test, with the critical region falling in either or both tails of the standard normal distribution, depending on the form of H_1.

The use of the Wilcoxon rank-sum test is not restricted to nonnormal populations. It can be used in place of the two-sample t test when the populations are normal, although the probability of committing a type II error will be larger. The Wilcoxon rank-sum test is always superior to the t test for decidedly nonnormal populations.

13.5

Kruskal–Wallis Test

The Kruskal–Wallis test, also called the Kruskal–Wallis H test, is a generalization of the Wilcoxon two-sample test to the case of $k > 2$ samples. It is used to test the null hypothesis H_0 that k independent samples are from identical populations. Introduced in 1952 by W. H. Kruskal and W. A. Wallis, the test is an alternative nonparametric procedure to the F test for testing the equality of means in the one-factor analysis of variance when the experimenter wishes to avoid the assumption that the samples were selected from normal populations.

Let n_i ($i = 1, 2, \ldots, k$) be the number of observations in the ith sample. First we combine all k samples and arrange the $n = n_1 + n_2 + \cdots + n_k$

observations in ascending order, substituting the appropriate rank from 1, 2, ..., n for each observation. In the case of ties (identical observations), we follow the usual procedure of replacing the observations by the means of the ranks that the observations would have if they were distinguishable. The sum of the ranks corresponding to the n_i observations in the ith sample is denoted by the random variable R_i. Now let us consider the statistic

$$H = \frac{12}{n(n+1)} \sum_{i=1}^{k} \frac{R_i^2}{n_i} - 3(n+1),$$

which is approximated very well by a chi-square distribution with $k - 1$ degrees of freedom when H_0 is true and if each sample consists of at least five observations. Note that the statistic H assumes the value h, where

$$h = \frac{12}{n(n+1)} \sum_{i=1}^{k} \frac{r_i^2}{n_i} - 3(n+1),$$

when R_1 assumes the value r_1, R_2 assumes the value r_2, and so forth. The fact that h is large when the independent samples come from populations that are not identical allows us to establish the following decision criterion for testing H_0:

Kruskal–Wallis Test. To test the null hypothesis H_0 that k independent samples are from identical populations, compute

$$h = \frac{12}{n(n+1)} \sum_{i=1}^{k} \frac{r_i^2}{n_i} - 3(n+1).$$

If h falls in the critical region $h > \chi_\alpha^2$ with $v = k - 1$ degrees of freedom, reject H_0 at the α level of significance; otherwise, accept H_0.

Example 6. In an experiment to determine which of three different missile systems is preferable, the propellant burning rate was measured. The data, after coding, are given in Table 13.3.

TABLE 13.3
Propellant Burning Rates

Missile System		
1	2	3
24.0	23.2	18.4
16.7	19.8	19.1
22.8	18.1	17.3
19.8	17.6	17.3
18.9	20.2	19.7
	17.8	18.9
		18.8
		19.3

Use the Kruskal–Wallis test and a significance level of $\alpha = 0.05$ to test the hypothesis that the propellant burning rates are the same for the three missile systems.

Solution

1. H_0: $\mu_1 = \mu_2 = \mu_3$.
2. H_1: the three means are not all equal.
3. $\alpha = 0.05$.
4. Critical region: $h > \chi^2_{0.05} = 5.991$.
5. Computations: In Table 13.4 we convert the 19 observations to ranks and sum the ranks for each missile system.

TABLE 13.4
Ranks for Propellant Burning Rates

Missile System		
1	2	3
19	18	7
1	14.5	11
17	6	2.5
14.5	4	2.5
9.5	16	13
$r_1 = \overline{61.0}$	5	9.5
	$r_2 = \overline{63.5}$	8
		12
		$r_3 = \overline{65.5}$

Now, substituting $n_1 = 5$, $n_2 = 6$, $n_3 = 8$, and $r_1 = 61.0$, $r_2 = 63.5$, $r_3 = 65.5$, our test statistic H assumes the value

$$h = \frac{12}{(19)(20)} \left[\frac{61.0^2}{5} + \frac{63.5^2}{6} + \frac{65.5^2}{8} \right] - (3)(20)$$

$$= 1.66.$$

6. Decision: Since $h = 1.66$ does not fall in the critical region $h > 5.991$, we have insufficient evidence to reject the hypothesis that the propellant burning rates are the same for the three missile systems.

EXERCISES

1. A cigarette manufacturer claims that the tar content of brand B cigarettes is lower than that of brand A. To test this claim, the following determinations of tar content, in milligrams, were recorded:

Brand A	12	9	13	11	14
Brand B	8	10	7		

Use the Wilcoxon rank-sum test with $\alpha = 0.05$ to test whether the claim is valid.

2. To find out whether a new serum will arrest leukemia, 9 patients, who have all reached an advanced stage of the disease, are selected. Five patients receive the treatment and 4 do not. The survival times, in years, from the time the experiment commenced are

Treatment	2.1	5.3	1.4	4.6	0.9
No treatment	1.9	0.5	2.8	3.1	

Use the Wilcoxon rank-sum test, at the 0.05 level of significance, to determine if the serum is effective.

3. The following data represent the number of hours that 2 different types of scientific pocket calculators operate before a recharge is required:

Calculator A	5.5	5.6	6.3	4.6	5.3	5.0	6.2	5.8	5.1
Calculator B	3.8	4.8	4.3	4.2	4.0	4.9	4.5	5.2	4.5

Use the Wilcoxon rank-sum test with $\alpha = 0.01$ to determine if calculator A operates longer than calculator B on a full battery charge.

4. The following data represent the weights of personal luggage carried on a large aircraft by the members of 2 baseball clubs:

Club A	34	39	41	28	33	
Club B	36	40	35	31	39	36

Use the Wilcoxon rank-sum test with $\alpha = 0.05$ to test the hypothesis that the two clubs carry the same amount of luggage on the average against the alternative hypothesis that the average weight of luggage for club B is greater than that of club A.

5. A fishing line is being manufactured by two processes. To determine if there is a difference in the mean breaking strength of the lines, 10 pieces by each process are selected and tested for breaking strength. The results are as follows:

Process 1	10.4	9.8	11.5	10.0	9.9	9.6	10.9	11.8	9.3	10.7
Process 2	8.7	11.2	9.8	10.1	10.8	9.5	11.0	9.8	10.5	9.9

Use the Wilcoxon rank-sum test with $\alpha = 0.1$ to determine if there is a difference between the mean breaking strengths of the lines manufactured by the two processes.

6. From a mathematics class of 12 equally capable students using programmed materials, 5 are selected at random and given additional instruction by the teacher. The results on the final examination were as follows:

	Grade						
Additional instruction	87	69	78	91	80		
No additional instruction	75	88	64	82	93	79	67

Use the Wilcoxon rank-sum test with $\alpha = 0.05$ to determine if the additional instruction affects the average grade.

7. The following data represent the operating times in hours for 3 types of scientific pocket calculators before a recharge is required:

Calculator		
A	B	C
4.9	5.5	6.4
6.1	5.4	6.8
4.3	6.2	5.6
4.6	5.8	6.5
5.3	5.5	6.3
	5.2	6.6
	4.8	

Use the Kruskal–Wallis test, at the 0.01 level of significance, to test the hypothesis that the operating times for all three calculators are equal.

8. Random samples of 4 brands of cigarettes were tested for tar content. The following figures show the milligrams of tar found in the 16 cigarettes tested:

Brand A	Brand B	Brand C	Brand D
14	16	16	17
10	18	15	20
11	14	14	19
13	15	12	21

Use the Kruskal–Wallis test, at the 0.05 level of significance, to test whether there is a significant difference in tar content among the 4 brands of cigarettes.

9. In Exercise 4, on page 401, use the Kruskal–Wallis test, at the 0.05 level of significance to determine if the grade distributions given by the 3 teachers differ significantly.

10. In Exercise 7, on page 401, use the Kruskal–Wallis test, at the 0.05 level of significance, to determine if the chemical analyses performed by the 4 laboratories give, on the average, the same results.

13.6

Runs Test

In applying the many statistical concepts that were discussed throughout this text, it was always assumed that our sample data had been collected by some randomization procedure. The **runs test,** based on the order in which the sample observations are obtained, is a useful technique for testing the null hypothesis H_0 that the observations have indeed been drawn at random.

To illustrate the runs test, let us suppose that 12 people have been polled to find out if they use a certain product. One would seriously question the assumed randomness of the sample if all 12 people were of the same sex. We shall designate a male and female by the symbols M and F, respectively, and record the outcomes according to their sex in the order in which they occur. A typical sequence for the experiment might be

$$M \ M \ \ F \ F \ F \ \ M \ \ F \ F \ \ M \ M \ M \ M \ ,$$

where we have grouped subsequences of similar symbols. Such groupings are called **runs.**

DEFINITION

_____

Run. A _run_ is a subsequence of one or more identical symbols representing a common property of the data.

_____

Regardless of whether our sample measurements represent qualitative or quantitative data, the runs test divides the data into two mutually exclusive categories: male or female; defective or nondefective; heads or tails; above or below the median; and so forth. Consequently, a sequence will always be limited to two distinct symbols. Let n_1 be the number of symbols associated with the category that occurs the least and n_2 be the number of symbols that belong to the other category. Then the sample size $n = n_1 + n_2$.

For the $n = 12$ symbols in our poll we have five runs with the first containing two M's, the second containing three F's, and so on. If the number of runs is larger or smaller than what we would expect by chance, the hypothesis that the sample was drawn at random should be rejected. Certainly, a sample resulting in only two runs,

$$M \ \ M \ \ M \ \ M \ \ M \ \ M \ \ M \ \ F \ \ F \ \ F \ \ F \ \ F,$$

or the reverse, is most unlikely to occur from a random selection process. Such a result indicates that the first seven people interviewed were all males followed by five females. Similarly, if the sample resulted in the maximum number of 12 runs, as in the alternating sequence

$$M \ \ F \ \ M \ \ F \ \ M \ \ F \ \ M \ \ F \ \ M \ \ F \ \ M \ \ F,$$

we would again be suspicious of the order in which the individuals were selected for the poll.

The runs test for randomness is based on the random variable V, the total number of runs that occur in the complete sequence of our experiment. In Table A.10, values of $P(V \leq v^*$ when H_0 is true) are given for $v^* = 2, 3, \ldots, 10$ runs, and values of n_1 and n_2 less than or equal to 10. Critical values for either tail of the distribution of V can be obtained using these tabled values. For example, the largest critical region of size not exceeding α for a two-tailed test of randomness is established by forming the inequalities

$$v \leq a \quad \text{and} \quad v \geq b,$$

where a is the largest value of v^* in Table A.10 for which

$$P(V \leq a \text{ when } H_0 \text{ is true}) \leq \frac{\alpha}{2}$$

and b is the smallest value of v^* in Table A.10 for which

$$P(V \geq b \text{ when } H_0 \text{ is true}) \geq \frac{\alpha}{2}.$$

In the poll taken above we exhibit a total of five F's and seven M's. Hence, with $n_1 = 5$ and $n_2 = 7$, we find from Table A.10, for α as close as possible but not exceeding 0.05, that

$$P(V \leq 3 \text{ when } H_0 \text{ is true}) = 0.015$$

and

$$P(V \geq 11 \text{ when } H_0 \text{ is true}) = 1 - P(V \leq 10 \text{ when } H_0 \text{ is true})$$
$$= 1 - 0.992 = 0.008.$$

Therefore, $a = 3$ and $b = 11$, and the hypothesis of randomness is rejected when $v \leq 3$ or $v \geq 11$. For our example, the value $v = 5$ falls in the acceptance region.

The runs test can also be used to detect departures in randomness of a sequence of quantitative measurements over time, caused by trends or periodicities. Replacing each measurement in the order in which they are collected by a *plus* symbol if it falls above the median, by a *minus* symbol if it falls below the median, and omitting all measurements that are exactly equal to the median, we generate a sequence of plus and minus symbols that are tested for randomness as illustrated in the following example.

Example 7. A machine is adjusted to dispense acrylic paint thinner into a container. Would you say that the amount of paint thinner being dispensed by this machine varies randomly if the contents of the next 15 containers are measured and found to be 3.6, 3.9, 4.1, 3.6, 3.8, 3.7, 3.4, 4.0, 3.8, 4.1, 3.9, 4.0, 3.8, 4.2, and 4.1 liters? Use a 0.1 level of significance.

Solution. Using the six-step procedure, we have

1. H_0: sequence is random.
2. H_1: sequence is not random.
3. $\alpha = 0.1$.
4. Critical region: $v \leq a$ and $v \geq b$, where v is the number of runs and a and b are found from Table A.10.
5. Computations: For the given sample we find $\tilde{x} = 3.9$. Replacing each measurement by the symbol '' $+$ '' if it falls above 3.9, by the symbol '' $-$ '' if it falls below 3.9, and omitting the two measurements that equal 3.9, we obtain the sequence

$$- \quad + \quad - \quad - \quad - \quad - \quad + \quad + \quad + \quad + \quad - \quad + \quad +$$

for which $n_1 = 6$, $n_2 = 7$, and $v = 6$.
6. Decision: Consulting Table A.10, we find that $a = 4$ and $b = 11$. Since $v = 6$ falls in the acceptance region, we accept the hypothesis that the sequence of measurements vary randomly.

The runs test, although less powerful, can also be used as an alternative to the Wilcoxon rank-sum test to test the claim that two random samples come from populations having the same distributions and therefore equal means. If the populations are symmetric, rejection of the claim of equal distributions is equivalent to accepting the alternative hypothesis that the means are not equal. In performing the test, we first combine the observations from both samples and arrange them in ascending order. Now assign the letter A to each observation taken from one of the populations and the letter B to each observation from the second population, thereby generating a sequence consisting of the symbols A and B. If observations from one population are tied with observations from the other population, the sequence of A and B symbols generated will not be unique and consequently the number of runs is unlikely to be unique. Procedures for breaking ties usually result in additional tedious computations, and for this reason one might prefer to apply the Wilcoxon rank-sum test whenever these situations occur.

To illustrate, consider the survival times of the leukemia patients of Exercise 2 on page 453 for which we have

$$0.5 \quad 0.9 \quad 1.4 \quad 1.9 \quad 2.1 \quad 2.8 \quad 3.1 \quad 4.6 \quad 5.3$$
$$B \quad A \quad A \quad B \quad A \quad B \quad B \quad A \quad A,$$

resulting in $v = 6$ runs. If the two symmetric populations have equal means the observations from the two samples will be intermingled, resulting in many runs. However, if the population means are significantly different, we would expect most of the observations for one of the two samples to be smaller than those for the other sample. In the extreme case where the populations do not overlap, we would obtain a sequence of the form

$$A \quad A \quad A \quad A \quad A \quad B \quad B \quad B \quad B \quad \text{or} \quad B \quad B \quad B \quad B \quad A \quad A \quad A \quad A \quad A,$$

and in either case there are only two runs. Consequently, the hypothesis of equal population means will be rejected when v is small and falls in the critical region $v \le a$, implying a one-tailed test.

Returning to the data of Exercise 2 on page 453, for which $n_1 = 4$, and $n_2 = 5$, we find from Table A.10, for $\alpha = 0.05$, that $a = 2$. Since the value $v = 6$ falls outside the critical region $v \le 2$, we conclude that the new serum does not prolong life by arresting leukemia.

When n_1 and n_2 increase in size, the sampling distribution of V approaches the normal distribution with mean

$$\mu_V = \frac{2n_1 n_2}{n_1 + n_2} + 1$$

and variance

$$\sigma_V^2 = \frac{2n_1 n_2 (2n_1 n_2 - n_1 - n_2)}{(n_1 + n_2)^2 (n_1 + n_2 - 1)}.$$

Consequently, when n_1 and n_2 are both greater than 10, one could use the statistic

$$Z = \frac{V - \mu_V}{\sigma_V}$$

to establish the critical region for the runs test.

13.7

Rank Correlation Coefficient

In Chapter 11 we used the sample correlation coefficient r to measure the linear relationship between two continuous variables X and Y. If ranks 1, 2, ..., n are assigned to the x observations in order of magnitude and similarly to the y observations, and if these ranks are then substituted for the actual numerical values into the formula for r, we obtain the nonparametric counterpart of the conventional correlation coefficient. A correlation coefficient calculated in this manner is known as the **Spearman rank correlation coefficient** and is denoted by r_S. When there are no ties among either set of measurements, the formula for r_S reduces to a much simpler expression, which we now state.

Rank Correlation Coefficient. A nonparametric measure of association between two variables X and Y is given by the *rank correlation coefficient*

$$r_S = 1 - \frac{6 \sum_{i=1}^{n} d_i^2}{n(n^2 - 1)},$$

where d_i is the difference between the ranks assigned to x_i and y_i, and n is the number of pairs of data.

In practice the preceding formula is also used when there are ties either among the x or y observations. The ranks for tied observations are assigned as in the Wilcoxon signed-rank test by averaging the ranks that would have been assigned if the observations were distinguishable.

The value of r_S will usually be close to the value obtained by finding r based on numerical measurements and is interpreted in much the same way. As before, the values of r_S will range from -1 to $+1$. A value of $+1$ or -1 indicates perfect association between X and Y, the plus sign occurring for identical rankings and the minus sign occurring for reverse rankings. When r_S is close to zero, we would conclude that the variables are uncorrelated.

Example 8. The following figures, released by the Federal Trade Commission, show the milligrams of tar and nicotine found in 10 brands of cigarettes.

Cigarette Brand	Tar Content	Nicotine Content
Viceroy	14	0.9
Marlboro	17	1.1
Chesterfield	28	1.6
Kool	17	1.3
Kent	16	1.0
Raleigh	13	0.8
Old Gold	24	1.5
Philip Morris	25	1.4
Oasis	18	1.2
Players	31	2.0

Calculate the rank correlation coefficient to measure the degree of relationship between tar and nicotine content in cigarettes.

Solution. Let X and Y represent the tar and nicotine contents, respectively. First we assign ranks to each set of measurements with the rank of 1 assigned to the lowest number in each set, the rank of 2 to the second lowest number in each set, and so forth, until the rank of 10 is assigned to the largest number. Table 13.5 shows the individual rankings of the measurements and the differences in ranks for the 10 pairs of observations.

TABLE 13.5
Rankings for Tar and Nicotine Contents

Cigarette Brand	x_i	y_i	d_i
Viceroy	2	2	0
Marlboro	4.5	4	0.5
Chesterfield	9	9	0
Kool	4.5	6	-1.5
Kent	3	3	0
Raleigh	1	1	0
Old Gold	7	8	-1
Philip Morris	8	7	1
Oasis	6	5	1
Players	10	10	0

Substituting into the formula for r_S, we find that

$$r_S = 1 - \frac{(6)(5.5)}{(10)(100 - 1)} = 0.97,$$

indicating a high positive correlation between the amount of tar and nicotine found in cigarettes.

There are some advantages to using r_S rather than r. For instance, we no longer assume the underlying relationship between X and Y to be linear and therefore, when the data possess a distinct curvilinear relationship, the rank correlation coefficient will likely be more reliable than the conventional measure. A second advantage in using the rank correlation coefficient is the fact that no assumptions of normality are made concerning the distributions of X and Y. Perhaps the greatest advantage occurs when one is unable to make meaningful numerical measurements but nevertheless can establish rankings. Such is the case, for example, when different judges rank a group of individuals according to some attribute. The rank correlation coefficient can be used in this situation as a measure of the consistency of the two judges.

To test the significance of the rank correlation coefficient, one needs to consider the distribution of the r_S values under the assumption that X and Y are independent. Critical values for $\alpha = 0.05, 0.025, 0.01$, and 0.005 have been calculated and are given in Table A.14. This table is set up similar to the table of critical values for the t distribution except for the left column, which now gives the number of pairs of observations rather than the degrees of freedom. Since the distribution of the r_S values is symmetric about $r_S = 0$, the r_S value that leaves an area of α to the left is equal to the negative of the r_S value that leaves an area of α to the right. For a two-sided alternative hypothesis, the critical region of size α falls equally in the two tails of the distribution. For a test in which the alternative hypothesis is negative, the critical region is entirely in the left tail of the distribution, and when the alternative is positive, the critical region is placed entirely in the right tail. In Example 8 the critical value for testing the null hypothesis H_0 that the rank correlation coefficient is zero against the alternative hypothesis H_1 that it is greater than zero, with $\alpha = 0.01$ and $n = 10$, is 0.745. That is, we reject H_0 if $r_S > 0.745$, and since our calculated value was $r_S = 0.97$, we conclude at the 0.01 level of significance that a high positive correlation does exist between the amount of tar and nicotine found in cigarettes.

When X and Y are independent, it can be shown that the distribution of the r_S values approaches a normal distribution with a mean of zero and a standard deviation of $1/\sqrt{n-1}$ as n increases. Consequently, when n exceeds the values given in Table A.14, one could test for a significant correlation by computing

$$ z = \frac{r_S - 0}{1/\sqrt{n-1}} = r_S\sqrt{n-1} $$

and comparing with critical values obtained from the standard normal curve. The test for independence of two *continuous variables* presented in this section is a simplified alternative to the more cumbersome chi-square procedure using contingency tables outlined in Section 10.9. Unfortunately, all too often the values of the two random variables must fall into certain established categories and therefore cannot be measured on a continuous scale, thereby necessitating the more complex calculations associated with contingency tables.

EXERCISES

1. A random sample of 15 adults living in a small town are selected to estimate the proportion of voters favoring a certain candidate for mayor. Each individual was also asked if he or she was a college graduate. By letting Y and N designate the responses of "yes" and "no" to the education question, the following sequence was obtained:

$$N \ N \ N \ N \ Y \ Y \ N \ Y \ Y \ Y \ N \ Y \ N \ N \ N \ N.$$

Use the runs test at the 0.1 level of significance to determine if the sequence supports the contention that the sample was selected at random.

2. A silver-plating process is being used to coat a certain type of serving tray. When the process is in control, the thickness of the silver on the trays will vary randomly following a normal distribution with a mean of 0.02 millimeter and a standard deviation of 0.005 millimeter. Suppose that the next 12 trays examined show the following thicknesses of silver: 0.019, 0.021, 0.020, 0.019, 0.020, 0.018, 0.023, 0.021, 0.024, 0.022, 0.023, 0.022. Use the runs test to determine if the fluctuations in thickness from one tray to another is random. Let $\alpha = 0.05$.

3. Use the runs test to test whether there is a difference in the average operating time for the two calculators of Exercise 3 on page 453.

4. In an industrial production line, items are inspected periodically for defectives. The following is a sequence of defective items, D, and nondefective items, N, produced by this production line:

$$D \ D \ N \ N \ N \ D \ N \ N \ D \ D \ N \ N \ N \ N$$
$$N \ D \ D \ D \ N \ N \ D \ N \ N \ N \ N \ D \ N \ D.$$

Use the large-sample theory for the runs test, with a significance level of 0.05, to determine whether the defectives are occurring at random or not.

5. Assuming that the measurements of Exercise 6 on page 65 were recorded in successive rows from left to right as they were collected, use the runs test, with $\alpha = 0.05$, to test the hypothesis that the data represent a random sequence.

6. The following table gives the recorded grades for 10 students on a midterm test and the final examination in a calculus course:

Student	Midterm Test	Final Examination
L.S.A.	84	73
W.P.B.	98	63
R.W.K.	91	87
J.R.L.	72	66
J.K.L.	86	78
D.L.P.	93	78
B.L.P.	80	91
D.W.M.	0	0
M.N.M.	92	88
R.H.S.	87	77

(a) Calculate the rank correlation coefficient.

(b) Test the hypothesis that the rank correlation coefficient is zero against the alternative that it is greater than zero. Use $\alpha = 0.025$.

7. With reference to the weights and chest sizes of infants in Exercise 6 on page 383,

(a) calculate the rank correlation coefficient;

(b) test the hypothesis at the 0.025 level of significance that the rank correlation coefficient is zero against the alternative that it is greater than zero.

8. Calculate the rank correlation coefficient for the daily rainfall and amount of particulate removed in Exercise 8 on page 352.

9. A consumer panel tested nine makes of microwave ovens for overall quality. The ranks assigned by the panel and the suggested retail prices were as follows:

Manufacturer	Panel Rating	Suggested Price
A	6	$480
B	9	395
C	2	575
D	8	550
E	5	510
F	1	545
G	7	400
H	4	465
I	3	420

Is there a significant relationship between the quality and the price of a microwave oven?

10. Two judges at a college homecoming parade ranked 8 floats in the following order:

	Float							
	1	2	3	4	5	6	7	8
Judge A	5	8	4	3	6	2	7	1
Judge B	7	5	4	2	8	1	6	3

   (a) Calculate the rank correlation coefficient.

   (b) Test the hypothesis that the rank correlation coefficient is zero against the alternative hypothesis that it is greater than zero. Use $\alpha = 0.05$.

11. Test the hypothesis that X and Y are independent against the alternative that they are dependent if for a sample of size $n = 50$ pairs of observations we find that $r_S = -0.29$. Use $\alpha = 0.05$.

References

ALDER, H. L., and E. B. ROESSLER. *Introduction to Probability and Statistics,* 6th ed. San Francisco: W. H. Freeman & Co., Publishers, 1977.

CHAO, L. L. *Introduction to Statistics*. Monterey, Calif.: Brooks/Cole Publishing Co., 1980.

DIXON, W. J., and F. J. MASSEY, Jr. *Introduction to Statistical Analysis,* 3rd ed. New York: McGraw-Hill Book Company, 1969.

EZEKIEL, M., and K. A. FOX. *Methods of Correlation and Regression Analysis,* 3rd ed. New York: John Wiley & Sons, Inc., 1959.

FREUND, J. E. *Modern Elementary Statistics,* 5th ed. Englewood Cliffs, N.J.: Prentice-Hall, Inc., 1979.

GUENTHER, W. C. *Analysis of Variance*. Englewood Cliffs, N.J.: Prentice-Hall, Inc., 1964.

HICKS, C. R. *Fundamental Concepts in the Design of Experiments*. New York: Holt, Rinehart and Winston, 1964.

HODGES, J. L., Jr., and E. L. LEHMANN. *Basic Concepts of Probability and Statistics,* 2nd ed. San Francisco: Holden-Day, Inc., 1970.

HOEL, P. G. *Elementary Statistics,* 4th ed. New York: John Wiley & Sons, Inc., 1976.

JOHNSON, R. R. *Elementary Statistics,* 3rd ed. North Scituate, Mass.: Duxbury Press, 1980.

LAPIN, L. L. *Statistics Meaning and Method,* 2nd ed. New York: Harcourt Brace Jovanovich, Inc., 1980.

LARSON, H. J. *Statistics: An Introduction*. New York: John Wiley & Sons, Inc., 1975.

LI, C. C. *Introduction to Experimental Statistics*. New York: McGraw-Hill Book Company, 1964.

LINDGREN, B. W. *Basic Ideas of Statistics*. New York: Macmillan Publishing Co., Inc., 1975.

MENDENHALL, W. *Introduction to Probability and Statistics,* 5th ed. North Scituate, Mass.: Duxbury Press, 1979.

MILLER, I., and J. E. FREUND. *Probability and Statistics for Engineers,* 2nd ed. Englewood Cliffs, N.J.: Prentice-Hall, Inc., 1977.

MOSTELLER, F., R. E. K. ROURKE, and G. B. THOMAS, Jr. *Probability with Statistical Applications*. Reading, Mass.: Addison-Wesley Publishing Co., Inc., 1970.

NETER, J., W. WASSERMAN, and G. A. WHITMORE. *Applied Statistics.* Boston: Allyn and Bacon, Inc., 1978.

NOETHER, G. E. *Introduction to Statistics: A Nonparametric Approach,* 2nd ed. Boston: Houghton Mifflin Company, 1976.

SCHEFFÉ, H. *Analysis of Variance.* New York: John Wiley & Sons, Inc., 1959.

SIEGEL, S. *Nonparametric Statistics for the Behavioral Sciences.* New York: McGraw-Hill Book Company, 1956.

TANUR, J. M., et al. *Statistics: A Guide to the Unknown,* 2nd ed. San Francisco: Holden-Day, Inc., 1978.

WALPOLE, R. E., and R. H. MYERS. *Probability and Statistics for Engineers and Scientists,* 2nd ed. New York: Macmillan Publishing Co., Inc., 1978.

WEINBERG, G., and J. SCHUMAKER. *Statistics: An Intuitive Approach,* 3rd ed. Monterey, Calif.: Brooks/Cole Publishing Co., 1974.

Appendix: Statistical Tables

469

TABLE A.1
Squares and Square Roots

n	n^2	$\sqrt{n}$	$\sqrt{10n}$	n	n^2	$\sqrt{n}$	$\sqrt{10n}$
1.0	1.00	1.000	3.162	5.5	30.25	2.345	7.416
1.1	1.21	1.049	3.317	5.6	31.36	2.366	7.483
1.2	1.44	1.095	3.464	5.7	32.49	2.387	7.550
1.3	1.69	1.140	3.606	5.8	33.64	2.408	7.616
1.4	1.96	1.183	3.742	5.9	34.81	2.429	7.681
1.5	2.25	1.225	3.873	6.0	36.00	2.449	7.746
1.6	2.56	1.265	4.000	6.1	37.21	2.470	7.810
1.7	2.89	1.304	4.123	6.2	38.44	2.490	7.874
1.8	3.24	1.342	4.243	6.3	39.69	2.510	7.937
1.9	3.61	1.378	4.359	6.4	40.96	2.530	8.000
2.0	4.00	1.414	4.472	6.5	42.25	2.550	8.062
2.1	4.41	1.449	4.583	6.6	43.56	2.569	8.124
2.2	4.84	1.483	4.690	6.7	44.89	2.588	8.185
2.3	5.29	1.517	4.796	6.8	46.24	2.608	8.246
2.4	5.76	1.549	4.899	6.9	47.61	2.627	8.307
2.5	6.25	1.581	5.000	7.0	49.00	2.646	8.367
2.6	6.76	1.612	5.099	7.1	50.41	2.665	8.426
2.7	7.29	1.643	5.196	7.2	51.84	2.683	8.485
2.8	7.84	1.673	5.292	7.3	53.29	2.702	8.544
2.9	8.41	1.703	5.385	7.4	54.76	2.720	8.602
3.0	9.00	1.732	5.477	7.5	56.25	2.739	8.660
3.1	9.61	1.761	5.568	7.6	57.76	2.757	8.718
3.2	10.24	1.789	5.657	7.7	59.29	2.775	8.775
3.3	10.89	1.817	5.745	7.8	60.84	2.793	8.832
3.4	11.56	1.844	5.831	7.9	62.41	2.811	8.888
3.5	12.25	1.871	5.916	8.0	64.00	2.828	8.944
3.6	12.96	1.897	6.000	8.1	65.61	2.846	9.000
3.7	13.69	1.924	6.083	8.2	67.24	2.864	9.055
3.8	14.44	1.949	6.164	8.3	68.89	2.881	9.110
3.9	15.21	1.975	6.245	8.4	70.56	2.898	9.165
4.0	16.00	2.000	6.325	8.5	72.25	2.915	9.220
4.1	16.81	2.025	6.403	8.6	73.96	2.933	9.274
4.2	17.64	2.049	6.481	8.7	75.69	2.950	9.327
4.3	18.49	2.074	6.557	8.8	77.44	2.966	9.381
4.4	19.36	2.098	6.633	8.9	79.21	2.983	9.434
4.5	20.25	2.121	6.708	9.0	81.00	3.000	9.487
4.6	21.16	2.145	6.782	9.1	82.81	3.017	9.539
4.7	22.09	2.168	6.856	9.2	84.64	3.033	9.592
4.8	23.04	2.191	6.928	9.3	86.49	3.050	9.644
4.9	24.01	2.214	7.000	9.4	88.36	3.066	9.695
5.0	25.00	2.236	7.071	9.5	90.25	3.082	9.747
5.1	26.01	2.258	7.141	9.6	92.16	3.098	9.798
5.2	27.04	2.280	7.211	9.7	94.09	3.114	9.849
5.3	28.09	2.302	7.280	9.8	96.04	3.130	9.899
5.4	29.16	2.324	7.348	9.9	98.01	3.146	9.950

TABLE A.2

Binomial Probability Sums $\sum_{x=0}^{r} b(x; n, p)$

						p					
n	r	.10	.20	.25	.30	.40	.50	.60	.70	.80	.90
1	0	.9000	.8000	.7500	.7000	.6000	.5000	.4000	.3000	.2000	.1000
	1	1.0000	1.0000	1.0000	1.0000	1.0000	1.0000	1.0000	1.0000	1.0000	1.0000
2	0	.8100	.6400	.5625	.4900	.3600	.2500	.1600	.0900	.0400	.0100
	1	.9900	.9600	.9375	.9100	.8400	.7500	.6400	.5100	.3600	.1900
	2	1.0000	1.0000	1.0000	1.0000	1.0000	1.0000	1.0000	1.0000	1.0000	1.0000
3	0	.7290	.5120	.4219	.3430	.2160	.1250	.0640	.0270	.0080	.0010
	1	.9720	.8960	.8438	.7840	.6480	.5000	.3520	.2160	.1040	.0280
	2	.9990	.9920	.9844	.9730	.9360	.8750	.7840	.6570	.4880	.2710
	3	1.0000	1.0000	1.0000	1.0000	1.0000	1.0000	1.0000	1.0000	1.0000	1.0000
4	0	.6561	.4096	.3164	.2401	.1296	.0625	.0256	.0081	.0016	.0001
	1	.9477	.8192	.7383	.6517	.4752	.3125	.1792	.0837	.0272	.0037
	2	.9963	.9728	.9492	.9163	.8208	.6875	.5248	.3483	.1808	.0523
	3	.9999	.9984	.9961	.9919	.9744	.9375	.8704	.7599	.5904	.3439
	4	1.0000	1.0000	1.0000	1.0000	1.0000	1.0000	1.0000	1.0000	1.0000	1.0000
5	0	.5905	.3277	.2373	.1681	.0778	.0312	.0102	.0024	.0003	.0000
	1	.9185	.7373	.6328	.5282	.3370	.1875	.0870	.0308	.0067	.0005
	2	.9914	.9421	.8965	.8369	.6826	.5000	.3174	.1631	.0579	.0086
	3	.9995	.9933	.9844	.9692	.9130	.8125	.6630	.4718	.2627	.0815
	4	1.0000	.9997	.9990	.9976	.9898	.9688	.9222	.8319	.6723	.4095
	5		1.0000	1.0000	1.0000	1.0000	1.0000	1.0000	1.0000	1.0000	1.0000
6	0	.5314	.2621	.1780	.1176	.0467	.0156	.0041	.0007	.0001	.0000
	1	.8857	.6554	.5339	.4202	.2333	.1094	.0410	.0109	.0016	.0001
	2	.9841	.9011	.8306	.7443	.5443	.3438	.1792	.0705	.0170	.0013
	3	.9987	.9830	.9624	.9295	.8208	.6563	.4557	.2557	.0989	.0158
	4	.9999	.9984	.9954	.9891	.9590	.8906	.7667	.5798	.3447	.1143
	5	1.0000	.9999	.9998	.9993	.9959	.9844	.9533	.8824	.7379	.4686
	6		1.0000	1.0000	1.0000	1.0000	1.0000	1.0000	1.0000	1.0000	1.0000
7	0	.4783	.2097	.1335	.0824	.0280	.0078	.0016	.0002	.0000	
	1	.8503	.5767	.4449	.3294	.1586	.0625	.0188	.0038	.0004	.0000
	2	.9743	.8520	.7564	.6471	.4199	.2266	.0963	.0288	.0047	.0002
	3	.9973	.9667	.9294	.8740	.7102	.5000	.2898	.1260	.0333	.0027
	4	.9998	.9953	.9871	.9712	.9037	.7734	.5801	.3529	.1480	.0257
	5	1.0000	.9996	.9987	.9962	.9812	.9375	.8414	.6706	.4233	.1497
	6		1.0000	.9999	.9998	.9984	.9922	.9720	.9176	.7903	.5217
	7			1.0000	1.0000	1.0000	1.0000	1.0000	1.0000	1.0000	1.0000

TABLE A.2 (*continued*)

Binomial Probability Sums $\sum_{x=0}^{r} b(x; n, p)$

n	r	.10	.20	.25	.30	.40	.50	.60	.70	.80	.90
									p		
8	0	.4305	.1678	.1001	.0576	.0168	.0039	.0007	.0001	.0000	
	1	.8131	.5033	.3671	.2553	.1064	.0352	.0085	.0013	.0001	
	2	.9619	.7969	.6785	.5518	.3154	.1445	.0498	.0113	.0012	.0000
	3	.9950	.9437	.8862	.8059	.5941	.3633	.1737	.0580	.0104	.0004
	4	.9996	.9896	.9727	.9420	.8263	.6367	.4059	.1941	.0563	.0050
	5	1.0000	.9988	.9958	.9887	.9502	.8555	.6846	.4482	.2031	.0381
	6		.9991	.9996	.9987	.9915	.9648	.8936	.7447	.4967	.1869
	7		1.0000	1.0000	.9999	.9993	.9961	.9832	.9424	.8322	.5695
	8				1.0000	1.0000	1.0000	1.0000	1.0000	1.0000	1.0000
9	0	.3874	.1342	.0751	.0404	.0101	.0020	.0003	.0000		
	1	.7748	.4362	.3003	.1960	.0705	.0195	.0038	.0004	.0000	
	2	.9470	.7382	.6007	.4628	.2318	.0898	.0250	.0043	.0003	.0000
	3	.9917	.9144	.8343	.7297	.4826	.2539	.0994	.0253	.0031	.0001
	4	.9991	.9804	.9511	.9012	.7334	.5000	.2666	.0988	.0196	.0009
	5	.9999	.9969	.9900	.9747	.9006	.7461	.5174	.2703	.0856	.0083
	6	1.0000	.9997	.9987	.9957	.9750	.9102	.7682	.5372	.2618	.0530
	7		1.0000	.9999	.9996	.9962	.9805	.9295	.8040	.5638	.2252
	8			1.0000	1.0000	.9997	.9980	.9899	.9596	.8658	.6126
	9					1.0000	1.0000	1.0000	1.0000	1.0000	1.0000
10	0	.3487	.1074	.0563	.0282	.0060	.0010	.0001	.0000		
	1	.7361	.3758	.2440	.1493	.0464	.0107	.0017	.0001	.0000	
	2	.9298	.6778	.5256	.3828	.1673	.0547	.0123	.0016	.0001	
	3	.9872	.8791	.7759	.6496	.3823	.1719	.0548	.0106	.0009	.0000
	4	.9984	.9672	.9219	.8497	.6331	.3770	.1662	.0474	.0064	.0002
	5	.9999	.9936	.9803	.9527	.8338	.6230	.3669	.1503	.0328	.0016
	6	1.0000	.9991	.9965	.9894	.9452	.8281	.6177	.3504	.1209	.0128
	7		.9999	.9996	.9984	.9877	.9453	.8327	.6172	.3222	.0702
	8		1.0000	1.0000	.9999	.9983	.9893	.9536	.8507	.6242	.2639
	9				1.0000	.9999	.9990	.9940	.9718	.8926	.6513
	10					1.0000	1.0000	1.0000	1.0000	1.0000	1.0000
11	0	.3138	.0859	.0422	.0198	.0036	.0005	.0000			
	1	.6974	.3221	.1971	.1130	.0302	.0059	.0007	.0000		
	2	.9104	.6174	.4552	.3127	.1189	.0327	.0059	.0006	.0000	
	3	.9815	.8369	.7133	.5696	.2963	.1133	.0293	.0043	.0002	
	4	.9972	.9496	.8854	.7897	.5328	.2744	.0994	.0216	.0020	.0000
	5	.9997	.9883	.9657	.9218	.7535	.5000	.2465	.0782	.0117	.0003
	6	1.0000	.9980	.9924	.9784	.9006	.7256	.4672	.2103	.0504	.0028
	7		.9998	.9988	.9957	.9707	.8867	.7037	.4304	.1611	.0185
	8		1.0000	.9999	.9994	.9941	.9673	.8811	.6873	.3826	.0896
	9			1.0000	1.0000	.9993	.9941	.9698	.8870	.6779	.3026
	10					1.0000	.9995	.9964	.9802	.9141	.6862
	11						1.0000	1.0000	1.0000	1.0000	1.0000

TABLE A.2 (*continued*)

Binomial Probability Sums $\sum_{x=0}^{r} b(x; n, p)$

						p					
n	*r*	.10	.20	.25	.30	.40	.50	.60	.70	.80	.90
12	0	.2824	.0687	.0317	.0138	.0022	.0002	.0000			
	1	.6590	.2749	.1584	.0850	.0196	.0032	.0003	.0000		
	2	.8891	.5583	.3907	.2528	.0834	.0193	.0028	.0002	.0000	
	3	.9744	.7946	.6488	.4925	.2253	.0730	.0153	.0017	.0001	
	4	.9957	.9274	.8424	.7237	.4382	.1938	.0573	.0095	.0006	.0000
	5	.9995	.9806	.9456	.8821	.6652	.3872	.1582	.0386	.0039	.0001
	6	.9999	.9961	.9857	.9614	.8418	.6128	.3348	.1178	.0194	.0005
	7	1.0000	.9994	.9972	.9905	.9427	.8062	.5618	.2763	.0726	.0043
	8		.9999	.9996	.9983	.9847	.9270	.7747	.5075	.2054	.0256
	9		1.0000	1.0000	.9998	.9972	.9807	.9166	.7472	.4417	.1109
	10				1.0000	.9997	.9968	.9804	.9150	.7251	.3410
	11					1.0000	.9998	.9978	.9862	.9313	.7176
	12						1.0000	1.0000	1.0000	1.0000	1.0000
13	0	.2542	.0550	.0238	.0097	.0013	.0001	.0000			
	1	.6213	.2336	.1267	.0637	.0126	.0017	.0001	.0000		
	2	.8661	.5017	.3326	.2025	.0579	.0112	.0013	.0001		
	3	.9658	.7473	.5843	.4206	.1686	.0461	.0078	.0007	.0000	
	4	.9935	.9009	.7940	.6543	.3530	.1334	.0321	.0040	.0002	
	5	.9991	.9700	.9198	.8346	.5744	.2905	.0977	.0182	.0012	.0000
	6	.9999	.9930	.9757	.9376	.7712	.5000	.2288	.0624	.0070	.0001
	7	1.0000	.9980	.9944	.9818	.9023	.7095	.4256	.1654	.0300	.0009
	8		.9998	.9990	.9960	.9679	.8666	.6470	.3457	.0991	.0065
	9		1.0000	.9999	.9993	.9922	.9539	.8314	.5794	.2527	.0342
	10			1.0000	.9999	.9987	.9888	.9421	.7975	.4983	.1339
	11				1.0000	.9999	.9983	.9874	.9363	.7664	.3787
	12					1.0000	.9999	.9987	.9903	.9450	.7458
	13						1.0000	1.0000	1.0000	1.0000	1.0000
14	0	.2288	.0440	.0178	.0068	.0008	.0001	.0000			
	1	.5846	.1979	.1010	.0475	.0081	.0009	.0001			
	2	.8416	.4481	.2811	.1608	.0398	.0065	.0006	.0000		
	3	.9559	.6982	.5213	.3552	.1243	.0287	.0039	.0002		
	4	.9908	.8702	.7415	.5842	.2793	.0898	.0175	.0017	.0000	
	5	.9985	.9561	.8883	.7805	.4859	.2120	.0583	.0083	.0004	
	6	.9998	.9884	.9617	.9067	.6925	.3953	.1501	.0315	.0024	.0000
	7	1.0000	.9976	.9897	.9685	.8499	.6047	.3075	.0933	.0116	.0002
	8		.9996	.9978	.9917	.9417	.7880	.5141	.2195	.0439	.0015
	9		1.0000	.9997	.9983	.9825	.9102	.7207	.4158	.1298	.0092
	10			1.0000	.9998	.9961	.9713	.8757	.6448	.3018	.0441
	11				1.0000	.9994	.9935	.9602	.8392	.5519	.1584
	12					.9999	.9991	.9919	.9525	.8021	.4154
	13					1.0000	.9999	.9992	.9932	.9560	.7712
	14						1.0000	1.0000	1.0000	1.0000	1.0000

TABLE A.2 (*continued*)

Binomial Probability Sums $\sum\limits_{x=0}^{r} b(x; n, p)$

							p				
n	r	.10	.20	.25	.30	.40	.50	.60	.70	.80	.90
15	0	.2059	.0352	.0134	.0047	.0005	.0000				
	1	.5490	.1671	.0802	.0353	.0052	.0005	.0000			
	2	.8159	.3980	.2361	.1268	.0271	.0037	.0003	.0000		
	3	.9444	.6482	.4613	.2969	.0905	.0176	.0019	.0001		
	4	.9873	.8358	.6865	.5155	.2173	.0592	.0094	.0007	.0000	
	5	.9978	.9389	.8516	.7216	.4032	.1509	.0338	.0037	.0001	
	6	.9997	.9819	.9434	.8689	.6098	.3036	.0951	.0152	.0008	
	7	1.0000	.9958	.9827	.9500	.7869	.5000	.2131	.0500	.0042	.0000
	8		.9992	.9958	.9848	.9050	.6964	.3902	.1311	.0181	.0003
	9		.9999	.9992	.9963	.9662	.8491	.5968	.2784	.0611	.0023
	10		1.0000	.9999	.9993	.9907	.9408	.7827	.4845	.1642	.0127
	11			1.0000	.9999	.9981	.9824	.9095	.7031	.3518	.0556
	12				1.0000	.9997	.9963	.9729	.8732	.6020	.1841
	13					1.0000	.9995	.9948	.9647	.8329	.4510
	14						1.0000	.9995	.9953	.9648	.7941
	15							1.0000	1.0000	1.0000	1.0000
16	0	.1853	.0281	.0100	.0033	.0003	.0000				
	1	.5147	.1407	.0635	.0261	.0033	.0003	.0000			
	2	.7892	.3518	.1971	.0994	.0183	.0021	.0001			
	3	.9316	.5981	.4050	.2459	.0651	.0106	.0009	.0000		
	4	.9830	.7982	.6302	.4499	.1666	.0384	.0049	.0003		
	5	.9967	.9183	.8103	.6598	.3288	.1051	.0191	.0016	.0000	
	6	.9995	.9733	.9204	.8247	.5272	.2272	.0583	.0071	.0002	
	7	.9999	.9930	.9729	.9256	.7161	.4018	.1423	.0257	.0015	.0000
	8	1.0000	.9985	.9925	.9743	.8577	.5982	.2839	.0744	.0070	.0001
	9		.9998	.9984	.9929	.9417	.7728	.4728	.1753	.0267	.0005
	10		1.0000	.9997	.9984	.9809	.8949	.6712	.3402	.0817	.0033
	11			1.0000	.9997	.9951	.9616	.8334	.5501	.2018	.0170
	12				1.0000	.9991	.9894	.9349	.7541	.4019	.0684
	13					.9999	.9979	.9817	.9006	.6482	.2108
	14					1.0000	.9997	.9967	.9739	.8593	.4853
	15						1.0000	.9997	.9967	.9719	.8147
	16							1.0000	1.0000	1.0000	1.0000

TABLE A.2 (*continued*)

Binomial Probability Sums $\sum_{x=0}^{r} b(x; n, p)$

						p					
n	*r*	.10	.20	.25	.30	.40	.50	.60	.70	.80	.90
17	0	.1668	.0225	.0075	.0023	.0002	.0000				
	1	.4818	.1182	.0501	.0193	.0021	.0001	.0000			
	2	.7618	.3096	.1637	.0774	.0123	.0012	.0001			
	3	.9174	.5489	.3530	.2019	.0464	.0064	.0005	.0000		
	4	.9779	.7582	.5739	.3887	.1260	.0245	.0025	.0001		
	5	.9953	.8943	.7653	.5968	.2639	.0717	.0106	.0007	.0000	
	6	.9992	.9623	.8929	.7752	.4478	.1662	.0348	.0032	.0001	
	7	.9999	.9891	.9598	.8954	.6405	.3145	.0919	.0127	.0005	
	8	1.0000	.9974	.9876	.9597	.8011	.5000	.1989	.0403	.0026	.0000
	9		.9995	.9969	.9873	.9081	.6855	.3595	.1046	.0109	.0001
	10		.9999	.9994	.9968	.9652	.8338	.5522	.2248	.0377	.0008
	11		1.0000	.9999	.9993	.9894	.9283	.7361	.4032	.1057	.0047
	12			1.0000	.9999	.9975	.9755	.8740	.6113	.2418	.0221
	13				1.0000	.9995	.9936	.9536	.7981	.4511	.0826
	14					.9999	.9988	.9877	.9226	.6904	.2382
	15					1.0000	.9999	.9979	.9807	.8818	.5182
	16						1.0000	.9998	.9977	.9775	.8332
	17							1.0000	1.0000	1.0000	1,0000
18	0	.1501	.0180	.0056	.0016	.0001	.0000				
	1	.4503	.0991	.0395	.0142	.0013	.0001				
	2	.7338	.2713	.1353	.0600	.0082	.0007	.0000			
	3	.9018	.5010	.3057	.1646	.0328	.0038	.0002			
	4	.9718	.7164	.5787	.3327	.0942	.0154	.0013	.0000		
	5	.9936	.8671	.7175	.5344	.2088	.0481	.0058	.0003		
	6	.9988	.9487	.8610	.7217	.3743	.1189	.0203	.0014	.0000	
	7	.9998	.9837	.9431	.8593	.5634	.2403	.0576	.0061	.0002	
	8	1.0000	.9957	.9807	.9404	.7368	.4073	.1347	.0210	.0009	
	9		.9991	.9946	.9790	.8653	.5927	.2632	.0596	.0043	.0000
	10		.9998	.9988	.9939	.9424	.7597	.4366	.1407	.0163	.0002
	11		1.0000	.9998	.9986	.9797	.8811	.6257	.2783	.0513	.0012
	12			1.0000	.9997	.9942	.9519	.7912	.4656	.1329	.0064
	13				1.0000	.9987	.9846	.9058	.6673	.2836	.0282
	14					.9998	.9962	.9672	.8354	.4990	.0982
	15					1.0000	.9993	.9918	.9400	.7287	.2662
	16						.9999	.9987	.9858	.9009	.5497
	17						1.0000	.9999	.9984	.9820	.8499
	18							1.0000	1.0000	1.0000	1.0000

TABLE A.2 (*continued*)

Binomial Probability Sums $\sum_{x=0}^{r} b(x; n, p)$

							p				
n	r	.10	.20	.25	.30	.40	.50	.60	.70	.80	.90
19	0	.1351	.0144	.0042	.0011	.0001					
	1	.4203	.0829	.0310	.0104	.0008	.0000				
	2	.7054	.2369	.1113	.0462	.0055	.0004	.0000			
	3	.8850	.4551	.2631	.1332	.0230	.0022	.0001			
	4	.9648	.6733	.4654	.2822	.0696	.0096	.0006	.0000		
	5	.9914	.8369	.6678	.4739	.1629	.0318	.0031	.0001		
	6	.9983	.9324	.8251	.6655	.3081	.0835	.0116	.0006		
	7	.9997	.9767	.9225	.8180	.4878	.1796	.0352	.0028	.0000	
	8	1.0000	.9933	.9713	.9161	.6675	.3238	.0885	.0105	.0003	
	9		.9984	.9911	.9674	.8139	.5000	.1861	.0326	.0016	
	10		.9997	.9977	.9895	.9115	.6762	.3325	.0839	.0067	.0000
	11		.9999	.9995	.9972	.9648	.8204	.5122	.1820	.0233	.0003
	12		1.0000	.9999	.9994	.9884	.9165	.6919	.3345	.0676	.0017
	13			1.0000	.9999	.9969	.9682	.8371	.5261	.1631	.0086
	14				1.0000	.9994	.9904	.9304	.7178	.3267	.0352
	15					.9999	.9978	.9770	.8668	.5449	.1150
	16					1.0000	.9996	.9945	.9538	.7631	.2946
	17						1.0000	.9992	.9896	.9171	.5797
	18							.9999	.9989	.9856	.8649
	19							1.0000	1.0000	1.0000	1.0000
20	0	.1216	.0115	.0032	.0008	.0000					
	1	.3917	.0692	.0243	.0076	.0005	.0000				
	2	.6769	.2061	.0913	.0355	.0036	.0002	.0000			
	3	.8670	.4114	.2252	.1071	.0160	.0013	.0001			
	4	.9568	.6296	.4148	.2375	.0510	.0059	.0003			
	5	.9887	.8042	.6172	.4164	.1256	.0207	.0016	.0000		
	6	.9976	.9133	.7858	.6080	.2500	.0577	.0065	.0003		
	7	.9996	.9679	.8982	.7723	.4159	.1316	.0210	.0013	.0000	
	8	.9999	.9900	.9591	.8867	.5956	.2517	.0565	.0051	.0001	
	9	1.0000	.9974	.9861	.9520	.7553	.4119	.1275	.0171	.0006	
	10		.9994	.9961	.9829	.8725	.5881	.2447	.0480	.0026	.0000
	11		.9999	.9991	.9949	.9435	.7483	.4044	.1133	.0100	.0001
	12		1.0000	.9998	.9987	.9790	.8684	.5841	.2277	.0321	.0004
	13			1.0000	.9997	.9935	.9423	.7500	.3920	.0867	.0024
	14				1.0000	.9984	.9793	.8744	.5836	.1958	.0113
	15					.9997	.9941	.9490	.7625	.3704	.0432
	16					1.0000	.9987	.9840	.8929	.5886	.1330
	17						.9998	.9964	.9645	.7939	.3231
	18						1.0000	.9995	.9924	.9308	.6083
	19							1.0000	.9992	.9885	.8784
	20								1.0000	1.0000	1.0000

TABLE A.3*

Poisson Probability Sums $\sum_{x=0}^{r} p(x; \mu)$

	μ								
r	0.1	0.2	0.3	0.4	0.5	0.6	0.7	0.8	0.9
0	0.9048	0.8187	0.7408	0.6730	0.6065	0.5488	0.4966	0.4493	0.4066
1	0.9953	0.9825	0.9631	0.9384	0.9098	0.8781	0.8442	0.8088	0.7725
2	0.9998	0.9989	0.9964	0.9921	0.9856	0.9769	0.9659	0.9526	0.9371
3	1.0000	0.9999	0.9997	0.9992	0.9982	0.9966	0.9942	0.9909	0.9865
4		1.0000	1.0000	0.9999	0.9998	0.9996	0.9992	0.9986	0.9977
5				1.0000	1.0000	1.0000	0.9999	0.9998	0.9997
6							1.0000	1.0000	1.0000

	μ								
r	1.0	1.5	2.0	2.5	3.0	3.5	4.0	4.5	5.0
0	0.3679	0.2231	0.1353	0.0821	0.0498	0.0302	0.0183	0.0111	0.0067
1	0.7358	0.5578	0.4060	0.2873	0.1991	0.1359	0.0916	0.0611	0.0404
2	0.9197	0.8088	0.6767	0.5438	0.4232	0.3208	0.2381	0.1736	0.1247
3	0.9810	0.9344	0.8571	0.7576	0.6472	0.5366	0.4335	0.3423	0.2650
4	0.9963	0.9814	0.9473	0.8912	0.8153	0.7254	0.6288	0.5321	0.4405
5	0.9994	0.9955	0.9834	0.9580	0.9161	0.8576	0.7851	0.7029	0.6160
6	0.9999	0.9991	0.9955	0.9858	0.9665	0.9347	0.8893	0.8311	0.7622
7	1.0000	0.9998	0.9989	0.9958	0.9881	0.9733	0.9489	0.9134	0.8666
8		1.0000	0.9998	0.9989	0.9962	0.9901	0.9786	0.9597	0.9319
9			1.0000	0.9997	0.9989	0.9967	0.9919	0.9829	0.9682
10				0.9999	0.9997	0.9990	0.9972	0.9933	0.9863
11				1.0000	0.9999	0.9997	0.9991	0.9976	0.9945
12					1.0000	0.9999	0.9997	0.9992	0.9980
13						1.0000	0.9999	0.9997	0.9993
14							1.0000	0.9999	0.9998
15								1.0000	0.9999
16									1.0000

*From E. C. Molina, *Poisson's Exponential Binomial Limit,* copyright 1942, Van Nostrand Reinhold Company, New York, by permission of the publisher.

TABLE A.3 (*continued*)

Poisson Probability Sums $\sum_{x=0}^{r} p(x; \mu)$

					μ				
r	5.5	6.0	6.5	7.0	7.5	8.0	8.5	9.0	9.5
0	0.0041	0.0025	0.0015	0.0009	0.0006	0.0003	0.0002	0.0001	0.0001
1	0.0266	0.0174	0.0113	0.0073	0.0047	0.0030	0.0019	0.0012	0.0008
2	0.0884	0.0620	0.0430	0.0296	0.0203	0.0138	0.0093	0.0062	0.0042
3	0.2017	0.1512	0.1118	0.0818	0.0591	0.0424	0.0301	0.0212	0.0149
4	0.3575	0.2851	0.2237	0.1730	0.1321	0.0996	0.0744	0.0550	0.0403
5	0.5289	0.4457	0.3690	0.3007	0.2414	0.1912	0.1496	0.1157	0.0885
6	0.6860	0.6063	0.5265	0.4497	0.3782	0.3134	0.2562	0.2068	0.1649
7	0.8095	0.7440	0.6728	0.5987	0.5246	0.4530	0.3856	0.3239	0.2687
8	0.8944	0.8472	0.7916	0.7291	0.6620	0.5925	0.5231	0.4557	0.3918
9	0.9462	0.9161	0.8774	0.8305	0.7764	0.7166	0.6530	0.5874	0.5218
10	0.9747	0.9574	0.9332	0.9015	0.8622	0.8159	0.7634	0.7060	0.6453
11	0.9890	0.9799	0.9661	0.9466	0.9208	0.8881	0.8487	0.8030	0.7520
12	0.9955	0.9912	0.9840	0.9730	0.9573	0.9362	0.9091	0.8758	0.8364
13	0.9983	0.9964	0.9929	0.9872	0.9784	0.9658	0.9486	0.9261	0.8981
14	0.9994	0.9986	0.9970	0.9943	0.9897	0.9827	0.9726	0.9585	0.9400
15	0.9998	0.9995	0.9988	0.9976	0.9954	0.9918	0.9862	0.9780	0.9665
16	0.9999	0.9998	0.9996	0.9990	0.9980	0.9963	0.9934	0.9889	0.9823
17	1.0000	0.9999	0.9998	0.9996	0.9992	0.9984	0.9970	0.9947	0.9911
18		1.0000	0.9999	0.9999	0.9997	0.9994	0.9987	0.9976	0.9957
19			1.0000	1.0000	0.9999	0.9997	0.9995	0.9989	0.9980
20					1.0000	0.9999	0.9998	0.9996	0.9991
21						1.0000	0.9999	0.9998	0.9996
22							1.0000	0.9999	0.9999
23								1.0000	0.9999
24									1.0000

TABLE A.3 (*continued*)

Poisson Probability Sums $\sum_{x=0}^{r} p(x; \mu)$

	μ								
r	10.0	11.0	12.0	13.0	14.0	15.0	16.0	17.0	18.0
0	0.0000	0.0000	0.0000						
1	0.0005	0.0002	0.0001	0.0000	0.0000				
2	0.0028	0.0012	0.0005	0.0002	0.0001	0.0000	0.0000		
3	0.0103	0.0049	0.0023	0.0010	0.0005	0.0002	0.0001	0.0000	0.0000
4	0.0293	0.0151	0.0076	0.0037	0.0018	0.0009	0.0004	0.0002	0.0001
5	0.0671	0.0375	0.0203	0.0107	0.0055	0.0028	0.0014	0.0007	0.0003
6	0.1301	0.0786	0.0458	0.0259	0.0142	0.0076	0.0040	0.0021	0.0010
7	0.2202	0.1432	0.0895	0.0540	0.0316	0.0180	0.0100	0.0054	0.0029
8	0.3328	0.2320	0.1550	0.0998	0.0621	0.0374	0.0220	0.0126	0.0071
9	0.4579	0.3405	0.2424	0.1658	0.1094	0.0699	0.0433	0.0261	0.0154
10	0.5830	0.4599	0.3472	0.2517	0.1757	0.1185	0.0774	0.0491	0.0304
11	0.6968	0.5793	0.4616	0.3532	0.2600	0.1848	0.1270	0.0847	0.0549
12	0.7916	0.6887	0.5760	0.4631	0.3585	0.2676	0.1931	0.1350	0.0917
13	0.8645	0.7813	0.6815	0.5730	0.4644	0.3632	0.2745	0.2009	0.1426
14	0.9165	0.8540	0.7720	0.6751	0.5704	0.4657	0.3675	0.2808	0.2081
15	0.9513	0.9074	0.8444	0.7636	0.6694	0.5681	0.4667	0.3715	0.2867
16	0.9730	0.9441	0.8987	0.8355	0.7559	0.6641	0.5660	0.4677	0.3750
17	0.9857	0.9678	0.9370	0.8905	0.8272	0.7489	0.6593	0.5640	0.4686
18	0.9928	0.9823	0.9626	0.9302	0.8826	0.8195	0.7423	0.6550	0.5622
19	0.9965	0.9907	0.9787	0.9573	0.9235	0.8752	0.8122	0.7363	0.6509
20	0.9984	0.9953	0.9884	0.9750	0.9521	0.9170	0.8682	0.8055	0.7307
21	0.9993	0.9977	0.9939	0.9859	0.9712	0.9469	0.9108	0.8615	0.7991
22	0.9997	0.9990	0.9970	0.9924	0.9833	0.9673	0.9418	0.9047	0.8551
23	0.9999	0.9995	0.9985	0.9960	0.9907	0.9805	0.9633	0.9367	0.8989
24	1.0000	0.9998	0.9993	0.9980	0.9950	0.9888	0.9777	0.9594	0.9317
25		0.9999	0.9997	0.9990	0.9974	0.9938	0.9869	0.9748	0.9554
26		1.0000	0.9999	0.9995	0.9987	0.9967	0.9925	0.9848	0.9718
27			0.9999	0.9998	0.9994	0.9983	0.9959	0.9912	0.9827
28			1.0000	0.9999	0.9997	0.9991	0.9978	0.9950	0.9897
29				1.0000	0.9999	0.9996	0.9989	0.9973	0.9941
30					0.9999	0.9998	0.9994	0.9986	0.9967
31					1.0000	0.9999	0.9997	0.9993	0.9982
32						1.0000	0.9999	0.9996	0.9990
33							0.9999	0.9998	0.9995
34							1.0000	0.9999	0.9998
35								1.0000	0.9999
36									0.9999
37									1.0000

TABLE A.4

Areas Under the Normal Curve

z	0.00	0.01	0.02	0.03	0.04	0.05	0.06	0.07	0.08	0.09
−3.4	0.0003	0.0003	0.0003	0.0003	0.0003	0.0003	0.0003	0.0003	0.0003	0.0002
−3.3	0.0005	0.0005	0.0005	0.0004	0.0004	0.0004	0.0004	0.0004	0.0004	0.0003
−3.2	0.0007	0.0007	0.0006	0.0006	0.0006	0.0006	0.0006	0.0005	0.0005	0.0005
−3.1	0.0010	0.0009	0.0009	0.0009	0.0008	0.0008	0.0008	0.0008	0.0007	0.0007
−3.0	0.0013	0.0013	0.0013	0.0012	0.0012	0.0011	0.0011	0.0011	0.0010	0.0010
−2.9	0.0019	0.0018	0.0017	0.0017	0.0016	0.0016	0.0015	0.0015	0.0014	0.0014
−2.8	0.0026	0.0025	0.0024	0.0023	0.0023	0.0022	0.0021	0.0021	0.0020	0.0019
−2.7	0.0035	0.0034	0.0033	0.0032	0.0031	0.0030	0.0029	0.0028	0.0027	0.0026
−2.6	0.0047	0.0045	0.0044	0.0043	0.0041	0.0040	0.0039	0.0038	0.0037	0.0036
−2.5	0.0062	0.0060	0.0059	0.0057	0.0055	0.0054	0.0052	0.0051	0.0049	0.0048
−2.4	0.0082	0.0080	0.0078	0.0075	0.0073	0.0071	0.0069	0.0068	0.0066	0.0064
−2.3	0.0107	0.0104	0.0102	0.0099	0.0096	0.0094	0.0091	0.0089	0.0087	0.0084
−2.2	0.0139	0.0136	0.0132	0.0129	0.0125	0.0122	0.0119	0.0116	0.0113	0.0110
−2.1	0.0179	0.0174	0.0170	0.0166	0.0162	0.0158	0.0154	0.0150	0.0146	0.0143
−2.0	0.0228	0.0222	0.0217	0.0212	0.0207	0.0202	0.0197	0.0192	0.0188	0.0183
−1.9	0.0287	0.0281	0.0274	0.0268	0.0262	0.0256	0.0250	0.0244	0.0239	0.0233
−1.8	0.0359	0.0352	0.0344	0.0336	0.0329	0.0322	0.0314	0.0307	0.0301	0.0294
−1.7	0.0446	0.0436	0.0427	0.0418	0.0409	0.0401	0.0392	0.0384	0.0375	0.0367
−1.6	0.0548	0.0537	0.0526	0.0516	0.0505	0.0495	0.0485	0.0475	0.0465	0.0455
−1.5	0.0668	0.0655	0.0643	0.0630	0.0618	0.0606	0.0594	0.0582	0.0571	0.0559
−1.4	0.0808	0.0793	0.0778	0.0764	0.0749	0.0735	0.0722	0.0708	0.0694	0.0681
−1.3	0.0968	0.0951	0.0934	0.0918	0.0901	0.0885	0.0869	0.0853	0.0838	0.0823
−1.2	0.1151	0.1131	0.1112	0.1093	0.1075	0.1056	0.1038	0.1020	0.1003	0.0985
−1.1	0.1357	0.1335	0.1314	0.1292	0.1271	0.1251	0.1230	0.1210	0.1190	0.1170
−1.0	0.1587	0.1562	0.1539	0.1515	0.1492	0.1469	0.1446	0.1423	0.1401	0.1379
−0.9	0.1841	0.1814	0.1788	0.1762	0.1736	0.1711	0.1685	0.1660	0.1635	0.1611
−0.8	0.2119	0.2090	0.2061	0.2033	0.2005	0.1977	0.1949	0.1922	0.1894	0.1867
−0.7	0.2420	0.2389	0.2358	0.2327	0.2296	0.2266	0.2236	0.2206	0.2177	0.2148
−0.6	0.2743	0.2709	0.2676	0.2643	0.2611	0.2578	0.2546	0.2514	0.2483	0.2451
−0.5	0.3085	0.3050	0.3015	0.2981	0.2946	0.2912	0.2877	0.2843	0.2810	0.2776
−0.4	0.3446	0.3409	0.3372	0.3336	0.3300	0.3264	0.3228	0.3192	0.3156	0.3121
−0.3	0.3821	0.3783	0.3745	0.3707	0.3669	0.3632	0.3594	0.3557	0.3520	0.3483
−0.2	0.4207	0.4168	0.4129	0.4090	0.4052	0.4013	0.3974	0.3936	0.3897	0.3859
−0.1	0.4602	0.4562	0.4522	0.4483	0.4443	0.4404	0.4364	0.4325	0.4286	0.4247
−0.0	0.5000	0.4960	0.4920	0.4880	0.4840	0.4801	0.4761	0.4721	0.4681	0.4641
0.0	0.5000	0.5040	0.5080	0.5120	0.5160	0.5199	0.5239	0.5279	0.5319	0.5359
0.1	0.5398	0.5438	0.5478	0.5517	0.5557	0.5596	0.5636	0.5675	0.5714	0.5753
0.2	0.5793	0.5832	0.5871	0.5910	0.5948	0.5987	0.6026	0.6064	0.6103	0.6141
0.3	0.6179	0.6217	0.6255	0.6293	0.6331	0.6368	0.6406	0.6443	0.6480	0.6517
0.4	0.6554	0.6591	0.6628	0.6664	0.6700	0.6736	0.6772	0.6808	0.6844	0.6879
0.5	0.6915	0.6950	0.6985	0.7019	0.7054	0.7088	0.7123	0.7157	0.7190	0.7224
0.6	0.7257	0.7291	0.7324	0.7357	0.7389	0.7422	0.7454	0.7486	0.7517	0.7549
0.7	0.7580	0.7611	0.7642	0.7673	0.7704	0.7734	0.7764	0.7794	0.7823	0.7852
0.8	0.7881	0.7910	0.7939	0.7967	0.7995	0.8023	0.8051	0.8078	0.8106	0.8133
0.9	0.8159	0.8186	0.8212	0.8238	0.8264	0.8289	0.8315	0.8340	0.8365	0.8389
1.0	0.8413	0.8438	0.8461	0.8485	0.8508	0.8531	0.8554	0.8577	0.8599	0.8621
1.1	0.8643	0.8665	0.8686	0.8708	0.8729	0.8749	0.8770	0.8790	0.8810	0.8830
1.2	0.8849	0.8869	0.8888	0.8907	0.8925	0.8944	0.8962	0.8980	0.8997	0.9015
1.3	0.9032	0.9049	0.9066	0.9082	0.9099	0.9115	0.9131	0.9147	0.9162	0.9177
1.4	0.9192	0.9207	0.9222	0.9236	0.9251	0.9265	0.9278	0.9292	0.9306	0.9319
1.5	0.9332	0.9345	0.9357	0.9370	0.9382	0.9394	0.9406	0.9418	0.9429	0.9441
1.6	0.9452	0.9463	0.9474	0.9484	0.9495	0.9505	0.9515	0.9525	0.9535	0.9545
1.7	0.9554	0.9564	0.9573	0.9582	0.9591	0.9599	0.9608	0.9616	0.9625	0.9633
1.8	0.9641	0.9649	0.9656	0.9664	0.9671	0.9678	0.9686	0.9693	0.9699	0.9706
1.9	0.9713	0.9719	0.9726	0.9732	0.9738	0.9744	0.9750	0.9756	0.9761	0.9767
2.0	0.9772	0.9778	0.9783	0.9788	0.9793	0.9798	0.9803	0.9808	0.9812	0.9817
2.1	0.9821	0.9826	0.9830	0.9834	0.9838	0.9842	0.9846	0.9850	0.9854	0.9857
2.2	0.9861	0.9864	0.9868	0.9871	0.9875	0.9878	0.9881	0.9884	0.9887	0.9890
2.3	0.9893	0.9896	0.9898	0.9901	0.9904	0.9906	0.9909	0.9911	0.9913	0.9916
2.4	0.9918	0.9920	0.9922	0.9925	0.9927	0.9929	0.9931	0.9932	0.9934	0.9936
2.5	0.9938	0.9940	0.9941	0.9943	0.9945	0.9946	0.9948	0.9949	0.9951	0.9952
2.6	0.9953	0.9955	0.9956	0.9957	0.9959	0.9960	0.9961	0.9962	0.9963	0.9964
2.7	0.9965	0.9966	0.9967	0.9968	0.9969	0.9970	0.9971	0.9972	0.9973	0.9974
2.8	0.9974	0.9975	0.9976	0.9977	0.9977	0.9978	0.9979	0.9979	0.9980	0.9981
2.9	0.9981	0.9982	0.9982	0.9983	0.9984	0.9984	0.9985	0.9985	0.9986	0.9986
3.0	0.9987	0.9987	0.9987	0.9988	0.9988	0.9989	0.9989	0.9989	0.9990	0.9990
3.1	0.9990	0.9991	0.9991	0.9991	0.9992	0.9992	0.9992	0.9992	0.9993	0.9993
3.2	0.9993	0.9993	0.9994	0.9994	0.9994	0.9994	0.9994	0.9995	0.9995	0.9995
3.3	0.9995	0.9995	0.9995	0.9996	0.9996	0.9996	0.9996	0.9996	0.9996	0.9997
3.4	0.9997	0.9997	0.9997	0.9997	0.9997	0.9997	0.9997	0.9997	0.9997	0.9998

TABLE A.5*
Critical Values of the *t* Distribution

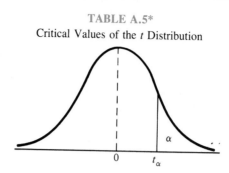

ν	0.10	0.05	0.025	0.01	0.005
1	3.078	6.314	12.706	31.821	63.657
2	1.886	2.920	4.303	6.965	9.925
3	1.638	2.353	3.182	4.541	5.841
4	1.533	2.132	2.776	3.747	4.604
5	1.476	2.015	2.571	3.365	4.032
6	1.440	1.943	2.447	3.143	3.707
7	1.415	1.895	2.365	2.998	3.499
8	1.397	1.860	2.306	2.896	3.355
9	1.383	1.833	2.262	2.821	3.250
10	1.372	1.812	2.228	2.764	3.169
11	1.363	1.796	2.201	2.718	3.106
12	1.356	1.782	2.179	2.681	3.055
13	1.350	1.771	2.160	2.650	3.012
14	1.345	1.761	2.145	2.624	2.977
15	1.341	1.753	2.131	2.602	2.947
16	1.337	1.746	2.120	2.583	2.921
17	1.333	1.740	2.110	2.567	2.898
18	1.330	1.734	2.101	2.552	2.878
19	1.328	1.729	2.093	2.539	2.861
20	1.325	1.725	2.086	2.528	2.845
21	1.323	1.721	2.080	2.518	2.831
22	1.321	1.717	2.074	2.508	2.819
23	1.319	1.714	2.069	2.500	2.807
24	1.318	1.711	2.064	2.492	2.797
25	1.316	1.708	2.060	2.485	2.787
26	1.315	1.706	2.056	2.479	2.779
27	1.314	1.703	2.052	2.473	2.771
28	1.313	1.701	2.048	2.467	2.763
29	1.311	1.699	2.045	2.462	2.756
inf.	1.282	1.645	1.960	2.326	2.576

*Table A.5 is taken from Table IV of R. A. Fisher, *Statistical Methods for Research Workers,* Oliver & Boyd Ltd., Edinburgh, by permission of the author and publishers.

TABLE A.6*
Critical Values of the Chi-Square Distribution

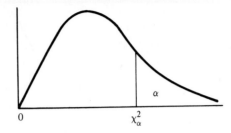

	α							
ν	0.995	0.99	0.975	0.95	0.05	0.025	0.01	0.005
1	0.0^4393	0.0^3157	0.0^3982	0.0^2393	3.841	5.024	6.635	7.879
2	0.0100	0.0201	0.0506	0.103	5.991	7.378	9.210	10.597
3	0.0717	0.115	0.216	0.352	7.815	9.348	11.345	12.838
4	0.207	0.297	0.484	0.711	9.488	11.143	13.277	14.860
5	0.412	0.554	0.831	1.145	11.070	12.832	15.086	16.750
6	0.676	0.872	1.237	1.635	12.592	14.449	16.812	18.548
7	0.989	1.239	1.690	2.167	14.067	16.013	18.475	20.278
8	1.344	1.646	2.180	2.733	15.507	17.535	20.090	21.955
9	1.735	2.088	2.700	3.325	16.919	19.023	21.666	23.589
10	2.156	2.558	3.247	3.940	18.307	20.483	23.209	25.188
11	2.603	3.053	3.816	4.575	19.675	21.920	24.725	26.757
12	3.074	3.571	4.404	5.226	21.026	23.337	26.217	28.300
13	3.565	4.107	5.009	5.892	22.362	24.736	27.688	29.819
14	4.075	4.660	5.629	6.571	23.685	26.119	29.141	31.319
15	4.601	5.229	6.262	7.261	24.996	27.488	30.578	32.801
16	5.142	5.812	6.908	7.962	26.296	28.845	32.000	34.267
17	5.697	6.408	7.564	8.672	27.587	30.191	33.409	35.718
18	6.265	7.015	8.231	9.390	28.869	31.526	34.805	37.156
19	6.844	7.633	8.907	10.117	30.144	32.852	36.191	38.582
20	7.434	8.260	9.591	10.851	31.410	34.170	37.566	39.997
21	8.034	8.897	10.283	11.591	32.671	35.479	38.932	41.401
22	8.643	9.542	10.982	12.338	33.924	36.781	40.289	42.796
23	9.260	10.196	11.689	13.091	35.172	38.076	41.638	44.181
24	9.886	10.856	12.401	13.848	36.415	39.364	42.980	45.558
25	10.520	11.524	13.120	14.611	37.652	40.646	44.314	46.928
26	11.160	12.198	13.844	15.379	38.885	41.923	45.642	48.290
27	11.808	12.879	14.573	16.151	40.113	43.194	46.963	49.645
28	12.461	13.565	15.308	16.928	41.337	44.461	48.278	50.993
29	13.121	14.256	16.047	17.708	42.557	45.722	49.588	52.336
30	13.787	14.953	16.791	18.493	43.773	46.979	50.892	53.672

*Abridged from Table 8 of *Biometrika Tables for Statisticians,* Vol. I, by permission of E. S. Pearson and the Biometrika Trustees.

TABLE A.7*
Critical Values of the F Distribution

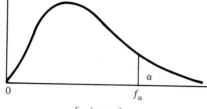

$$f_{0.05}(\nu_1, \nu_2)$$

ν_2					ν_1				
	1	2	3	4	5	6	7	8	9
1	161.4	199.5	215.7	224.6	230.2	234.0	236.8	238.9	240.5
2	18.51	19.00	19.16	19.25	19.30	19.33	19.35	19.37	19.38
3	10.13	9.55	9.28	9.12	9.01	8.94	8.89	8.85	8.81
4	7.71	6.94	6.59	6.39	6.26	6.16	6.09	6.04	6.00
5	6.61	5.79	5.41	5.19	5.05	4.95	4.88	4.82	4.77
6	5.99	5.14	4.76	4.53	4.39	4.28	4.21	4.15	4.10
7	5.59	4.74	4.35	4.12	3.97	3.87	3.79	3.73	3.68
8	5.32	4.46	4.07	3.84	3.69	3.58	3.50	3.44	3.39
9	5.12	4.26	3.86	3.63	3.48	3.37	3.29	3.23	3.18
10	4.96	4.10	3.71	3.48	3.33	3.22	3.14	3.07	3.02
11	4.84	3.98	3.59	3.36	3.20	3.09	3.01	2.95	2.90
12	4.75	3.89	3.49	3.26	3.11	3.00	2.91	2.85	2.80
13	4.67	3.81	3.41	3.18	3.03	2.92	2.83	2.77	2.71
14	4.60	3.74	3.34	3.11	2.96	2.85	2.76	2.70	2.65
15	4.54	3.68	3.29	3.06	2.90	2.79	2.71	2.64	2.59
16	4.49	3.63	3.24	3.01	2.85	2.74	2.66	2.59	2.54
17	4.45	3.59	3.20	2.96	2.81	2.70	2.61	2.55	2.49
18	4.41	3.55	3.16	2.93	2.77	2.66	2.58	2.51	2.46
19	4.38	3.52	3.13	2.90	2.74	2.63	2.54	2.48	2.42
20	4.35	3.49	3.10	2.87	2.71	2.60	2.51	2.45	2.39
21	4.32	3.47	3.07	2.84	2.68	2.57	2.49	2.42	2.37
22	4.30	3.44	3.05	2.82	2.66	2.55	2.46	2.40	2.34
23	4.28	3.42	3.03	2.80	2.64	2.53	2.44	2.37	2.32
24	4.26	3.40	3.01	2.78	2.62	2.51	2.42	2.36	2.30
25	4.24	3.39	2.99	2.76	2.60	2.49	2.40	2.34	2.28
26	4.23	3.37	2.98	2.74	2.59	2.47	2.39	2.32	2.27
27	4.21	3.35	2.96	2.73	2.57	2.46	2.37	2.31	2.25
28	4.20	3.34	2.95	2.71	2.56	2.45	2.36	2.29	2.24
29	4.18	3.33	2.93	2.70	2.55	2.43	2.35	2.28	2.22
30	4.17	3.32	2.92	2.69	2.53	2.42	2.33	2.27	2.21
40	4.08	3.23	2.84	2.61	2.45	2.34	2.25	2.18	2.12
60	4.00	3.15	2.76	2.53	2.37	2.25	2.17	2.10	2.04
120	3.92	3.07	2.68	2.45	2.29	2.17	2.09	2.02	1.96
∞	3.84	3.00	2.60	2.37	2.21	2.10	2.01	1.94	1.88

TABLE A.7 (*continued*)

Critical Values of the F Distribution

$$f_{0.05}(v_1, v_2)$$

v_2	v_1									
	10	12	15	20	24	30	40	60	120	∞
1	241.9	243.9	245.9	248.0	249.1	250.1	251.1	252.2	253.3	254.3
2	19.40	19.41	19.43	19.45	19.45	19.46	19.47	19.48	19.49	19.50
3	8.79	8.74	8.70	8.66	8.64	8.62	8.59	8.57	8.55	8.53
4	5.96	5.91	5.86	5.80	5.77	5.75	5.72	5.69	5.66	5.63
5	4.74	4.68	4.62	4.56	4.53	4.50	4.46	4.43	4.40	4.36
6	4.06	4.00	3.94	3.87	3.84	3.81	3.77	3.74	3.70	3.67
7	3.64	3.57	3.51	3.44	3.41	3.38	3.34	3.30	3.27	3.23
8	3.35	3.28	3.22	3.15	3.12	3.08	3.04	3.01	2.97	2.93
9	3.14	3.07	3.01	2.94	2.90	2.86	2.83	2.79	2.75	2.71
10	2.98	2.91	2.85	2.77	2.74	2.70	2.66	2.62	2.58	2.54
11	2.85	2.79	2.72	2.65	2.61	2.57	2.53	2.49	2.45	2.40
12	2.75	2.69	2.62	2.54	2.51	2.47	2.43	2.38	2.34	2.30
13	2.67	2.60	2.53	2.46	2.42	2.38	2.34	2.30	2.25	2.21
14	2.60	2.53	2.46	2.39	2.35	2.31	2.27	2.22	2.18	2.13
15	2.54	2.48	2.40	2.33	2.29	2.25	2.20	2.16	2.11	2.07
16	2.49	2.42	2.35	2.28	2.24	2.19	2.15	2.11	2.06	2.01
17	2.45	2.38	2.31	2.23	2.19	2.15	2.10	2.06	2.01	1.96
18	2.41	2.34	2.27	2.19	2.15	2.11	2.06	2.02	1.97	1.92
19	2.38	2.31	2.23	2.16	2.11	2.07	2.03	1.98	1.93	1.88
20	2.35	2.28	2.20	2.12	2.08	2.04	1.99	1.95	1.90	1.84
21	2.32	2.25	2.18	2.10	2.05	2.01	1.96	1.92	1.87	1.81
22	2.30	2.23	2.15	2.07	2.03	1.98	1.94	1.89	1.84	1.78
23	2.27	2.20	2.13	2.05	2.01	1.96	1.91	1.86	1.81	1.76
24	2.25	2.18	2.11	2.03	1.98	1.94	1.89	1.84	1.79	1.73
25	2.24	2.16	2.09	2.01	1.96	1.92	1.87	1.82	1.77	1.71
26	2.22	2.15	2.07	1.99	1.95	1.90	1.85	1.80	1.75	1.69
27	2.20	2.13	2.06	1.97	1.93	1.88	1.84	1.79	1.73	1.67
28	2.19	2.12	2.04	1.96	1.91	1.87	1.82	1.77	1.71	1.65
29	2.18	2.10	2.03	1.94	1.90	1.85	1.81	1.75	1.70	1.64
30	2.16	2.09	2.01	1.93	1.89	1.84	1.79	1.74	1.68	1.62
40	2.08	2.00	1.92	1.84	1.79	1.74	1.69	1.64	1.58	1.51
60	1.99	1.92	1.84	1.75	1.70	1.65	1.59	1.53	1.47	1.39
120	1.91	1.83	1.75	1.66	1.61	1.55	1.50	1.43	1.35	1.25
∞	1.83	1.75	1.67	1.57	1.52	1.46	1.39	1.32	1.22	1.00

TABLE A.7 (*continued*)

Critical Values of the F Distribution

$f_{0.01}(v_1, v_2)$

v_2	v_1								
	1	2	3	4	5	6	7	8	9
1	4052	4999.5	5403	5625	5764	5859	5928	5981	6022
2	98.50	99.00	99.17	99.25	99.30	99.33	99.36	99.37	99.39
3	34.12	30.82	29.46	28.71	28.24	27.91	27.67	27.49	27.35
4	21.20	18.00	16.69	15.98	15.52	15.21	14.98	14.80	14.66
5	16.26	13.27	12.06	11.39	10.97	10.67	10.46	10.29	10.16
6	13.75	10.92	9.78	9.15	8.75	8.47	8.26	8.10	7.98
7	12.25	9.55	8.45	7.85	7.46	7.19	6.99	6.84	6.72
8	11.26	8.65	7.59	7.01	6.63	6.37	6.18	6.03	5.91
9	10.56	8.02	6.99	6.42	6.06	5.80	5.61	5.47	5.35
10	10.04	7.56	6.55	5.99	5.64	5.39	5.20	5.06	4.94
11	9.65	7.21	6.22	5.67	5.32	5.07	4.89	4.74	4.63
12	9.33	6.93	5.95	5.41	5.06	4.82	4.64	4.50	4.39
13	9.07	6.70	5.74	5.21	4.86	4.62	4.44	4.30	4.19
14	8.86	6.51	5.56	5.04	4.69	4.46	4.28	4.14	4.03
15	8.68	6.36	5.42	4.89	4.56	4.32	4.14	4.00	3.89
16	8.53	6.23	5.29	4.77	4.44	4.20	4.03	3.89	3.78
17	8.40	6.11	5.18	4.67	4.34	4.10	3.93	3.79	3.68
18	8.29	6.01	5.09	4.58	4.25	4.01	3.84	3.71	3.60
19	8.18	5.93	5.01	4.50	4.17	3.94	3.77	3.63	3.52
20	8.10	5.85	4.94	4.43	4.10	3.87	3.70	3.56	3.46
21	8.02	5.78	4.87	4.37	4.04	3.81	3.64	3.51	3.40
22	7.95	5.72	4.82	4.31	3.99	3.76	3.59	3.45	3.35
23	7.88	5.66	4.76	4.26	3.94	3.71	3.54	3.41	3.30
24	7.82	5.61	4.72	4.22	3.90	3.67	3.50	3.36	3.26
25	7.77	5.57	4.68	4.18	3.85	3.63	3.46	3.32	3.22
26	7.72	5.53	4.64	4.14	3.82	3.59	3.42	3.29	3.18
27	7.68	5.49	4.60	4.11	3.78	3.56	3.39	3.26	3.15
28	7.64	5.45	4.57	4.07	3.75	3.53	3.36	3.23	3.12
29	7.60	5.42	4.54	4.04	3.73	3.50	3.33	3.20	3.09
30	7.56	5.39	4.51	4.02	3.70	3.47	3.30	3.17	3.07
40	7.31	5.18	4.31	3.83	3.51	3.29	3.12	2.99	2.89
60	7.08	4.98	4.13	3.65	3.34	3.12	2.95	2.82	2.72
120	6.85	4.79	3.95	3.48	3.17	2.96	2.79	2.66	2.56
∞	6.63	4.61	3.78	3.32	3.02	2.80	2.64	2.51	2.41

TABLE A.7 (*continued*)

Critical Values of the F Distribution

$$f_{0.01}(\nu_1, \nu_2)$$

ν_2	ν_1									
	10	12	15	20	24	30	40	60	120	∞
1	6056	6106	6157	6209	6235	6261	6287	6313	6339	6366
2	99.40	99.42	99.43	99.45	99.46	99.47	99.47	99.48	99.49	99.50
3	27.23	27.05	26.87	26.69	26.60	26.50	26.41	26.32	26.22	26.13
4	14.55	14.37	14.20	14.02	13.93	13.84	13.75	13.65	13.56	13.46
5	10.05	9.89	9.72	9.55	9.47	9.38	9.29	9.20	9.11	9.02
6	7.87	7.72	7.56	7.40	7.31	7.23	7.14	7.06	6.97	6.88
7	6.62	6.47	6.31	6.16	6.07	5.99	5.91	5.82	5.74	5.65
8	5.81	5.67	5.52	5.36	5.28	5.20	5.12	5.03	4.95	4.86
9	5.26	5.11	4.96	4.81	4.73	4.65	4.57	4.48	4.40	4.31
10	4.85	4.71	4.56	4.41	4.33	4.25	4.17	4.08	4.00	3.91
11	4.54	4.40	4.25	4.10	4.02	3.94	3.86	3.78	3.69	3.60
12	4.30	4.16	4.01	3.86	3.78	3.70	3.62	3.54	3.45	3.36
13	4.10	3.96	3.82	3.66	3.59	3.51	3.43	3.34	3.25	3.17
14	3.94	3.80	3.66	3.51	3.43	3.35	3.27	3.18	3.09	3.00
15	3.80	3.67	3.52	3.37	3.29	3.21	3.13	3.05	2.96	2.87
16	3.69	3.55	3.41	3.26	3.18	3.10	3.02	2.93	2.84	2.75
17	3.59	3.46	3.31	3.16	3.08	3.00	2.92	2.83	2.75	2.65
18	3.51	3.37	3.23	3.08	3.00	2.92	2.84	2.75	2.66	2.57
19	3.43	3.30	3.15	3.00	2.92	2.84	2.76	2.67	2.58	2.49
20	3.37	3.23	3.09	2.94	2.86	2.78	2.69	2.61	2.52	2.42
21	3.31	3.17	3.03	2.88	2.80	2.72	2.64	2.55	2.46	2.36
22	3.26	3.12	2.98	2.83	2.75	2.67	2.58	2.50	2.40	2.31
23	3.21	3.07	2.93	2.78	2.70	2.62	2.54	2.45	2.35	2.26
24	3.17	3.03	2.89	2.74	2.66	2.58	2.49	2.40	2.31	2.21
25	3.13	2.99	2.85	2.70	2.62	2.54	2.45	2.36	2.27	2.17
26	3.09	2.96	2.81	2.66	2.58	2.50	2.42	2.33	2.23	2.13
27	3.06	2.93	2.78	2.63	2.55	2.47	2.38	2.29	2.20	2.10
28	3.03	2.90	2.75	2.60	2.52	2.44	2.35	2.26	2.17	2.06
29	3.00	2.87	2.73	2.57	2.49	2.41	2.33	2.23	2.14	2.03
30	2.98	2.84	2.70	2.55	2.47	2.39	2.30	2.21	2.11	2.01
40	2.80	2.66	2.52	2.37	2.29	2.20	2.11	2.02	1.92	1.80
60	2.63	2.50	2.35	2.20	2.12	2.03	1.94	1.84	1.73	1.60
120	2.47	2.34	2.19	2.03	1.95	1.86	1.76	1.66	1.53	1.38
∞	2.32	2.18	2.04	1.88	1.79	1.70	1.59	1.47	1.32	1.00

TABLE A.8*
Critical Values for the Wilcoxon Signed-Rank Test

n	One-sided $\alpha = 0.01$ Two-sided $\alpha = 0.02$	One-sided $\alpha = 0.025$ Two-sided $\alpha = 0.05$	One-sided $\alpha = 0.05$ Two-sided $\alpha = 0.10$
5			1
6		1	2
7	0	2	4
8	2	4	6
9	3	6	8
10	5	8	11
11	7	11	14
12	10	14	17
13	13	17	21
14	16	21	26
15	20	25	30
16	24	30	36
17	28	35	41
18	33	40	47
19	38	46	54
20	43	52	60
21	49	59	68
22	56	66	75
23	62	73	83
24	69	81	92
25	77	90	101
26	85	98	110
27	93	107	120
28	102	117	130
29	111	127	141
30	120	137	152

*Reproduced from F. Wilcoxon and R. A. Wilcox, *Some Rapid Approximate Statistical Procedures*, American Cyanamid Company, Pearl River, N.Y., 1964, by permission of the American Cyanamid Company.

TABLE A.9*
Critical Values for the Wilcoxon Rank-Sum Test
One-Tailed Test at $\alpha = 0.001$ or Two-Tailed Test at $\alpha = 0.002$

n_1	6	7	8	9	10	11	12	13	14	15	16	17	18	19	20
1															
2															
3												0	0	0	0
4					0	0	0	1	1	1	2	2	3	3	3
5		0	0	1	1	2	2	3	3	4	5	5	6	7	7
6	0	1	2	2	3	4	4	5	6	7	8	9	10	11	12
7		2	3	3	5	6	7	8	9	10	11	13	14	15	16
8			5	5	6	8	9	11	12	14	15	17	18	20	21
9				7	8	10	12	14	15	17	19	21	23	25	26
10					10	12	14	17	19	21	23	25	27	29	32
11						15	17	20	22	24	27	29	32	34	37
12							20	23	25	28	31	34	37	40	42
13								26	29	32	35	38	42	45	48
14									32	36	39	43	46	50	54
15										40	43	47	51	55	59
16											48	52	56	60	65
17												57	61	66	70
18													66	71	76
19														77	82
20															88

One-Tailed Test at $\alpha = 0.01$ or Two-Tailed Test at $\alpha = 0.02$

n_1	5	6	7	8	9	10	11	12	13	14	15	16	17	18	19	20
1																
2									0	0	0	0	0	0	1	1
3			0	0	1	1	1	2	2	2	3	3	4	4	4	5
4	0	1	1	2	3	3	4	5	5	6	7	7	8	9	9	10
5	1	2	3	4	5	6	7	8	9	10	11	12	13	14	15	16
6		3	4	6	7	8	9	11	12	13	15	16	18	19	20	22
7			6	8	9	11	12	14	16	17	19	21	23	24	26	28
8				10	11	13	15	17	20	22	24	26	28	30	32	34
9					14	16	18	21	23	26	28	31	33	36	38	40
10						19	22	24	27	30	33	36	38	41	44	47
11							25	28	31	34	37	41	44	47	50	53
12								31	35	38	42	46	49	53	56	60
13									39	43	47	51	55	59	63	67
14										47	51	56	60	65	69	73
15											56	61	66	70	75	80
16												66	71	76	82	87
17													77	82	88	93
18														88	94	100
19															101	107
20																114

*Based in part on Tables 1, 3, 5, and 7 of D. Auble, "Extended tables for the Mann–Whitney statistic," *Bulletin of the Institute of Educational Research at Indiana University*, vol. 1, no. 2 (1953), by permission of the director.

TABLE A.9 (*continued*)
Critical Values for the Wilcoxon Rank-Sum Test
One-Tailed Test at $\alpha = 0.025$ or Two-Tailed Test at $\alpha = 0.05$

n_1	4	5	6	7	8	9	10	11	12	13	14	15	16	17	18	19	20
1																	
2					0	0	0	0	1	1	1	1	1	2	2	2	2
3		0	1	1	2	2	3	3	4	4	5	5	6	6	7	7	8
4	0	1	2	3	4	4	5	6	7	8	9	10	11	11	12	13	13
5		2	3	5	6	7	8	9	11	12	13	14	15	17	18	19	20
6			5	6	8	10	11	13	14	16	17	19	21	22	24	25	27
7				8	10	12	14	16	18	20	22	24	26	28	30	32	34
8					13	15	17	19	22	24	26	29	31	34	36	38	41
9						17	20	23	26	28	31	34	37	39	42	45	48
10							23	26	29	33	36	39	42	45	48	52	55
11								30	33	37	40	44	47	51	55	58	62
12									37	41	45	49	53	57	61	65	69
13										45	50	54	59	63	67	72	76
14											55	59	64	67	74	78	83
15												64	70	75	80	85	90
16													75	81	86	92	98
17														87	93	99	105
18															99	106	112
19																113	119
20																	127

One-Tailed Test at $\alpha = 0.05$ or Two-Tailed Test at $\alpha = 0.10$

n_1	3	4	5	6	7	8	9	10	11	12	13	14	15	16	17	18	19	20
1																	0	0
2			0	0	0	1	1	1	1	2	2	3	3	3	3	4	4	4
3	0	0	1	2	2	3	4	4	5	5	6	7	7	8	9	9	10	11
4		1	2	3	4	5	6	7	8	9	10	11	12	14	15	16	17	18
5			4	5	6	8	9	11	12	13	15	16	18	19	20	22	23	25
6				7	8	10	12	14	16	17	19	21	23	25	26	28	30	32
7					11	13	15	17	19	21	24	26	28	30	33	35	37	39
8						15	18	20	23	26	28	31	33	36	39	41	44	47
9							21	24	27	30	33	36	39	42	45	48	51	54
10								27	31	34	37	41	44	48	51	55	58	62
11									34	38	42	46	50	54	57	61	65	69
12										42	47	51	55	60	64	68	72	77
13											51	56	61	65	70	75	80	84
14												61	66	71	77	82	87	92
15													72	77	83	88	94	100
16														83	89	95	101	107
17															96	102	109	115
18																109	116	123
19																	123	130
20																		138

TABLE A.10*

$P(V \leq a$ when H_0 is true) in the Runs Test

(n_1, n_2)	a								
	2	3	4	5	6	7	8	9	10
(2, 3)	0.200	0.500	0.900	1.000					
(2, 4)	0.133	0.400	0.800	1.000					
(2, 5)	0.095	0.333	0.714	1.000					
(2, 6)	0.071	0.286	0.643	1.000					
(2, 7)	0.056	0.250	0.583	1.000					
(2, 8)	0.044	0.222	0.533	1.000					
(2, 9)	0.036	0.200	0.491	1.000					
(2, 10)	0.030	0.182	0.455	1.000					
(3, 3)	0.100	0.300	0.700	0.900	1.000				
(3, 4)	0.057	0.200	0.543	0.800	0.971	1.000			
(3, 5)	0.036	0.143	0.429	0.714	0.929	1.000			
(3, 6)	0.024	0.107	0.345	0.643	0.881	1.000			
(3, 7)	0.017	0.083	0.283	0.583	0.833	1.000			
(3, 8)	0.012	0.067	0.236	0.533	0.788	1.000			
(3, 9)	0.009	0.055	0.200	0.491	0.745	1.000			
(3, 10)	0.007	0.045	0.171	0.455	0.706	1.000			
(4, 4)	0.029	0.114	0.371	0.629	0.886	0.971	1.000		
(4, 5)	0.016	0.071	0.262	0.500	0.786	0.929	0.992	1.000	
(4, 6)	0.010	0.048	0.190	0.405	0.690	0.881	0.976	1.000	
(4, 7)	0.006	0.033	0.142	0.333	0.606	0.833	0.954	1.000	
(4, 8)	0.004	0.024	0.109	0.279	0.533	0.788	0.929	1.000	
(4, 9)	0.003	0.018	0.085	0.236	0.471	0.745	0.902	1.000	
(4, 10)	0.002	0.014	0.068	0.203	0.419	0.706	0.874	1.000	
(5, 5)	0.008	0.040	0.167	0.357	0.643	0.833	0.960	0.992	1.000
(5, 6)	0.004	0.024	0.110	0.262	0.522	0.738	0.911	0.976	0.998
(5, 7)	0.003	0.015	0.076	0.197	0.424	0.652	0.854	0.955	0.992
(5, 8)	0.002	0.010	0.054	0.152	0.347	0.576	0.793	0.929	0.984
(5, 9)	0.001	0.007	0.039	0.119	0.287	0.510	0.734	0.902	0.972
(5, 10)	0.001	0.005	0.029	0.095	0.239	0.455	0.678	0.874	0.958
(6, 6)	0.002	0.013	0.067	0.175	0.392	0.608	0.825	0.933	0.987
(6, 7)	0.001	0.008	0.043	0.121	0.296	0.500	0.733	0.879	0.966
(6, 8)	0.001	0.005	0.028	0.086	0.226	0.413	0.646	0.821	0.937
(6, 9)	0.000	0.003	0.019	0.063	0.175	0.343	0.566	0.762	0.902
(6, 10)	0.000	0.002	0.013	0.047	0.137	0.288	0.497	0.706	0.864
(7, 7)	0.001	0.004	0.025	0.078	0.209	0.383	0.617	0.791	0.922
(7, 8)	0.000	0.002	0.015	0.051	0.149	0.296	0.514	0.704	0.867
(7, 9)	0.000	0.001	0.010	0.035	0.108	0.231	0.427	0.622	0.806
(7, 10)	0.000	0.001	0.006	0.024	0.080	0.182	0.355	0.549	0.743
(8, 8)	0.000	0.001	0.009	0.032	0.100	0.214	0.405	0.595	0.786
(8, 9)	0.000	0.001	0.005	0.020	0.069	0.157	0.319	0.500	0.702
(8, 10)	0.000	0.000	0.003	0.013	0.048	0.117	0.251	0.419	0.621
(9, 9)	0.000	0.000	0.003	0.012	0.044	0.109	0.238	0.399	0.601
(9, 10)	0.000	0.000	0.002	0.008	0.029	0.077	0.179	0.319	0.510
(10, 10)	0.000	0.000	0.001	0.004	0.019	0.051	0.128	0.242	0.414

*Reproduced from C. Eisenhart and F. Swed, "Tables for testing randomness of grouping in a sequence of alternatives," *Ann. Math. Stat.*, vol. 14 (1943), by permission of the editor.

TABLE A.10 (*continued*)

$P(V \leq a$ when H_0 is true) in the Runs Test

(n_1, n_2)	a									
	11	12	13	14	15	16	17	18	19	20
(2, 3)										
(2, 4)										
(2, 5)										
(2, 6)										
(2, 7)										
(2, 8)										
(2, 9)										
(2, 10)										
(3, 3)										
(3, 4)										
(3, 5)										
(3, 6)										
(3, 7)										
(3, 8)										
(3, 9)										
(3, 10)										
(4, 4)										
(4, 5)										
(4, 6)										
(4, 7)										
(4, 8)										
(4, 9)										
(4, 10)										
(5, 5)										
(5, 6)	1.000									
(5, 7)	1.000									
(5, 8)	1.000									
(5, 9)	1.000									
(5, 10)	1.000									
(6, 6)	0.998	1.000								
(6, 7)	0.992	0.999	1.000							
(6, 8)	0.984	0.998	1.000							
(6, 9)	0.972	0.994	1.000							
(6, 10)	0.958	0.990	1.000							
(7, 7)	0.975	0.996	0.999	1.000						
(7, 8)	0.949	0.988	0.998	1.000	1.000					
(7, 9)	0.916	0.975	0.994	0.999	1.000					
(7, 10)	0.879	0.957	0.990	0.998	1.000					
(8, 8)	0.900	0.968	0.991	0.999	1.000	1.000				
(8, 9)	0.843	0.939	0.980	0.996	0.999	1.000	1.000			
(8, 10)	0.782	0.903	0.964	0.990	0.998	1.000	1.000			
(9, 9)	0.762	0.891	0.956	0.988	0.997	1.000	1.000	1.000		
(9, 10)	0.681	0.834	0.923	0.974	0.992	0.999	1.000	1.000	1.000	
(10, 10)	0.586	0.758	0.872	0.949	0.981	0.996	0.999	1.000	1.000	1.000

TABLE A.11*
Least-Significant Studentized Ranges r_p
$\alpha = 0.05$

					p				
ν	2	3	4	5	6	7	8	9	10
1	17.97	17.97	17.97	17.97	17.97	17.97	17.97	17.97	17.97
2	6.085	6.085	6.085	6.085	6.085	6.085	6.085	6.085	6.085
3	4.501	4.516	4.516	4.516	4.516	4.516	4.516	4.516	4.516
4	3.927	4.013	4.033	4.033	4.033	4.033	4.033	4.033	4.033
5	3.635	3.749	3.797	3.814	3.814	3.814	3.814	3.814	3.814
6	3.461	3.587	3.649	3.680	3.694	3.697	3.697	3.697	3.697
7	3.344	3.477	3.548	3.588	3.611	3.622	3.626	3.626	3.626
8	3.261	3.399	3.475	3.521	3.549	3.566	3.575	3.579	3.579
9	3.199	3.339	3.420	3.470	3.502	3.523	3.536	3.544	3.547
10	3.151	3.293	3.376	3.430	3.465	3.489	3.505	3.516	3.522
11	3.113	3.256	3.342	3.397	3.435	3.462	3.480	3.493	3.501
12	3.082	3.225	3.313	3.370	3.410	3.439	3.459	3.474	3.484
13	3.055	3.200	3.289	3.348	3.389	3.419	3.442	3.458	3.470
14	3.033	3.178	3.268	3.329	3.372	3.403	3.426	3.444	3.457
15	3.014	3.160	3.250	3.312	3.356	3.389	3.413	3.432	3.446
16	2.998	3.144	3.235	3.298	3.343	3.376	3.402	3.422	3.437
17	2.984	3.130	3.222	3.285	3.331	3.366	3.392	3.412	3.429
18	2.971	3.118	3.210	3.274	3.321	3.356	3.383	3.405	3.421
19	2.960	3.107	3.199	3.264	3.311	3.347	3.375	3.397	3.415
20	2.950	3.097	3.190	3.255	3.303	3.339	3.368	3.391	3.409
24	2.919	3.066	3.160	3.226	3.276	3.315	3.345	3.370	3.390
30	2.888	3.035	3.131	3.199	3.250	3.290	3.322	3.349	3.371
40	2.858	3.006	3.102	3.171	3.224	3.266	3.300	3.328	3.352
60	2.829	2.976	3.073	3.143	3.198	3.241	3.277	3.307	3.333
120	2.800	2.947	3.045	3.116	3.172	3.217	3.254	3.287	3.314
∞	2.772	2.918	3.017	3.089	3.146	3.193	3.232	3.265	3.294

*Abridged from H. L. Harter, "Critical values for Duncan's new multiple range test," *Biometrics*, vol. 16, no. 4 (1960), by permission of the author and the editor.

TABLE A.11 (*continued*)
Least-Significant Studentized Ranges r_p
$\alpha = 0.01$

ν	\multicolumn{9}{c}{p}								
	2	3	4	5	6	7	8	9	10
1	90.03	90.03	90.03	90.03	90.03	90.03	90.03	90.03	90.03
2	14.04	14.04	14.04	14.04	14.04	14.04	14.04	14.04	14.04
3	8.261	8.321	8.321	8.321	8.321	8.321	8.321	8.321	8.321
4	6.512	6.677	6.740	6.756	6.756	6.756	6.756	6.756	6.756
5	5.702	5.893	5.989	6.040	6.065	6.074	6.074	6.074	6.074
6	5.243	5.439	5.549	5.614	5.655	5.680	5.694	5.701	5.703
7	4.949	5.145	5.260	5.334	5.383	5.416	5.439	5.454	5.464
8	4.746	4.939	5.057	5.135	5.189	5.227	5.256	5.276	5.291
9	4.596	4.787	4.906	4.986	5.043	5.086	5.118	5.142	5.160
10	4.482	4.671	4.790	4.871	4.931	4.975	5.010	5.037	5.058
11	4.392	4.579	4.697	4.780	4.841	4.887	4.924	4.952	4.975
12	4.320	4.504	4.622	4.706	4.767	4.815	4.852	4.883	4.907
13	4.260	4.442	4.560	4.644	4.706	4.755	4.793	4.824	4.850
14	4.210	4.391	4.508	4.591	4.654	4.704	4.743	4.775	4.802
15	4.168	4.347	4.463	4.547	4.610	4.660	4.700	4.733	4.760
16	4.131	4.309	4.425	4.509	4.572	4.622	4.663	4.696	4.724
17	4.099	4.275	4.391	4.475	4.539	4.589	4.630	4.664	4.693
18	4.071	4.246	4.362	4.445	4.509	4.560	4.601	4.635	4.664
19	4.046	4.220	4.335	4.419	4.483	4.534	4.575	4.610	4.639
20	4.024	4.197	4.312	4.395	4.459	4.510	4.552	4.587	4.617
24	3.956	4.126	4.239	4.322	4.386	4.437	4.480	4.516	4.546
30	3.889	4.056	4.168	4.250	4.314	4.366	4.409	4.445	4.477
40	3.825	3.988	4.098	4.180	4.244	4.296	4.339	4.376	4.408
60	3.762	3.922	4.031	4.111	4.174	4.226	4.270	4.307	4.340
120	3.702	3.858	3.965	4.044	4.107	4.158	4.202	4.239	4.272
∞	3.643	3.796	3.900	3.978	4.040	4.091	4.135	4.172	4.205

TABLE A.12
Random Numbers

Line	1–5	6–10	11–15	16–20	21–25	26–30	31–35	36–40
				Column				
1	62956	95735	70988	86027	27648	65155	46301	27217
2	17143	50118	41681	87224	75674	43371	09846	83403
3	99285	01369	94610	71099	69207	01999	23931	34711
4	12940	81308	40436	82916	74245	70324	88555	82182
5	28089	80216	08681	83524	00583	55179	31911	68484
6	78079	74747	17626	74930	41300	04858	85634	42398
7	36009	01306	33858	96930	71087	11354	85891	52644
8	95695	52933	39459	84218	34670	91542	02186	86134
9	89221	34158	16364	16532	50070	78159	18445	05884
10	91937	35854	13168	24642	22369	87396	64367	89259
11	07339	63159	94886	51002	85834	94109	56843	03769
12	73238	34352	81008	95682	13029	76288	22054	54849
13	87940	32625	44838	39920	57188	41771	43185	74236
14	46904	92456	64675	66930	54980	11631	54596	50563
15	02580	92653	33907	54380	00763	60452	18860	48829
16	86983	20150	78561	97095	15990	45947	88542	86519
17	92608	22144	67209	88807	82087	06616	16605	95621
18	26988	49617	87118	28108	13110	40766	21216	01567
19	75370	38794	51939	20879	30221	73593	76238	85702
20	18826	84055	91391	78487	07594	74994	64239	00808
21	20198	45182	09914	45305	97352	00516	56804	10931
22	74784	75807	79881	45290	56117	39798	62617	26912
23	08050	25691	87922	75747	55031	82704	97667	03734
24	63096	27123	94686	39205	68047	12108	62144	31291
25	23099	48428	16697	82597	74983	22452	46283	97317

TABLE A.12
Random Numbers (*continued*)

Line	1–5	6–10	11–15	16–20	21–25	26–30	31–35	36–40
26	84827	81473	19453	95401	01363	40795	86600	78317
27	97965	30432	92410	42482	31448	78558	55152	27863
28	96097	51256	61546	93683	46277	30115	37682	15694
29	77733	98610	86615	19007	29402	26348	96477	97154
30	73159	81085	96957	48358	90944	58155	73014	79515
31	19074	14518	91372	73333	42832	17500	91049	74510
32	83098	95483	17986	79141	92419	36887	65473	05675
33	10416	60700	37527	26169	07315	08340	31597	05568
34	08693	25225	54798	60498	32060	60310	36587	30579
35	50451	52350	37860	40950	14377	16485	62250	96104
36	73128	88097	01832	19463	28038	00222	83868	74422
37	89677	39620	49118	49660	96852	71822	66195	28204
38	67828	36965	63617	60332	10525	78030	06835	59222
39	30001	63542	05680	12956	96058	80149	79950	39309
40	14283	75479	39727	79075	87995	74464	49102	93185
41	84051	28694	03885	97247	43578	48213	97929	49951
42	80815	60959	58747	50798	47455	18738	58154	95800
43	28515	30696	23612	87285	96888	25681	65597	50837
44	17402	25186	12526	19012	42374	47886	43367	61815
45	66814	38016	61219	14760	99030	38070	81369	94157
46	49751	96432	63666	47760	70192	10367	17197	95801
47	35597	97760	47288	34700	25569	91920	02045	24344
48	03026	00712	49279	10272	30083	61603	26715	89026
49	96637	00092	97446	75109	53899	93915	37789	13073
50	34324	90440	76224	71230	92581	06794	39559	05362

TABLE A.13*
Critical Values for Bartlett's Test
$b_k(0.01; n)$

	Number of Populations, k								
n	2	3	4	5	6	7	8	9	10
3	.1411	.1672	*	*	*	*	*	*	*
4	.2843	.3165	.3475	.3729	.3937	.4110	*	*	*
5	.3984	.4304	.4607	.4850	.5046	.5207	.5343	.5458	.5558
6	.4850	.5149	.5430	.5653	.5832	.5978	.6100	.6204	.6293
7	.5512	.5787	.6045	.6248	.6410	.6542	.6652	.6744	.6824
8	.6031	.6282	.6518	.6704	.6851	.6970	.7069	.7153	.7225
9	.6445	.6676	.6892	.7062	.7197	.7305	.7395	.7471	.7536
10	.6783	.6996	.7195	.7352	.7475	.7575	.7657	.7726	.7786
11	.7063	.7260	.7445	.7590	.7703	.7795	.7871	.7935	.7990
12	.7299	.7483	.7654	.7789	.7894	.7980	.8050	.8109	.8160
13	.7501	.7672	.7832	.7958	.8056	.8135	.8201	.8256	.8303
14	.7674	.7835	.7985	.8103	.8195	.8269	.8330	.8382	.8426
15	.7825	.7977	.8118	.8229	.8315	.8385	.8443	.8491	.8532
16	.7958	.8101	.8235	.8339	.8421	.8486	.8541	.8586	.8625
17	.8076	.8211	.8338	.8436	.8514	.8576	.8627	.8670	.8707
18	.8181	.8309	.8429	.8523	.8596	.8655	.8704	.8745	.8780
19	.8275	.8397	.8512	.8601	.8670	.8727	.8773	.8811	.8845
20	.8360	.8476	.8586	.8671	.8737	.8791	.8835	.8871	.8903
21	.8437	.8548	.8653	.8734	.8797	.8848	.8890	.8926	.8956
22	.8507	.8614	.8714	.8791	.8852	.8901	.8941	.8975	.9004
23	.8571	.8673	.8769	.8844	.8902	.8949	.8988	.9020	.9047
24	.8630	.8728	.8820	.8892	.8948	.8993	.9030	.9061	.9087
25	.8684	.8779	.8867	.8936	.8990	.9034	.9069	.9099	.9124
26	.8734	.8825	.8911	.8977	.9029	.9071	.9105	.9134	.9158
27	.8781	.8869	.8951	.9015	.9065	.9105	.9138	.9166	.9190
28	.8824	.8909	.8988	.9050	.9099	.9138	.9169	.9196	.9219
29	.8864	.8946	.9023	.9083	.9130	.9167	.9198	.9224	.9246
30	.8902	.8981	.9056	.9114	.9159	.9195	.9225	.9250	.9271
40	.9175	.9235	.9291	.9335	.9370	.9397	.9420	.9439	.9455
50	.9339	.9387	.9433	.9468	.9496	.9518	.9536	.9551	.9564
60	.9449	.9489	.9527	.9557	.9580	.9599	.9614	.9626	.9637
80	.9586	.9617	.9646	.9668	.9685	.9699	.9711	.9720	.9728
100	.9669	.9693	.9716	.9734	.9748	.9759	.9769	.9776	.9783

* Reproduced from D. D. Dyer and J. P. Keating, "On the determination of critical values for Bartlett's test," *J. Am. Stat. Assoc.*, vol. 75 (1980), by permission of the Board of Directors.

TABLE A.13 *(continued)*
Critical Values for Bartlett's Test
$b_k(0.05; n)$

n	Number of Populations, k								
	2	3	4	5	6	7	8	9	10
3	.3123	.3058	.3173	.3299	*	*	*	*	*
4	.4780	.4699	.4803	.4921	.5028	.5122	.5204	.5277	.5341
5	.5845	.5762	.5850	.5952	.6045	.6126	.6197	.6260	.6315
6	.6563	.6483	.6559	.6646	.6727	.6798	.6860	.6914	.6961
7	.7075	.7000	.7065	.7142	.7213	.7275	.7329	.7376	.7418
8	.7456	.7387	.7444	.7512	.7574	.7629	.7677	.7719	.7757
9	.7751	.7686	.7737	.7798	.7854	.7903	.7946	.7984	.8017
10	.7984	.7924	.7970	.8025	.8076	.8121	.8160	.8194	.8224
11	.8175	.8118	.8160	.8210	.8257	.8298	.8333	.8365	.8392
12	.8332	.8280	.8317	.8364	.8407	.8444	.8477	.8506	.8531
13	.8465	.8415	.8450	.8493	.8533	.8568	.8598	.8625	.8648
14	.8578	.8532	.8564	.8604	.8641	.8673	.8701	.8726	.8748
15	.8676	.8632	.8662	.8699	.8734	.8764	.8790	.8814	.8834
16	.8761	.8719	.8747	.8782	.8815	.8843	.8868	.8890	.8909
17	.8836	.8796	.8823	.8856	.8886	.8913	.8936	.8957	.8975
18	.8902	.8865	.8890	.8921	.8949	.8975	.8997	.9016	.9033
19	.8961	.8926	.8949	.8979	.9006	.9030	.9051	.9069	.9086
20	.9015	.8980	.9003	.9031	.9057	.9080	.9100	.9117	.9132
21	.9063	.9030	.9051	.9078	.9103	.9124	.9143	.9160	.9175
22	.9106	.9075	.9095	.9120	.9144	.9165	.9183	.9199	.9213
23	.9146	.9116	.9135	.9159	.9182	.9202	.9219	.9235	.9248
24	.9182	.9153	.9172	.9195	.9217	.9236	.9253	.9267	.9280
25	.9216	.9187	.9205	.9228	.9249	.9267	.9283	.9297	.9309
26	.9246	.9219	.9236	.9258	.9278	.9296	.9311	.9325	.9336
27	.9275	.9249	.9265	.9286	.9305	.9322	.9337	.9350	.9361
28	.9301	.9276	.9292	.9312	.9330	.9347	.9361	.9374	.9385
29	.9326	.9301	.9316	.9336	.9354	.9370	.9383	.9396	.9406
30	.9348	.9325	.9340	.9358	.9376	.9391	.9404	.9416	.9426
40	.9513	.9495	.9506	.9520	.9533	.9545	.9555	.9564	.9572
50	.9612	.9597	.9606	.9617	.9628	.9637	.9645	.9652	.9658
60	.9677	.9665	.9672	.9681	.9690	.9698	.9705	.9710	.9716
80	.9758	.9749	.9754	.9761	.9768	.9774	.9779	.9783	.9787
100	.9807	.9799	.9804	.9809	.9815	.9819	.9823	.9827	.9830

TABLE A.14*

Critical Values of Spearman's Rank Correlation Coefficient

n	$\alpha = 0.05$	$\alpha = 0.025$	$\alpha = 0.01$	$\alpha = 0.005$
5	0.900	—	—	—
6	0.829	0.886	0.943	—
7	0.714	0.786	0.893	—
8	0.643	0.738	0.833	0.881
9	0.600	0.683	0.783	0.833
10	0.564	0.648	0.745	0.794
11	0.523	0.623	0.736	0.818
12	0.497	0.591	0.703	0.780
13	0.475	0.566	0.673	0.745
14	0.457	0.545	0.646	0.716
15	0.441	0.525	0.623	0.689
16	0.425	0.507	0.601	0.666
17	0.412	0.490	0.582	0.645
18	0.399	0.476	0.564	0.625
19	0.388	0.462	0.549	0.608
20	0.377	0.450	0.534	0.591
21	0.368	0.438	0.521	0.576
22	0.359	0.428	0.508	0.562
23	0.351	0.418	0.496	0.549
24	0.343	0.409	0.485	0.537
25	0.336	0.400	0.475	0.526
26	0.329	0.392	0.465	0.515
27	0.323	0.385	0.456	0.505
28	0.317	0.377	0.448	0.496
29	0.311	0.370	0.440	0.487
30	0.305	0.364	0.432	0.478

* Reproduced from E. G. Olds. "Distribution of sums of squares of rank differences for small samples," *Ann. Math. Stat.*, vol. 9 (1938), by permission of the editor.

Answers to Selected Exercises

Page 17

1. (a) Statistical inference.
   (b) Descriptive statistics.
   (c) Descriptive statistics.
   (d) Statistical inference.
   (e) Statistical inference.
3. (a) Responses of all people in Richmond who have telephones.
   (b) Outcomes for a large or infinite number of tosses of a coin.
   (c) Length of life of such tennis shoes when worn on the professional tour.
   (d) All possible time intervals for this lawyer to drive from her home to her office.
5. Bivens, Grayson, Wright, and Moorman.
7. (a) 26, 13, 201, 221, 256.   (b) 215, 102, 214, 151, 297.
8. (a) $w_6^2 + w_7^2 + w_8^2 + w_9^2 + w_{10}^2$.   (b) $x_2 + x_3 + x_4 + 9$.
   (c) $3(v_1 + v_2 + v_3 + v_4 + v_5) - 30$.
9. (a) $12x^2 + 36x + 29$.   (b) $4x^3 + 18x^2 + 42x + 36$.
10. (a) 66.   (b) 53.   (c) $\frac{5}{3}$.
11. (a) -7.   (b) -3.   (c) -14.

Page 27

1. (a) $\bar{x} = 2.4$.   (b) $\tilde{x} = 2$.   (c) Mode = 3.
2. (a) $\mu = 6$.   (b) $\tilde{\mu} = 6$.   (c) Mode = 7.
3. (a) $\bar{x} = 3.2$ seconds.   (b) $\tilde{x} = 3.1$ seconds.
5. (a) $\bar{x} = 35.7$ grams.   (b) $\tilde{x} = 32.5$ grams.   (c) Mode = 29 grams.
7. $\bar{x} = 10.56$ kilograms.
10. 121.
11. Disneyworld.
12. (a) 3.3 seconds.   (b) $27.50.
13. (a) 80.   (b) 41.15 cents per liter.   (c) 10.76%.
14. (a) 79.4   (b) 15 boys.

499

15. (a) 8.   (b) $1010.45.   (c) 8.119%.
16. (a) 86.3 kilometers per hour.   (b) $5.79.   (c) $4.80.

Page 42
2. (a) Range = 15.   (b) $\sigma^2 = 16$.
3. (a) Range = 2.0.   (b) $s^2 = 0.498$.
6. $s = 0.585$.
7. (a) 45.9.   (b) 5.1.
9. $s^2 = 1.018 \times 10^7$.
11. 2.
12. (a) From 105 to 141.   (b) Less than 96 and greater than 150.
13. (a) At least 75%.   (b) At least 93.75%.   (c) At least 84%.
16. Subcompact model.
18. $25,600.
19. (a) 82%.   (b) Yes.

Page 64
1.

	Class Boundaries	Class Mark	Class Width
(a)	6.5–13.5	10	7
(b)	(-5.5)–(-0.5)	-3	5
(c)	10.35–18.75	14.55	8.4
(d)	0.3455–0.4185	0.382	0.073
(e)	(-2.755)–1.355	-0.70	4.11
(f)	74.485–86.605	82.605	8.24

2. (a) $c = 2$.
   (b) 5.5–7.5, 7.5–9.5, 9.5–11.5, 11.5–13.5, 13.5–15.5.
   (c) 6–7, 8–9, 10–11, 12–13, 14–15.
4. (a) Frequencies are 3, 2, 3, 4, 5, 11, 14, 14, 4.
   (b) Cumulative frequencies are 0, 3, 5, 8, 12, 17, 28, 42, 56, 60.
5. (a) Percentages are 10, 18, 30, 20, 12, 4, 2, 4.
   (b) Cumulative percentages are 0, 10, 28, 58, 78, 90, 94, 96, 100.
7. Frequencies are 12, 5, 8, 2, 3.
12. (a) $\mu = 65.48$.   (b) $\tilde{\mu} = 71.5$.   (c) $\sigma = 20.96$.   (d) SK $= -0.86$.
13. (a) $\bar{x} = 2.40$.   (b) $\tilde{x} = 2.2$.   (c) $= 1.36$.   (d) SK $= 0.44$.
15. (a) 680.   (b) 25.   (c) 1 or 2.
17. (a) 1360.   (b) 1900.   (c) 1994.
19. 37 measurements in the interval from 1.04 to 3.76;
    47 measurements in the interval from -0.32 to 5.12.
20. (a) $\mu = 66$.   (b) $\bar{x} = 2.42$.
21. $s^2 = 0.49$.
23. (a) $P_{89} = 3.9$; $Q_3 = 3.1$; $D_4 = 1.95$.
    (b) $P_{89} = 3.98$; $Q_3 = 3.13$; $D_4 = 1.97$.
25. $P_{56} = 16.1$; $Q_1 = 13.0$; $D_2 = 12.5$.
26. 0.53.
27. (a) 0.75.   (b) 4.6.

Page 78

1. (a) {8, 16, 24, 32, 40, 48}.
   (b) {−5, 1}.
   (c) {*T, HT, HHT, HHH*}.
   (d) {N. America, S. America, Europe, Asia, Africa, Australia, Antarctica}.
   (e) ∅.

3. (a)

			Red			
Green	1	2	3	4	5	6
1	(1, 1)	(1, 2)	(1, 3)	(1, 4)	(1, 5)	(1, 6)
2	(2, 1)	(2, 2)	(2, 3)	(2, 4)	(2, 5)	(2, 6)
3	(3, 1)	(3, 2)	(3, 3)	(3, 4)	(3, 5)	(3, 6)
4	(4, 1)	(4, 2)	(4, 3)	(4, 4)	(4, 5)	(4, 6)
5	(5, 1)	(5, 2)	(5, 3)	(5, 4)	(5, 5)	(5, 6)
6	(6, 1)	(6, 2)	(6, 3)	(6, 4)	(6, 5)	(6, 6)

   (b) $S = \{(x, y) \mid x = 1, 2, ..., 6; y = 1, 2, ..., 6\}$.

4. $S = \{1HH, 1HT, 1TH, 1TT, 2H, 2T, 3HH, 3HT, 3TH, 3TT, 4H, 4T, 5HH, 5HT, 5TH,$
   $5TT, 6H, 6T\}$.

6. $S = \{(x, y) \mid x^2 + y^2 \leq 9\}$.

8. (a) $A = \{1HH, 1HT, 1TH, 1TT, 2H, 2T\}$.
   (b) $B = \{1TT, 3TT, 5TT\}$.
   (c) $A' = \{3HH, 3HT, 3TH, 3TT, 4H, 4T, 5HH, 5HT, 5TH, 5TT, 6H, 6T\}$.
   (d) $A' \cap B = \{3TT, 5TT\}$.
   (e) $A \cup B = \{1HH, 1HT, 1TH, 1TT, 2H, 2T, 3TT, 5TT\}$.

10. (a) $S = \{M_1M_2, M_1F_1, M_1F_2, M_2M_1, M_2F_1, M_2F_2, F_1M_1, F_1M_2, F_1F_2, F_2M_1, F_2M_2, F_2F_1\}$.
    (b) $A = \{M_1M_2, M_1F_1, M_1F_2, M_2M_1, M_2F_1, M_2F_2\}$.
    (c) $B = \{M_1F_1, M_1F_2, M_2F_1, M_2F_2, F_1M_1, F_1M_2, F_2M_1, F_2M_2\}$.
    (d) $C = \{F_1F_2, F_2F_1\}$.
    (e) $A \cap B = \{M_1F_1, M_1F_2, M_2F_1, M_2F_2\}$.
    (f) $A \cup C = \{M_1M_2, M_1F_1, M_1F_2, M_2M_1, M_2F_1, M_2F_2, F_1F_2, F_2F_1\}$.
    (g)

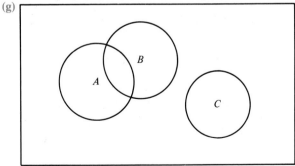

11. (a) $S = \{T, NT, NNT, NNNT, ...\}$, where $T = 3$ and $N =$ not a 3.
    (b) $E = \{T, NT, NNT, NNNT\}$.

13. (a) $A' \cup C = \{1, 2, 3, 4, 5, 6, 8\}$.
    (b) $B \cap C' = \{1, 7, 9\}$.

(c) $(S \cap B')' = \{1, 3, 5, 7, 9\}$.

(d) $(C' \cap D) \cup B = \{1, 3, 5, 6, 7, 9\}$.

(e) $(B \cap C') \cup A = \{1, 2, 4, 7, 9\}$.

(f) $A \cap C \cap D' = \{2, 4\}$.

14. (a) $M \cup N = \{x \mid 0 < x < 9\}$.

(b) $M \cap N = \{x \mid 1 < x < 5\}$.

(c) $M' \cap N' = \{x \mid 9 < x < 12\}$.

16.

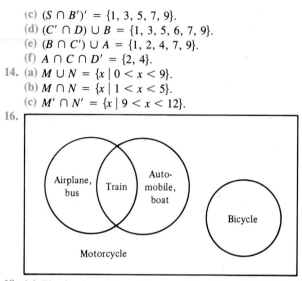

18. (a) The family will experience mechanical problems but will receive no ticket for a traffic violation and will not arrive at a campsite having no vacancies.

(b) The family will receive a traffic ticket and arrive at a campsite having no vacancies but will not experience mechanical problems.

(c) The family will experience mechanical problems and will arrive at a campsite having no vacancies.

(d) The family will receive a traffic ticket but will not arrive at a campsite having no vacancies.

(e) The family will not experience mechanical problems.

19. (a) 6.   (b) 2.   (c) 2, 5, 6.   (d) 4, 5, 7, 8.

Page 87

1. 18.

3. 156.

5. 48.

7. (a) 1024.   (b) 243.

9. 560.

10. (a) 720.   (b) 144.   (c) 480.

12. (a) 180.   (b) 75.   (c) 105.

14. (a) 40,320   (b) 384.   (c) 576.

16. 360.

17. 59,280.

19. 5040.

20. 1260.

22. 4400.

23. 56.

25. (a) 84.   (b) 40.   (c) 15.

27. 800.

Page 95

1. (a) Sum of the probabilities exceeds 1.
   (b) Sum of the probabilities is less than 1.
   (c) A negative probability.
   (d) Probability of both a heart and a black card is zero.
3. (a) $\frac{1}{4}$.   (b) $\frac{3}{4}$.
5. $\frac{9}{19}$.
6. (a) 0.8.   (b) 0.7.   (c) 0.5.
8. (a) $\frac{5}{26}$.   (b) $\frac{9}{26}$.   (c) $\frac{19}{26}$.
9. (a) $\frac{3}{5}$.   (b) $\frac{2}{5}$.   (c) $\frac{1}{10}$.
11. (a) $\frac{5}{36}$.   (b) $\frac{5}{18}$.
13. (a) $\frac{2}{3}$.   (b) $\frac{5}{14}$.
15. (a) $\frac{25}{108}$.   (b) $\frac{125}{1296}$.
17. (a) $\frac{22}{125}$.   (b) $\frac{31}{500}$.   (c) $\frac{171}{500}$.
19. (a) 0.75.   (b) 0.25.
21. (a) 0.55.   (b) 0.83.   (c) 0.27.
22. (a) $\frac{3}{4}$.   (b) 7 to 1 odds.

Page 105

1. (a) The probability that a convict who pushed dope also committed armed robbery.
   (b) The probability that a convict who committed armed robbery also pushed dope.
   (c) The probability that a convict who did not push dope also did not commit armed robbery.
3. (a) $\frac{14}{39}$.   (b) $\frac{95}{112}$.
4. (a) $\frac{5}{34}$.   (b) $\frac{3}{8}$.
5. (a) $\frac{2}{11}$.   (b) $\frac{5}{11}$.
7. (a) 0.56.   (b) 0.35.
9. (a) $\frac{9}{28}$.   (b) $\frac{3}{4}$.   (c) 0.91.
11. 0.27.
13. $\frac{5}{8}$.
15. 0.03.
17. (a) $\frac{91}{323}$.   (b) $\frac{91}{323}$.
19. (a) $\frac{1}{4}$.   (b) $\frac{1}{16}$.

Page 113

1. 0.0744.
3. 0.2097.
5. (a) $\frac{1}{3}$.   (b) $\frac{3}{10}$.
6. (a) 0.054.   (b) $\frac{4}{9}$.
8. 0.174.

Page 124

1. Discrete; continuous; continuous; discrete; discrete; continuous.

3.

Sample Space	w
HHH	3
HHT	1
HTH	1
THH	1
HTT	−1
THT	−1
TTH	−1
TTT	−3

5.

t	20	25	30
$P(T = t)$	$\frac{1}{5}$	$\frac{3}{5}$	$\frac{1}{5}$

7.

w	−3	−1	1	3
$P(W = w)$	$\frac{1}{27}$	$\frac{2}{9}$	$\frac{4}{9}$	$\frac{8}{27}$

8. $f(x) = \dfrac{\binom{5}{x}\binom{5}{4-x}}{\binom{10}{4}}$, for $x = 0, 1, 2, 3, 4$.

11.

x	0	1	2	3
$P(X = x)$	$\frac{703}{1700}$	$\frac{741}{1700}$	$\frac{117}{850}$	$\frac{11}{850}$

13. (a) $\frac{16}{27}$.   (b) $\frac{1}{3}$.
14. (a) 0.68.

Page 131

1. (a) $\frac{1}{36}$.   (b) $\frac{1}{15}$.
2. (a) $\frac{1}{5}$.   (b) $\frac{7}{30}$.   (c) $\frac{3}{5}$.   (d) $\frac{4}{15}$.
3. (a)   (b) $\frac{1}{2}$.

f(x, y)		0	1	2	3
	0		$\frac{3}{70}$	$\frac{9}{70}$	$\frac{3}{70}$
y	1	$\frac{2}{70}$	$\frac{18}{70}$	$\frac{18}{70}$	$\frac{2}{70}$
	2	$\frac{3}{70}$	$\frac{9}{70}$	$\frac{3}{70}$	

5.

f(x,y)		x 0	1	2	3
	−3	$\frac{1}{8}$			
y	−1		$\frac{3}{8}$		
	1			$\frac{3}{8}$	
	3				$\frac{1}{8}$

7. (a)

x	0	1	2	3
g(x)	$\frac{1}{10}$	$\frac{1}{5}$	$\frac{3}{10}$	$\frac{4}{10}$

(b)

y	0	1	2
h(y)	$\frac{1}{5}$	$\frac{1}{3}$	$\frac{7}{15}$

9. (a)

y	0	1	2
f(y|2)	$\frac{3}{10}$	$\frac{3}{5}$	$\frac{1}{10}$

(b) $\frac{3}{10}$.

11. (a)

x	2	4
g(x)	0.40	0.60

(b)

y	1	3	5
h(y)	0.25	0.50	0.25

13. Independent.

Page 150

1. $\frac{6}{7}$.
3. 2.
5. 1.
7. $1.23.
8. $2100.
10. 209.
11. $\frac{9}{8}$.
12. 35.2.
15. $\mu_X = \frac{65}{36}$; $\mu_Y = \frac{197}{90}$.
17. $\frac{20}{49}$.
19. −0.293 to 2.543.
20. 0.74.
22. 119.
23. (a) 10.33. (b) 10.33.
25. (a) $\frac{175}{12}$. (b) $\frac{175}{6}$.
27. (b) −0.1244.

Page 165

1. $f(x) = \frac{1}{12}$, for $x = 1, 2, \ldots, 12$; $P(X < 4) = \frac{1}{4}$.
3. $f(x) = \frac{1}{15}$, for $x = 1, 2, \ldots, 15$.
4. 0.3955.
6. 0.3134.

8. 0.1631.
10. $\frac{63}{64}$.
12. Four-engine plane.
15. 32; 24–40.
16. $\frac{15}{128}$.
17. 0.0095.
19. $\frac{21}{256}$.
20. 0.0077.

Page 173

1. (a) 0.3246.　　(b) 0.4773.

3. $f(x) = \dfrac{\binom{4}{x}\binom{2}{3-x}}{\binom{6}{3}}$, for $x = 1, 2, 3$; $P(2 \le X \le 3) = \frac{4}{5}$.

5. $\frac{5}{14}$.
7. 3.25; between 0.52 and 5.98.
9. 0.9453.
11. 0.7758.
13. (a) $\frac{4}{33}$.　(b) $\frac{8}{165}$.
15. 0.0308.

Page 180

1. 0.0515.
3. (a) 0.1638.　　(b) 0.032.
5. $\frac{63}{64}$.
6. (a) $\frac{2}{243}$.　　(b) $\frac{16}{81}$.
7. (a) 0.1008.　　(b) 0.4232.　　(c) 0.8009.
9. (a) 0.1512.　　(b) 0.4015.
11. (a) 0.3840.　　(b) 0.1395.　　(c) 0.0553.
13. 0.2657.
14. (a) 0.1321.　　(b) 0.3376.
16. At least $\frac{8}{9}$ of the time from 1 to 19 of every 10,000 forms examined are in error.

Page 197

1. (a) 0.0918.　　(b) 0.9850.　　(c) 0.3371.　　(d) 39.22.　　(e) 46.78.
3. (a) 0.1151.　　(b) 0.5403.　　(c) 16.375.　　(d) 20.55.
4. (a) 0.0548.　　(b) 0.4514.　　(c) 23.　　(d) 189.95 milliliters.
6. (a) 0.0571.　　(b) 99.11%.　　(c) 0.3974.　　(d) 27.952 minutes.　　(e) 0.0092.
7. (a) 64.　　(b) 86.　　(c) 78.
9. (a) 16.　　(b) 549.　　(c) 28.　　(d) 27.
11. (a) 0.0427.　　(b) 0.7642.　　(c) 0.6964.
13. (a) 19.36%.　　(b) 39.70%.
14. 26.
16. 6.24 years.

Page 206

1. 0.0020.
3. (a) 0.8643. (b). 0.2978. (c) 0.0796.
5. (a) 0.9966. (b) 0.1841.
6. (a) 0.0838. (b) 0.1635.
8. 0.1357.
10. 0.1737.

Page 230

1. (a)

$\bar{x}$	2.0	2.5	3.0	3.5	4.0	4.5	5.0	5.5	6.0
$f(\bar{x})$	$\frac{1}{81}$	$\frac{4}{81}$	$\frac{10}{81}$	$\frac{16}{81}$	$\frac{19}{81}$	$\frac{16}{81}$	$\frac{10}{81}$	$\frac{4}{81}$	$\frac{1}{81}$

   (d) Between 3.18 and 4.82.
2. 0.1912.

3. (a)

$\bar{x}$	2	3	4	5	6
$f(\bar{x})$	$\frac{4}{25}$	$\frac{4}{25}$	$\frac{9}{25}$	$\frac{4}{25}$	$\frac{4}{25}$

5. (a) Reduced from 0.7 to 0.4.
   (b) Increased from 0.2 to 0.8.
7. (a) 0.866. (b) 0.995. (c) 0.998.
9. Yes.
11. (a) $\mu = 5.3$; $\sigma^2 = 0.81$. (b) $\mu_{\bar{x}} = 5.3$; $\sigma_{\bar{x}}^2 = 0.0225$. (c) 0.9082.
13. (a) 2.145. (b) -1.372. (c) -3.499.
14. (a) 0.975. (b) 0.10. (c) 0.875. (d) 0.99.
15. (a) 0.985. (b) 0.975.
16. (a) 2.500. (b) 1.319. (c) 1.714.
17. $t = -2.000$; valid claim.
19. $t = 1.64$; yes.

20. (a)

$\bar{x}_1 - \bar{x}_2$	-3	-2	-1	0	1	2	3	4	5
$f(\bar{x}_1 - \bar{x}_2)$	$\frac{1}{64}$	$\frac{4}{64}$	$\frac{8}{64}$	$\frac{12}{64}$	$\frac{14}{64}$	$\frac{12}{64}$	$\frac{8}{64}$	$\frac{4}{64}$	$\frac{1}{64}$

22. 0.1940.
23. 0.7070.
25. (a) 0.0768. (b) 0.2812.

Page 239

2. 0, 1, 3, 0, 1.
3. 1, 1, 2, 4, 2, 1, 1, 3, 2, 3.
5. 3, 8, 7, 7, 1, 5, 6, 4, 1, 6, 7, 3, 5, 2, 2.
6. (a) Sample means are

4.4	3.2	3.8	3.4	3.0	4.0	3.4	4.6
3.0	3.6	3.8	3.4	3.6	4.0	2.4	3.0
2.6	3.2	3.4	3.4	3.8	3.0	4.6	3.6
3.6	3.6	3.0	2.8	2.6	3.6	2.6	4.4
3.4	4.4	3.2	3.0	4.6	3.2	1.6	3.6

(b) Frequencies are 1, 1, 10, 10, 12, 3, 3.

(c) 3.435 and 0.648; $\mu_{\bar{x}} = 3.43$ and $\sigma_{\bar{x}} = 0.664$.

7. 3, 2, 5, 8, 5; 5, 4, 7, 3, 10; 8, 6, 3, 8, 7; 5, 9, 12, 5, 6.

9. 11 whites, 3 blacks, 1 oriental.

10. (a) 64 dormitory, 19 fraternity, 117 private residence students.

(b) 12 whites, 2 blacks, 1 oriental.

Page 262

1. (a) $\mu = \frac{10}{3}$; $\sigma^2 = \frac{14}{9}$.

3. $765 < \mu < 795$.

5. (a) $172.23 < \mu < 176.77$.   (b) $e < 2.27$.

7. 68.

9. 28.

10. $10.15 < \mu < 12.45$.

11. $9.81 < \mu < 10.31$.

13. $\$7.09 < \mu < \$8.91$.

15. $6.56 < \mu_1 - \mu_2 < 11.24$.

17. $0.3 < \mu_1 - \mu_2 < 9.7$.

19. $-6522 < \mu_1 - \mu_2 < 2922$.

21. $-0.7 < \mu_D < 6.3$.

23. $0.99 < \mu_D < 6.12$.

Page 273

1. (a) $0.529 < p < 0.671$.   (b) $e < 0.071$.

3. $0.5924 < p < 0.6636$.

5. $0.739 < p < 0.961$.   (b) No.

7. 508.

9. 16.577.

11. $0.016 < p_A - p_B < 0.164$.

12. $-0.030 < p_M - p_F < 0.250$.

Page 282

1. (a) 34.805.   (b) 16.047.   (c) 13.277.

2. (a) 0.05.   (b) 0.94.

3. $0.293 < \sigma^2 < 6.736$; valid claim.

5. $1.863 < \sigma < 3.578$.

7. $1.410 < \sigma < 6.385$.

9. (a) 2.71.   (b) 3.51.   (c) 2.92.   (d) 0.47.   (e) 0.34.

11. 0.99.

12. $0.600 < \sigma_1/\sigma_2 < 2.819$; yes.

13. $0.238 < \sigma_1^2/\sigma_2^2 < 1.895$.

Page 290

1. $p^* = 0.173$.

2. (a)

p	0.05	0.10	0.15
$f(p\mid x = 2)$	0.12	0.55	0.33

(b) $p^* = 0.111$.

4. $8.077 < \mu < 8.692$.

5. (a) $\$7.04$.　　(b) $\$6.65 < \mu < \$7.43$.　　(c) 0.6532.

7. $R(\hat{P}; p) = pq/n$.

8. $R(\Theta_1; \theta) = \begin{cases} 0 \text{ for } \theta = 0 \\ \frac{2}{3} \text{ for } \theta = 1 \\ \frac{2}{3} \text{ for } \theta = 2 \\ 0 \text{ for } \theta = 3. \end{cases}$

10. $\hat{\Theta}_1$

11. $\hat{\Theta}_2$

Page 307

1. (a) $\alpha = 0.0853$.　　(b) $\beta = 0.8287$; $\beta = 0.7817$.　　(c) No.

2. (a) $\alpha = 0.0536$.　　(b) $\beta = 0.0918$; $\beta = 0.1401$.　　(c) Fair.

5. (a) $\alpha = 0.0466$.　　(b) $\beta = 0.0022$.

7. (a) $\alpha = 0.0793$.　　(b) $\beta = 0.0793$; $\beta = 0.5$.

8. (a) $\alpha = 0.0718$.　　(b) $\beta = 0.1151$.

10. (a) $H_0: \mu = 21.8$, $H_1: \mu \neq 21.8$; critical region in both tails.
　(b) $H_0: p = 0.2$, $H_1: p > 0.2$; critical region in right tail.
　(c) $H_0: \mu = 6.2$, $H_1: \mu > 6.2$; critical region in right tail.
　(d) $H_0: p = 0.7$, $H_1: p < 0.7$; critical region in left tail.
　(e) $H_0: p = 0.58$, $H_1: p \neq 0.58$; critical region in both tails.
　(f) $H_0: \mu = 340$, $H_1: \mu < 340$; critical region in left tail.

11.

Value of μ	Probability of Accepting H_0
184	0.08
188	0.27
192	0.58
196	0.84
200	0.93
204	0.84
208	0.58
212	0.27
216	0.08

Page 315

1. $z = -1.64$; accept H_0.

2. $z = -2.76$; reject H_0.

4. $z = 8.97$; reject H_0, $\mu > 20,000$.

5. $t = 0.77$; accept H_0.

6. $t = 4.38$; reject H_0, $\mu > 220$.

8. $t = 1.78$; accept H_0.

9. (a) 271.　　(b) 21.

11. $z = -2.60$; reject H_0, $\mu_A - \mu_B < 12$.

13. $t = 6.58$; reject H_0, $\mu_1 > \mu_2$.

15. $t = -0.92$; yes.

17. $t' = 0.22$; accept H_0.

19. $t = -2.16$; accept H_0.
20. $t = 2.45$; yes.
21. (a) 114.   (b) 12.

Page 323

1. $\chi^2 = 18.12$; accept H_0.
3. $\chi^2 = 17.45$; reject H_0.
5. $\chi^2 = 42.37$; machine is out of control.
6. (a) $z = 2.64$; reject H_0.   (b) $z = -1.92$; no.
7. $f = 1.33$; accept H_0.
10. $f = 1.18$; accept H_0.

Page 331

1. Critical region: $x = 0$ and $x \geq 9$; valid claim.
2. Critical region: $x \geq 10$; no.
5. $z = -3.27$; $p \neq \frac{2}{3}$.
7. $z = 2.40$; yes.
9. $z = 1.11$; no.
11. (a) $z = 2.18$; reject H_0.   (b) $z = -0.29$; accept H_0.

Page 341

1. $\chi^2 = 4.47$; yes.
3. $\chi^2 = 10.14$; reject H_0.
5. $\chi^2 = 2.33$; accept H_0.
7. $\chi^2 = 2.57$; accept H_0.
10. $\chi^2 = 12.62$; not normal.
12. $\chi^2 = 14.60$; not independent.
15. $\chi^2 = 22.32$; not independent.
17. $\chi^2 = 10.19$; not equal.
18. $\chi^2 = 10.20$; not equal.

Page 349

1. (a) $\hat{y} = 5.799 - 0.514x$.   (c) $\hat{y} = 3.743$.
3. (a) $\hat{y} = 6.414 + 1.809x$.   (b) $\hat{y} = 9.580$.
4. (a) $\hat{y} = 5.811 + 0.568x$.   (c) $\hat{y} = 34.211$.
6. (b) $\hat{y} = 343.699 + 3.221x$.   (c) $\hat{y} = \$456$.
7. (a) $\hat{y} = 2.776 - 0.180x$.   (b) $\hat{y} = 2.24$.

Page 364

1. (a) $s_e^2 = 1.344$.   (b) $t = -1.86$; accept $\beta = 0$.
2. $t = -0.19$; accept $\alpha = 6$.
4. (a) $s_x^2 = 0.11$; $s_y^2 = 0.72$; $s_e^2 = 0.40$.
    (b) $4.323 < \alpha < 8.505$.
    (c) $0.445 < \beta < 3.173$.
5. (a) $2.747 < \alpha < 8.875$.   (b) $0.501 < \beta < 0.635$.
7. $0.975 < \beta < 5.467$.
8. (a) $\hat{y} = 0.349 + 1.929x$.   (b) $t = 1.41$; accept $\alpha = 0$.
10. $7.808 < y_0 < 10.808$.

11. (a) $32.285 < \mu_{Y|50} < 36.137$.　　(b) $26.630 < y_0 < 41.792$.
13. (a) $\$452.97 < \mu_{Y|45} < \$524.32$.　　(b) $\$390.90 < y_0 < \$586.38$.

Page 371

1. $f = 1.36$; regression is linear.
3. $f = 52.02$; regression is nonlinear.
4. (a) $\hat{y} = 2760(0.804)^x$.　　(b) $\hat{y} = \$1153$.
5. (a) $\gamma = 2.660$; $C = 2.63 \times 10^6$.　　(b) $P = 22.9$ kg/cm.
7. (a) $\hat{y} = 20.1084 + 0.4136x_1 + 2.0253x_2$.　　(b) $\hat{y} = 57.5$ kg.
8. (a) $\hat{y} = 1.2522 + 0.0028x_1 - 0.1837x_2$.　　(b) $\hat{y} = 2.31$.
9. (a) $\hat{y} = 8.697 - 2.341x + 0.288x^2$.　　(b) $\hat{y} = 5.2$.
11. (a) $\hat{d} = 13.3587 - 0.3394v + 0.011825v^2$.　　(b) $\hat{d} = 47.54$.

Page 382

1. $r = -0.526$.
3. $r = 0.240$.
5. $z = 0.42$; accept H_0.
6. (a) $r = 0.784$.　　(b) $z = 2.59$; $\rho > 0$.　　(c) 61.5%.
8. $t = -1.38$; accept H_0.
9. $R^2_{Y.12} = 0.98$.
11. $r_{Y1.2} = 0.986$.　　(b) $r_{Y2.1} = 0.940$.
13. $r_{Y1.2} = 0.855$.
14. (a) $R_{Y.12} = 0.934$.　　(b) 87.3%　　(c) 67%.

Page 400

2. $f = 3.59$; significant.
4. $f = 0.46$; not significant.
6. $b = 0.86$; variances are equal.
7. (a) $b = 0.77$; variances are equal.
   (b) $f = 13.33$; significant.
8.

$\bar{x}_2$	$\bar{x}_1$	$\bar{x}_5$	$\bar{x}_4$	$\bar{x}_3$
26	30	33	37	42

$\bar{x}_3$	$\bar{x}_1$	$\bar{x}_4$	$\bar{x}_2$
56.52	59.66	61.12	61.96

Page 418

3. (a) $f = 0.15$; not significant.　　(b) $f = 4.37$; significant.
5. f(subjects) $= 4.86$; significant. f(treatments) $= 3.33$; not significant.
7. (a) $f = 7.70$; significant.　　(b) $f = 1.40$; not significant.
8. (a) $f = 0.32$; accept H_0''.　　(b) $f = 15.38$; reject H_0''.　　(c) $f = 1.67$; accept H_0'''.
9. (a) $f = 9.53$; reject H_0'.　　(b) $f = 15.18$; reject H_0'.　　(c) $f = 1.28$; accept H_0'''.
11. (a) $f = 14.81$; reject H_0'.　　(b) $f = 9.04$; reject H_0'.　　(c) $f = 0.61$; accept H_0'''.

Page 429

1. f (varieties) $= 1.79$; no difference in the yielding capabilities of the different varieties.
   f (locations) $= 8.14$; yes, it was necessary to plant each variety at each location.
3. (a) $f = 1.78$; no difference in the percent of foreign additives due to different analysts.
   (b) $f = 5.99$; percent of foreign additives is not the same for all three brands of jam.
4. (a) $f = 3.30$; no difference in the grades due to different time periods.
   (b) $f = 1.76$; courses are of equal difficulty.
   (c) $f = 5.03$; grades are affected by different professors.
5. $f = 1.29$; color additives have no effect on setting time.

Page 443

1. $x = 7$; accept H_0.
3. $z = -2.14$; accept H_0.
5. $x = 2$; accept H_0.
7. $z = 2.67$; reject H_0.
8. $w_- = 12.5$; accept H_0.
10. $w_+ = 3.5$; accept H_0.
11. $w_+ = 7.5$; reject H_0.

Page 453

1. $u_1 = 1$; claim is valid.
3. $u_2 = 5$; A operates longer.
5. $u = 43.5$; $\mu_1 = \mu_2$.
7. $h = 10.47$; reject H_0.
9. $h = 1.07$; no significant difference.

Page 463

1. $v = 7$; sample is random.
3. $v = 6$; no significant difference.
5. $z = 1.12$; random sequence.
7. (a) $r_S = 0.72$.   (b) Reject H_0.
9. $r_S = -0.47$; no significant relationship.
10. (a) $r_S = 0.71$.   (b) Reject H_0.
11. $z = -2.03$; reject H_0.

Index

513

Frequently Used Formulas (*Continued*)

Confidence interval for p

$$\hat{p} \pm z_{\alpha/2} \sqrt{\frac{\hat{p}\hat{q}}{n}}$$

Sample size for estimating p

$$n = \frac{z_{\alpha/2}^2}{4e^2}$$

Confidence interval for $p_1 - p_2$

$$(\hat{p}_1 - \hat{p}_2) \pm z_{\alpha/2} \sqrt{\frac{\hat{p}_1\hat{q}_1}{n_1} + \frac{\hat{p}_2\hat{q}_2}{n_2}}$$

Confidence interval for σ^2

$$\frac{(n-1)s^2}{\chi_{\alpha/2}^2} < \sigma^2 < \frac{(n-1)s^2}{\chi_{1-\alpha/2}^2}$$

Confidence interval for $\dfrac{\sigma_1^2}{\sigma_2^2}$

$$\frac{s_1^2}{s_2^2} \frac{1}{f_{\alpha/2}(v_1, v_2)} < \frac{\sigma_1^2}{\sigma_2^2} < \frac{s_1^2}{s_2^2} f_{\alpha/2}(v_2, v_1)$$

Testing $\mu = \mu_0$ (σ known)

$$z = \frac{\bar{x} - \mu_0}{\sigma/\sqrt{n}}$$

Testing $\mu = \mu_0$ (σ unknown)

$$t = \frac{\bar{x} - \mu_0}{s/\sqrt{n}}$$

Testing $\mu_1 - \mu_2 = d_0$
(σ_1 and σ_2 known)

$$z = \frac{(\bar{x}_1 - \bar{x}_2) - d_0}{\sqrt{\dfrac{\sigma_1^2}{n_1} + \dfrac{\sigma_2^2}{n_2}}}$$

Testing $\mu_1 - \mu_2 = d_0$
($\sigma_1 = \sigma_2$ but unknown)

$$t = \frac{(\bar{x}_1 - \bar{x}_2) - d_0}{s_p \sqrt{\dfrac{1}{n_1} + \dfrac{1}{n_2}}}$$

Testing $\mu_D = d_0$

$$t = \frac{\bar{d} - d_0}{s_d/\sqrt{n}}$$

Testing $\sigma^2 = \sigma_0^2$	$\chi^2 = \dfrac{(n-1)s^2}{\sigma_0^2}$
Testing $\sigma_1^2 = \sigma_2^2$	$f = \dfrac{s_1^2}{s_2^2}$
Testing $p = p_0$	$z = \dfrac{x - np_0}{\sqrt{np_0q_0}}$
Pooled estimate of p	$\hat{p} = \dfrac{x_1 + x_2}{n_1 + n_2}$
Testing $p_1 = p_2$	$z = \dfrac{\hat{p}_1 - \hat{p}_2}{\sqrt{\hat{p}\hat{q}[(1/n_1) + (1/n_2)]}}$
Chi-square tests using contingency tables	$\chi^2 = \Sigma \dfrac{(o - e)^2}{e}$
Linear regression line	$\hat{y} = a + bx$
Slope of regression line	$b = \dfrac{n\Sigma xy - (\Sigma x)(\Sigma y)}{n\Sigma x^2 - (\Sigma x)^2}$
y intercept of regression line	$a = \bar{y} - b\bar{x}$
Estimate of σ^2 in regression	$s_e^2 = \dfrac{n-1}{n-2}(s_y^2 - b^2 s_x^2)$
Testing $\beta = \beta_0$	$t = \dfrac{s_x\sqrt{n-1}\,(b - \beta_0)}{s_e}$
Correlation coefficient	$r = \dfrac{n\Sigma xy - (\Sigma x)(\Sigma y)}{\sqrt{[n\Sigma x^2 - (\Sigma x)^2][n\Sigma y^2 - (\Sigma y)^2]}}$
Rank correlation coefficient	$r_S = 1 - \dfrac{6\Sigma d^2}{n(n^2 - 1)}$